Handbook of
INDUSTRIAL ROBOTICS
Technology and Applications

Handbook of
INDUSTRIAL ROBOTICS
Technology and Applications

Contributors :
Alberto Sanfeliu,
Juan Andrade-Cetto, *et al.*

AURIS REFERENCE LTD.
London, UK

Handbook of Industrial Robotics : *Technology and Applications*
Contributors : Alberto Sanfeliu *and* Juan Andrade-Cetto, *et al.*

Auris Reference Ltd., UK

www.aurisreference.com

United Kingdom

Copyright 2016

Printed in 2017 for Sale in the Indian Subcontinent

The information in this book has been obtained from highly regarded resources. The copyrights for individual articles remain with the authors, as indicated. All chapters are distributed under the terms of the Creative Commons Attribution License, which permit unrestricted use, distribution, and reproduction in any medium, provided the original author and source are credited.

Notice

Contributors, whose names have been given on the book cover, are not associated with the Publisher. The editors and the Publisher have attempted to trace the copyright holders of all material reproduced in this publication and apologise to copyright holders if permission has not been obtained. If any copyright holder has not been acknowledged, please write to us so we may rectify.

Reasonable efforts have been made to publish reliable data. The views articulated in the chapters are those of the individual contributors, and not necessarily those of the editors or the Publisher. Editors and/or the Publisher are not responsible for the accuracy of the information in the published chapters or consequences from their use. The Publisher accepts no responsibility for any damage or grievance to individual(s) or property arising out of the use of any material(s), instruction(s), methods or thoughts in the book.

No part of this publication maybe reproduced, stored in a retrieval system or transmitted in any form or by any means, electronic, mechanical, photocopying, recording, scanning or otherwise without prior written permission of the publisher.

Handbook of Industrial Robotics : *Technology and Applications*

ISBN: 978-1-78154-499-0

British Library Cataloguing in Publication Data
A CIP record for this book is available from the British Library

Exclusively distributed by CBS Publishers & Distributors Pvt. Ltd.

Sales & Distribution Rights only for India, Pakistan, Bangladesh, Sri Lanka, Nepal and Bhutan.This book is not to be sold outside these territories.

PREFACE

An industrial robot is an automatically controlled, reprogrammable, multi-purpose manipulator programmable in three or more axes. The field of industrial robotics may be more practically defined as the study, design and use of robot systems for manufacturing (a top-level definition relying on the prior definition of robot). Typical applications of industrial robots include welding, painting, ironing, assembly, pick and place, palletizing, product inspection, and testing, all accomplished with high endurance, speed, and precision. The most commonly used robot configurations for industrial automation, include articulated robots, SCARA robots and gantry robots. In the context of general robotics, most types of industrial robots would fall into the category of robot arms.

This book is a springboard for readers of all backgrounds including students taking robotics as a subject, graduate students preparing to specialize in robotics.

This page left intentionally blank.

Contents

Preface	*v*
1. Robot	**1-60**
Summary	2
History of Robots	2
Modern Robots	11
Robots in Society	42
Roboethics	42
Contemporary Uses	45
Structure of Industrial Robots or Manipulators	53
Automatic Type Robot	55
The Brains and Body of a Robot	56
How Does a Robot's Voice Recognition System Works?	58
Position and Orientation of the Objects in Robotic Automation	59
2. Industrial Robot	**61-89**
Types and Features	62
History of Industrial Robotics	62
Technical Description	63
Recent and Future Developments	69
Market Structure	70
Dealing with Decommissioned Industrial Robots	71
Types of Industrial Robots	72
3. Robotics	**90-110**
Etymology	90
History of Robotics	91
Robotic Aspects	92
Components	94
Locomotion	99

Robotics Research	107
Education and Training	108
Employment	110

4. Artificial Intelligence — 111-131

History	112
Goals	113
Approaches	119
Tools	122
Evaluating Progress	125
Applications	126
Philosophy	127
Predictions and Ethics	128
Robots and Artificial Intelligence	129

5. Machine Vision — 132-136

Applications	132
Methods	132
Vision Guided Robotic Systems	134

6. Welding Robot Applications — 137-185

Robot arc Welding	137
Welding Safety	148
Welding Processes	148
Things to Know About Robotic Welding Systems	149
What You Must Know About Robotic Welding	153
Troubleshooting Robotic Welding	158
Welding in Today's Automotive Industry	163
Consumables for Robotic Welding	166
Handling, Installing, and Maintaining GMAW Consumables	171
Understanding MIG Welding Nozzles	175
Preventive Robotic MIG Gun Maintenance: The Whos, Whens, Whys and Hows	178
How Peripherals Can Maximize Your Robotic Welding Performance	180

7. Robot end Effector — 186-190

Mechanism of Gripping	186
Examples	187
Robotic Arm	188

8. Robotics Technology — 191-224

Levels of Processing	192
Switch Sensors	192
Light Sensors	193
Polarized Light	194
Resistive Position Sensors	194
Potentiometers	194
Biological Analogs	194
Break-beam Sensors	196
Effectors	203
Actuators	206
Controllers	211
Arms	217
Artificial Intelligence	219
Mobility	221

9. Robot Software — 225-232

Introduction	225
Industrial Robot Software	226
Examples of Programming Languages for Industrial Robots	226
Other Robot Programming Languages	227
Robot Application Software	227
Robot Programming Methods	228
Programmable Devices	230

10. Robot Programming Languages — 233-250

Which Language to Pick?	233
Writing Your Software	234
Robot Programming Languages	235
Action Language	236
Action Description Language	237
Enchanting	240
EUSLISP Robot Programming Language	240
LEJOS	241
ROBOTML	243
Next Byte Codes	243
Urbiscript	243
Variable Assembly Language	247

11. Robot Controllers — 251-256
- Programmable Controllers — 251
- Robot Controller — 252
- Different Levels of Robot Controller — 252
- Industrial Robot Control System — 255

12. Sensors and Actuators — 257-266
- Uses of sensors in Robotics — 257
- Electric Motors – AC Servomotors and Stepper Motors — 258
- Electric Motors – DC Servomotors — 259
- Robotic Actuators – Hydraulic and Pneumatic — 260
- Position Sensors — 261
- Velocity Sensors — 263
- Components Used for Robot Actuation and Feedback — 264
- Sensors Used in Robotics — 265

13. Grippers — 267-276
- Considerations in Robot Gripper Selection and Design — 267
- Miscellaneous Robot Gripper Types — 268
- Mechanical Gripper Mechanisms — 269
- Vacuum Grippers — 270
- Robot Magnetic Grippers — 271
- Mechanical Grippers — 273
- End Effector – Robot's Hand — 274
- Tools as End Effectors — 275

14. Safety Principles for Industrial Robots — 277-284
- Why Are Industrial Robots Dangerous? — 277
- Safety Goals for the Construction and Use of Industrial Robots — 281

15. Robots : Reshaping Manufacturing — 285-293
- Industrial Robots are Redefining Industry, Products and Working Practices — 285
- Robot Work Volume and its Comparisons — 288
- Machine Loading and Unloading by Industrial Robots — 289
- Types of Costs Required for Developing a Robot — 290
- Three Methods to Develop a Robot with Profit — 291
- Factors for Selecting a Robot Based on the Application — 292

16. ROBOT OPERATION **294-304**
   Robot Operating Techniques   294
   Robot Links and Joints   295
   Robot Drive Systems   296
   Robot Mechanical Transmission Systems   297
   Robotics/Feedback Sensors/Encoders   300

17. HUMAN–ROBOT INTERACTION **305-309**
   Origins   305
   The Goal of Friendly Human–Robot Interactions   306
   General HRI Research   307
   Application-oriented HRI Research   308

18. DECENTRALIZED SENSOR FUSION FOR UBIQUITOUS NETWORKING ROBOTICS IN URBAN AREAS **310-350**
   Alberto Sanfeliu, Juan Andrade-Cetto, Marco Barbosa, Richard Bowden, Jesus Capitan, Andreu Corominas, Andrew Gilbert, John Illingworth, Luis Merino, Josep M. Mirats, Plínio Moreno, Aníbal Ollero, João Sequeira and Matthijs T.J. Spaan
   Introduction   311
   The URUS Project   312
   The URUS Architecture   315
   Sensors in the Urban Site   317
   Sensors Included in Urban Robots   322
   Decentralized Sensor Fusion for Robotic Services   328
   Software Architecture to Manage Sensors Networks   330
   Some Results in the URUS Project   331
   Lessons Learned   345
   Conclusions   346
   References   347

This page left intentionally blank.

LIST OF CONTRIBUTORS

Alberto Sanfeliu
Institut de Robotica i Informatica Industrial, CSIC-UPC, Barcelona, Spain; E-Mails: cetto@iri.upc.edu (J.A.-C); acoromin@iri.upc.edu (A.C.); jmirats@iri.upc.edu (J.M.M.)

Juan Andrade-Cetto
Institut de Robotica i Informatica Industrial, CSIC-UPC, Barcelona, Spain; E-Mails: cetto@iri.upc.edu (J.A.-C); acoromin@iri.upc.edu (A.C.); jmirats@iri.upc.edu (J.M.M.)

Marco Barbosa
Instituto Superior Tecnico & Institute for Systems and Robotics, Lisbon, Portugal; E-Mails: mafb@isr.ist.utl.pt (M.B.); plinio@irs.ist.utl.pt (P.M.); jseq@isr.ist.utl.pt (J.S.); mtjspaan@isr.ist.utl.pt (M.TJ.S.)

Richard Bowden
Centre for Vision Speech and Signal Processing, University of Surrey, Guildford, UK; E-Mails: r.bowden@surrey.ac.uk (R.B.); a.gilbert@surrey.ac.uk (A.G.); j.illingworth@surrey.ac.uk (J.I.)

Jesus Capitan
Robotics, Vision and Control Group, University of Seville, Seville, Spain; E-Mail: aollero@cartuja.us.es

Andreu Corominas
Institut de Robotica i Informatica Industrial, CSIC-UPC, Barcelona, Spain; E-Mails: cetto@iri.upc.edu (J.A.-C); acoromin@iri.upc.edu (A.C.); jmirats@iri.upc.edu (J.M.M.)

Andrew Gilbert

Centre for Vision Speech and Signal Processing, University of Surrey, Guildford, UK; E-Mails: r.bowden@surrey.ac.uk (R.B.); a.gilbert@surrey.ac.uk (A.G.); j.illingworth@surrey.ac.uk (J.I.)

John Illingworth

Centre for Vision Speech and Signal Processing, University of Surrey, Guildford, UK; E-Mails: r.bowden@surrey.ac.uk (R.B.); a.gilbert@surrey.ac.uk (A.G.); j.illingworth@surrey.ac.uk (J.I.)

Luis Merino

Pablo de Olavide University, Seville, Spain; E-Mail: lmercab@upo.es

Josep M. Mirats

Institut de Robotica i Informatica Industrial, CSIC-UPC, Barcelona, Spain; E-Mails: cetto@iri.upc.edu (J.A.-C); acoromin@iri.upc.edu (A.C.); jmirats@iri.upc.edu (J.M.M.)

Plínio Moreno

Instituto Superior Tecnico & Institute for Systems and Robotics, Lisbon, Portugal; E-Mails: mafb@isr.ist.utl.pt (M.B.); plinio@irs.ist.utl.pt (P.M.); jseq@isr.ist.utl.pt (J.S.); mtjspaan@isr.ist.utl.pt (M.TJ.S.)

Aníbal Ollero

Robotics, Vision and Control Group, University of Seville, Seville, Spain; E-Mail: aollero@cartuja.us.es

and

Center for Advanced Aerospace Technology, Seville, Spain

João Sequeira

Instituto Superior Tecnico & Institute for Systems and Robotics, Lisbon, Portugal; E-Mails: mafb@isr.ist.utl.pt (M.B.); plinio@irs.ist.utl.pt (P.M.); jseq@isr.ist.utl.pt (J.S.); mtjspaan@isr.ist.utl.pt (M.TJ.S.)

Matthijs T.J. Spaan

Instituto Superior Tecnico & Institute for Systems and Robotics, Lisbon, Portugal; E-Mails: mafb@isr.ist.utl.pt (M.B.); plinio@irs.ist.utl.pt (P.M.); jseq@isr.ist.utl.pt (J.S.); mtjspaan@isr.ist.utl.pt (M.TJ.S.)

Chapter 1
ROBOT

A **robot** is a mechanical or virtual artificial agent, usually an electro-mechanical machine that is guided by a computer program or electronic circuitry. Robots can be autonomous or semi-autonomous and range from humanoids such as Honda's *Advanced Step in Innovative Mobility* (ASIMO) and TOSY's *TOSY Ping Pong Playing Robot* (TOPIO) to industrial robots, collectively programmed *swarm* robots, and even microscopic nano robots. By mimicking a lifelike appearance or automating movements, a robot may convey a sense of intelligence or thought of its own.

The branch of technology that deals with the design, construction, operation, and application of robots, as well as computer systems for their control, sensory feedback, and information processing is robotics. These technologies deal with automated machines that can take the place of humans in dangerous environments or manufacturing processes, or resemble humans in appearance, behavior, and/or cognition. Many of today's robots are inspired by nature contributing to the field of bio-inspired robotics. These robots have also created a newer branch of robotics : Soft robotics.

From the time of ancient civilization there have been many accounts of user-configurable automated devices and even automata resembling animals and humans, designed primarily as entertainment. As mechanical techniques developed through the Industrial age, there appeared more practical applications such as automated machines, remote-control and wireless remote-control. Electronics evolved into the driving force of development with the advent of the first electronic autonomous robots created by William Grey Walter in Bristol, England in 1948. The first digital and programmable robot was invented by George Devol in 1954 and was named the Unimate. It was sold to General Motors in 1961 where it was used to lift pieces of hot metal from die casting machines at the Inland Fisher Guide Plant in the West Trenton section of Ewing Township, New Jersey.

Robots have replaced humans in the assistance of performing those repetitive and dangerous tasks which humans prefer not to do, or are unable to do due to size limitations, or even those such as in outer space or at the bottom of the sea where humans could not survive the extreme environments.

There are concerns about the increasing use of robots and their role in society. Robots are blamed for rising unemployment as they replace workers in increasing numbers of functions. The use of robots in military combat raises ethical concerns. The possibilities of robot autonomy and potential repercussions have been addressed in fiction and may be a realistic concern in the future.

SUMMARY

The word *robot* can refer to both physical robots and virtual software agents, but the latter are usually referred to as bots. There is no consensus on which machines qualify as robots but there is general agreement among experts, and the public, that robots tend to do some or all of the following : move around, operate a mechanical limb, sense and manipulate their environment, and exhibit intelligent behavior — especially behavior which mimics humans or other animals. In practical terms, "robot" usually refers to a machine which can be electronically programmed to carry out a variety of physical tasks or actions.

There is no one definition of *robot* that satisfies everyone and many people have their own. For example Joseph Engelberger, a pioneer in industrial robotics, once remarked : "I can't define a robot, but I know one when I see one." The two ways that robots differ from actual beings are, simply stated, in the domain of cognition, and in the domain of biological form. The general consensus is that a "robot" is a machine and not a being simply because it is not intelligent (it requires programming to function), regardless of how human-like it may appear. In contrast, an imaginary "machine" or "artificial life form" (as in science fiction) that could think near or above human intelligence, and had a sensory body, would no longer be a "robot" but would be some kind of "artificial being" or "cognitive robot".

According to the Encyclopaedia Britannica a robot is "any automatically operated machine that replaces human effort, though it may not resemble human beings in appearance or perform functions in a humanlike manner." Merriam-Webster describes a robot as a "machine that looks like a human being and performs various complex acts (as walking or talking) of a human being", or a "device that automatically performs complicated often repetitive tasks", or a "mechanism guided by automatic controls".

HISTORY OF ROBOTS

The **history of robots** has its origins in the ancient world. The modern concept began to be developed with the onset of the Industrial Revolution which allowed for the use of complex mechanics and the subsequent introduction of electricity. This made it possible to power machines with small compact motors. In the early 20th century, the modern formulation of a humanoid machine was developed.

Today, it is now possible to envisage human sized robots with the capacity for near human thoughts and movement.

The first uses of modern robots were in factories as industrial robots – simple fixed machines capable of manufacturing tasks which allowed production without the need for human assistance. Digitally controlled industrial robots and robots making use of artificial intelligence have been built since the 1960s.

Ancient Mythology

Concepts of artificial servants and companions date at least as far back as the ancient legends of Cadmus, who sowed dragon teeth that turned into soldiers, and the myth of Pygmalion whose statue of Galatea came to life. Many ancient mythologies included artificial people, such as the talking mechanical handmaidens built by the Greek god Hephaestus (Vulcan to the Romans) out of gold, the clay golems of Jewish legend and clay giants of Norse legend. Chinese legend relates that in the 10th century BC, Yan Shi made an automaton resembling a human in an account from the *Lie Zi* text.

In Greek mythology, Hephaestus created utilitarian three-legged tables that could move about under their own power and a bronze man, Talos, that defended Crete. Talos was eventually destroyed by Media who cast a lightning bolt at his single vein of lead. To take the golden fleece Jason was also required to tame two fire breathing bulls with bronze hooves; and like Cadmus he sowed the teeth of a dragon into soldiers.

The Indian *Lokapannatti* (11th/12th century) tells the story of King Ajatasatru of Magadha who gathered the Buddhas relics and hid them in an underground stupa. The Buddhas relics were protected by mechanical robots, from the kingdom of Roma visaya; until they were disarmed by King Ashoka.

In the Islamic legend of Rocail, the younger brother of Seth created a palace and a sepulcher containing autonomous statues that lived out the lives of men so realistically they were mistaken for having souls.

According to Christian legends, the Doctor of the Church Albertus Magnus created a man of brass in the 13th century, that would only work under certain constellations. The brass man could respond to complicated questions and was employed as a domestic servant. The bronze man's capacity for intelligent speech deeply disturbed the theologian Thomas Aquinas, Magnus's pupil, and in a fit of rage, he beat the bronze man to pieces with a hammer.

Another famous medieval automaton legend is that of Roger Bacon. Bacon built statues that could move and was able to draw articulate sounds from a brass head, not with magic but through knowledge of the natural sciences. This story served as the basis for a popular English legend. Friar Bacon and Father Bungy sought to enclose England in a wall in order to make it inaccessible to invaders; summoning the devil they were told to construct a brazen head with all the internal structures and organs of a human head. This would take some time and they would have to wait for the head to learn speech. Eventually the head would answer any question asked of it; however after seven years of construction one night the head spoke "Time is" "Time was" and finally "Time has passed" before a bolt of lightning from a storm shattered the head into a thousand pieces.

Automata were popular in the imaginary worlds of medieval literature. For instance, the Middle Dutch tale *Roman van Walewein* ("The Romance of Walewein", early 13th century) described mechanical birds and angels producing sound by means of systems of pipes.

Early Beginnings

Concepts akin to a robot can be found as long ago as the 4th century BC, when the Greek mathematician Archytas of Tarentum postulated a mechanical bird he called "The Pigeon" which was propelled by steam. Yet another early automaton was the clepsydra, made in 250 BC by Ctesibius of Alexandria, a physicist and inventor from Ptolemaic Egypt. Hero of Alexandria made numerous innovations in the field of automata, including one that allegedly could speak.

Taking up the earlier reference in Homer's Iliad, Aristotle speculated in his *Politics* that automatons could someday bring about human equality by making possible the abolition of slavery :

- There is only one condition in which we can imagine managers not needing subordinates, and masters not needing slaves. This condition would be that each instrument could do its own work, at the word of command or by intelligent anticipation, like the statues of Daedalus or the tripods made by Hephaestus, of which Homer relates that "Of their own motion they entered the conclave of Gods on Olympus", as if a shuttle should weave of itself, and a plectrum should do its own harp playing.

In ancient China, an account on automata is found in the *Lie Zi* text, written in the 3rd century BC, in which King Mu of Zhou is presented with a life-size, human-shaped mechanical figure by Yan Shi, an "artificer".

The Cosmic Engine, a 10-metre (33 ft) clock tower built by Su Song in Kaifeng, China, in 1088, featured mechanical mannequins that chimed the hours, ringing gongs or bells among other devices.

Al-Jazari, a Muslim inventor during the Artuqid dynasty, designed and constructed a number of automatic machines, including kitchen appliances, musical automata powered by water, and the first programmable humanoid robot in 1206. Al-Jazari's robot was a boat with four automatic musicians that floated on a lake to entertain guests at royal drinking parties. His mechanism had a programmable drum machine with pegs (cams) that bump into little levers that operate the percussion. The drummer could be made to play different rhythms and different drum patterns by moving the pegs to different locations.

The early 13th century artist-engineer Villard de Honnecourt sketched plans for several automata. At the end of the thirteenth century, Robert II, Count of Artois, built a pleasure garden at his castle at Hesdin that incorporated a number of robots, humanoid and animal.

One of the first recorded designs of a humanoid robot was made by Leonardo da Vinci in around 1495. Da Vinci's notebooks, rediscovered in the 1950s, contain detailed drawings of a mechanical knight in armour which was able to sit up, wave its arms and move its head and jaw. The design is likely to be based on his anatomical research recorded in the *Vitruvian Man* but it is not known whether he attempted to build the robot. In 1533, Johannes Müller von Königsberg created an automaton eagle and fly made of iron; both could fly. John Dee is also known for creating a wooden beetle, capable of flying.

Around 1700, many automatons were built including ones capable of acting, drawing, flying, and playing music; some of the most famous works of the period were created by Jacques de Vaucanson in 1737, including an automaton flute player, tambourine player, and his most famous work, "The Digesting Duck". Vaucanson's duck was powered by weights and was capable of imitating a real duck by flapping its wings (over 400 parts were in each of the wings alone), eat grain, digest it, and defecate by excreting matter stored in a hidden compartment.

The Japanese craftsman Hisashige Tanaka, known as "Japan's Edison", created an array of extremely complex mechanical toys, some of which were capable of serving tea, firing arrows drawn from a quiver, or even painting a Japanese *kanji* character. The landmark text *Karakuri Zui* was published in 1796.

Remote-controlled Systems

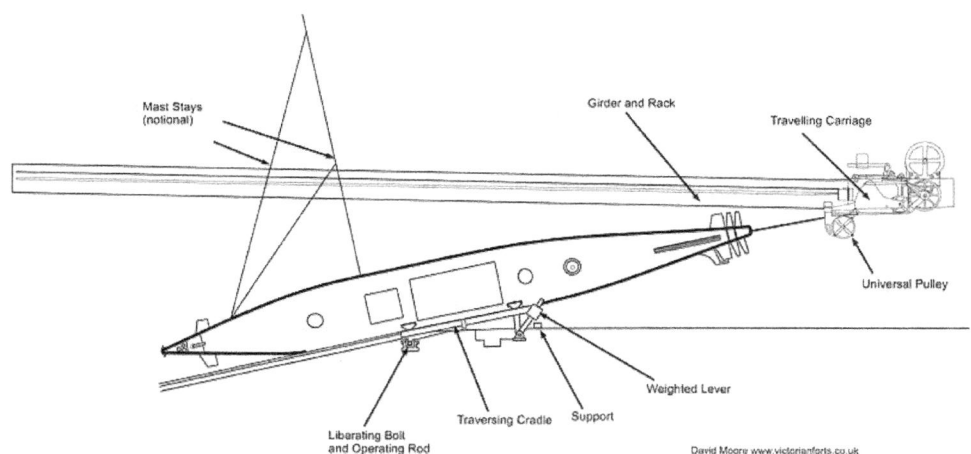

The Brennan torpedo, one of the earliest "guided missiles".

Remotely operated vehicles were demonstrated in the late 19th century in the form of several types of remotely controlled torpedos. The early 1870s saw remotely controlled torpedos by John Ericsson (pneumatic), John Louis Lay (electric wire guided), and Victor von Scheliha (electric wire guided).

The Brennan torpedo, invented by Louis Brennan in 1877 was powered by two contra-rotating propellors that were spun by rapidly pulling out wires from drums wound inside the torpedo. Differential speed on the wires connected to the shore station allowed the torpedo to be guided to its target, making it "the world's first *practical* guided missile". In 1898 Nikola Tesla publicly demonstrated a "wireless" radio-controlled torpedo that he hoped to sell to the U.S. Navy.

Archibald Low was known as the "father of radio guidance systems" for his pioneering work on guided rockets and planes during the First World War. In 1917, he demonstrated a remote controlled aircraft to the Royal Flying Corps and in the same year built the first wire-guided rocket.

Humanoid Robots

The term "robot" was first used to denote fictional automata in the 1921 play *R.U.R.* (Rossum's Universal Robots) by the Czech writer, Karel Čapek. According to Čapek, the word was created by his brother Josef from the Czech "robota", meaning servitude. In 1927, Fritz Lang's Metropolis was released; the Maschinenmensch ("machine-human"), a gynoid humanoid robot, also called "Parody", "Futura", "Robotrix", or the "Maria impersonator" (played by German actress Brigitte Helm), was the first robot ever to be depicted on film.

Many robots were constructed before the dawn of computer-controlled servomechanisms, for the public relations purposes of major firms. These were essentially machines that could perform a few stunts, like the automatons of the 18th century. In 1928, one of the first humanoid robots was exhibited at the annual exhibition of the Model Engineers Society in London. Invented by W. H. Richards, the robot Eric's frame consisted of an aluminium body of armour with eleven electromagnets and one motor powered by a twelve-volt power source. The robot could move its hands and head and could be controlled through remote control or voice control.

Westinghouse Electric Corporation built Televox in 1926 – it was a cardboard cutout connected to various devices which users could turn on and off. In 1939, the humanoid robot known as Elektro was debuted at the World's Fair. Seven feet tall (2.1 m) and weighing 265 pounds (120.2 kg), it could walk by voice command, speak about 700 words (using a 78-rpm record player), smoke cigarettes, blow up balloons, and move its head and arms. The body consisted of a steel gear cam and motor skeleton covered by an aluminum skin. In 1928, Japan's first robot, Gakutensoku, was designed and constructed by biologist Makoto Nishimura.

Modern Autonomous Robots

In 1941 and 1942, Isaac Asimov formulated the Three Laws of Robotics, and in the process of doing so, coined the word "robotics". In 1948, Norbert Wiener formulated the principles of cybernetics, the basis of practical robotics.

The first electronic autonomous robots with complex behaviour were created by William Grey Walter of the Burden Neurological Institute at Bristol, England in 1948 and 1949. He wanted to prove that rich connections between a small number of brain cells could give rise to very complex behaviors - essentially that the secret of how the brain worked lay in how it was wired up. His first robots, named *Elmer* and *Elsie*, were constructed between 1948 and 1949 and were often described as *tortoises* due to their shape and slow rate of movement. The three-wheeled tortoise robots were capable of phototaxis, by which they could find their way to a recharging station when they ran low on battery power.

Walter stressed the importance of using purely analogue electronics to simulate brain processes at a time when his contemporaries such as Alan Turing and John von Neumann were all turning towards a view of mental processes in terms of digital computation. His work inspired subsequent generations of robotics researchers such as Rodney Brooks, Hans Moravec and Mark Tilden. Modern incarnations of Walter's *turtles* may be found in the form of BEAM robotics.

The Turing test was proposed by British mathematician Alan Turing in his 1950 paper *Computing Machinery and Intelligence*, which opens with the words : "I propose to consider the question, 'Can machines think?'" The term "Artificial Intelligence was created at a conference held at Dartmouth College in 1956. Allen Newell, J. C. Shaw, and Herbert A. Simon pioneered the newly created artificial intelligence field with the Logic Theory Machine, and the General Problem Solver in 1957. In 1958, John McCarthy and Marvin Minsky started the MIT Artificial Intelligence lab with $50,000. John McCarthy also created LISP in the summer of 1958, a programming language still important in artificial intelligence research.

The first digitally operated and programmable robot was invented by George Devol in 1954 and was ultimately called the Unimate. This ultimately laid the foundations of the modern robotics industry. Devol sold the first Unimate to General Motors in 1960, and it was installed in 1961 in a plant in Trenton, New Jersey to lift hot pieces of metal from a die casting machine and stack them. Devol's patent for the first digitally operated programmable robotic arm represents the foundation of the modern robotics industry.

The Rancho Arm was developed as a robotic arm to help handicapped patients at the Rancho Los Amigos Hospital in Downey, California; this computer controlled arm was bought by Stanford University in 1963. IBM announced its IBM System/360 in 1964. The system was heralded as being more powerful, faster, and more capable than its predecessors.

The film 2001 : A Space Odyssey was released in 1968; the movie prominently features HAL 9000, a malevolent artificial intelligence unit which controls a spacecraft. Marvin Minsky created the Tentacle Arm in 1968; the arm was computer controlled and its 12 joints were powered by hydraulics. Mechanical Engineering student Victor Scheinman created the Stanford Arm in 1969; the Stanford Arm is recognized as the first electronic computer controlled robotic arm (Unimate's instructions were stored on a magnetic drum). The first mobile robot capable of reasoning about its surroundings, Shakey was built in 1970 by the Stanford Research Institute (now SRI International). Shakey combined multiple sensor inputs, including TV cameras, laser rangefinders, and "bump sensors" to navigate. In the winter of 1970, the Soviet Union explored the surface of the moon with the lunar vehicle Lunokhod 1, the first roving remote-controlled robot to land on another world.

1970s

Artificial intelligence critic Hubert Dreyfuss published his influential book *What Computers Cannot Do* in 1972. Freddy and Freddy II, both built in the United Kingdom, were robots capable of assembling wooden blocks in a period of several hours. German based company KUKA built the world's first industrial robot with six electromechanically driven axes, known as FAMULUS. In 1974, David Silver designed The Silver Arm; the Silver Arm was capable of fine movements replicating human hands. Feedback was provided by touch and pressure sensors and analyzed by a computer. Marvin Minsky published his landmark paper "A Framework for Representing Knowledge" on artificial intelligence.

Joseph Weizenbaum (creator of ELIZA, a program capable of simulating a Rogerian psychotherapist) published Computer Power and Human Reason, presenting an argument against the creation of artificial intelligence. The SCARA, Selective Compliance Assembly Robot Arm, was created in 1978 as an efficient, 4-axis robotic arm. Best used for picking up parts and placing them in another location, the SCARA was introduced to assembly lines in 1981. XCON, an expert system designed to customize orders for industrial use, was released in 1979. The Stanford Cart successfully crossed a room full of chairs in 1979. The Stanford Cart relied primarily on stereo vision to navigate and determine distances. The Robotics Institute at Carnegie Mellon University was founded in 1979 by Raj Reddy.

1980s

Takeo Kanade created the first "direct drive arm" in 1981. The first of its kind, the arm's motors were contained within the robot itself, eliminating long transmissions. Cyc, a project to create a database of common sense for artificial intelligence, was started in 1984 by Douglas Leant. The program attempts to deal with ambiguity in language, and is still underway. The first program to publish a book, the expert system Racter, programmed by William Chamberlain and Thomas Etter, wrote the book "The Policeman's Beard is Half-Constructed" in 1983. It is now thought that a system of complex templates were used.

In 1984 Wabot-2 was revealed; capable of playing the organ, Wabot-2 had 10 fingers and two feet. Wabot-2 was able to read a score of music and accompany a person. Chess playing programs HiTech and Deep Thought defeated chess masters in 1989. Both were developed by Carnegie Mellon University; Deep Thought development paved the way for the Deep Blue.

In 1986, Honda began its humanoid research and development program to create robots capable of interacting successfully with humans. A hexapodal robot named Genghis was revealed by MIT in 1989. Genghis was famous for being made quickly and cheaply due to construction methods; Genghis used 4 microprocessors, 22 sensors, and 12 servo motors. Rodney Brooks and Anita M. Flynn published "Fast, Cheap, and Out of Control : A Robot Invasion of The Solar System". The paper advocated creating smaller cheaper robots in greater numbers to increase production time and decrease the difficulty of launching robots into space.

1990s

The biomimetic robot RoboTuna was built by doctoral student David Barrett at the Massachusetts Institute of Technology in 1996 to study how fish swim in water. RoboTuna is designed to swim and resemble a blue fin tuna. Invented by Dr. John Adler, in 1994, the Cyberknife (a stereotactic radiosurgery performing robot) offered an alternative treatment of tumors with a comparable accuracy to surgery performed by human doctors.

Honda's P2 humanoid robot was first shown in 1996. Standing for "Prototype Model 2", P2 was an integral part of Honda's humanoid development project; over 6 feet tall, P2 was smaller than its predecessors and appeared to be more human-like in its motions. Expected to only operate for seven days, the Sojourner rover finally shuts down after 83 days of operation in 1997. This small robot (only weighing 23 lbs) performed semi-autonomous operations on the surface of Mars as part of the Mars Pathfinder mission; equipped with an obstacle avoidance program, Sojourner was capable of planning and navigating routes to study the surface of the planet. Sojourner's ability to navigate with little data about its environment and nearby surroundings allowed the robot to react to unplanned events and objects. Also in 1997, IBM's chess playing program Deep Blue beat the then current World Chess Champion Garry Kasparov playing at the "Grandmaster" level. The super computer was a specialized version of a framework produced by IBM, and was capable of processing twice as many moves per second as it had during the first match, reportedly 200,000,000 moves per second. The event was broadcast live over the internet and received over 74 million hits.

The P3 humanoid robot was revealed by Honda in 1998 as a part of the company's continuing humanoid project. In 1999, Sony introduced the AIBO, a robotic dog capable of interacting with humans, the first models released in Japan sold out in 20 minutes. Honda revealed the most advanced result of their humanoid project in 2000, named ASIMO. ASIMO is capable of running, walking, communication with humans, facial and environmental recognition, voice and posture recognition, and interacting with its environment. Sony also revealed its Sony Dream Robots, small humanoid robots in development for entertainment. In October 2000, the United Nations estimated that there were 742,500 industrial robots in the world, with more than half of the robots being used in Japan.

2001

In April 2001, the Canadarm2 was launched into orbit and attached to the International Space Station. The Canadarm2 is a larger, more capable version of the arm used by the Space Shuttle and is hailed as being "smarter." Also in April, the Unmanned Aerial Vehicle Global Hawk made the first autonomous non-stop flight over the Pacific Ocean from Edwards Air Force Base in California to RAAF Base Edinburgh in Southern Australia. The flight was made in 22 hours. The popular Roomba, a robotic vacuum cleaner, was first released in 2002 by the company iRobot.

In 2004, Cornell University revealed a robot capable of self-replication; a set of cubes capable of attaching and detaching, the first robot capable of building copies of itself. On 3 and 24 January the Mars rovers Spirit and Opportunity land on the surface of Mars. Launched in 2003, the two robots will drive many times the distance originally expected, and Opportunity is still operating as of mid 2012.

Self-driving cars had made their appearance by the middle of the first decade of the 21st century, but there was room for improvement. All 15 teams competing in the 2004 DARPA Grand Challenge failed to complete the course, with no robot successfully navigating more than five percent of the 150 mile off road course, leaving the $1 million prize unclaimed. In 2005, Honda revealed a new version of its ASIMO robot, updated with new behaviors and capabilities. In 2006, Cornell University revealed its "Starfish" robot, a 4-legged robot capable of self modeling and learning to walk after having been damaged. In 2007, TOMY launched the entertainment robot, i-sobot, which is a humanoid bipedal robot that can walk like a human beings and performs kicks and punches and also some entertaining tricks and special actions under "Special Action Mode".

Robonaut 2, the latest generation of the astronaut helpers, launched to the space station aboard Space Shuttle Discovery on the STS-133 mission. It is the first humanoid robot in space, and although its primary job for now is teaching engineers how dexterous robots behave in space, the hope is that through upgrades and advancements, it could one day venture outside the station to help spacewalkers make repairs or additions to the station or perform scientific work.

Commercial and industrial robots are now in widespread use performing jobs more cheaply or with greater accuracy and reliability than humans. They are also employed for jobs which are too dirty, dangerous or dull to be suitable for humans. Robots are widely used in manufacturing, assembly and packing, transport, earth and space exploration, surgery, weaponry, laboratory research, and mass production of consumer and industrial goods.

MODERN ROBOTS

Mobile Robot

A **mobile robot** is an automatic machine that is capable of locomotion.

A spying robot is an example of a mobile robot capable of movement in a given environment.

Mobile robots have the capability to move around in their environment and are not fixed to one physical location. By contrast, industrial robots are usually more-or-less stationary, consisting of a jointed arm (multi-linked manipulator) and gripper assembly (or end effector), attached to a fixed surface.

Mobile robots are a major focus of current research and almost every major university has one or more labs that focus on mobile robot research. Mobile robots are also found in industrial, military and security settings. Domestic robots are consumer products, including entertainment robots and those that perform certain household tasks such as vacuuming or gardening.

Classification

Mobile robots may be classified by :
- The environment in which they travel :
  - Land or home robots are usually referred to as Unmanned Ground Vehicles (UGVs). They are most commonly wheeled or tracked, but also include legged robots with two or more legs (humanoid, or resembling animals or insects).
  - Aerial robots are usually referred to as Unmanned Aerial Vehicles (UAVs)

- o Underwater robots are usually called autonomous underwater vehicles (AUVs)
- o Polar robots, designed to navigate icy, crevasse filled environments
- The device they use to move, mainly :
  - o Legged robot : human-like legs (*i.e.* an android) or animal-like legs.
  - o Wheeled robot.
  - o Tracks.

Mobile Robot Navigation

There are many types of mobile robot navigation :

Manual Remote or Tele-op

A manually teleoperated robot is totally under control of a driver with a joystick or other control device. The device may be plugged directly into the robot, may be a wireless joystick, or may be an accessory to a wireless computer or other controller. A tele-op'd robot is typically used to keep the operator out of harm's way. Examples of manual remote robots include Robotics Design's ANATROLLER ARI-100 and ARI-50, Foster-Miller's Talon, iRobot's PackBot, and KumoTek's MK-705 Roosterbot.

Guarded Tele-op

A guarded tele-op robot has the ability to sense and avoid obstacles but will otherwise navigate as driven, like a robot under manual tele-op. Few if any mobile robots offer only guarded tele-op.

Line-following Car

Some of the earliest Automated Guided Vehicles (AGVs) were line following mobile robots. They might follow a visual line painted or embedded in the floor or ceiling or an electrical wire in the floor. Most of these robots operated a simple "keep the line in the center sensor" algorithm. They could not circumnavigate obstacles; they just stopped and waited when something blocked their path. Many examples of such vehicles are still sold, by Transbotics, FMC, Egemin, HK Systems and many other companies.

Autonomously Randomized Robot

Autonomous robots with random motion basically bounce off walls, whether those walls are sensed

Autonomously Guided Robot

An autonomously guided robot knows at least some information about where it is and how to reach various goals and or waypoints along the way. "Localization" or knowledge of its current location, is calculated by one or more means, using sensors such motor encoders, vision, Stereopsis, lasers and global positioning systems. Positioning systems often use triangulation, relative position and/or Monte-Carlo/Markov localization to determine the location and orientation of the platform, from which it can plan a path to its next waypoint or goal. It can gather sensor readings that are time- and location-stamped, so that a hospital, for instance, can know exactly when and where radiation levels exceeded permissible levels. Such robots are often part of the wireless enterprise network, interfaced with other sensing and control systems in the building. For instance, the PatrolBot security robot responds to alarms, operates elevators and notifies the command center when an incident arises. Other autonomously guided robots include the SpeciMinder and the Tug delivery robots for hospital labs, though the latter actually has people at the ready to drive the robot remotely when its autonomy fails. The Tug sends a letter to its tech support person, who then takes the helm and steers it over the Internet by looking through a camera low in the base of the robot.In 2013, Autonomous movement controlled by plants was achieved by artist Elizabeth Demaray and engineer Dr. Qingze during the IndaPlant Project, and act of trans-species giving. They successfully created a part-robot, part-plant entity that allows a potted-plant to freely seek sunlight and water.

Sliding Autonomy

More capable robots combine multiple levels of navigation under a system called sliding autonomy. Most autonomously guided robots, such as the HelpMate hospital robot, also offer a manual mode. The Motivity autonomous robot operating system, which is used in the ADAM, PatrolBot, SpeciMinder, MapperBot and a number of other robots, offers full sliding autonomy, from manual to guarded to autonomous modes.

History

Date	Developments
1939–1945	During World War II the first mobile robots emerged as a result of technical advances on a number of relatively new research fields like computer science and cybernetics. They were mostly flying bombs. Examples are smart bombs that only detonate within a certain range of the target, the use of guiding systems and radar control. The V1 and V2 rockets had a crude 'autopilot' and automatic detonation systems. They were the predecessors of modern cruise missiles.
1948–1949	W. Grey Walter builds Elmer and Elsie, two autonomous robots called Machina Speculatrix because these robots liked to explore their environment. Elmer and Elsie were each equipped with a light sensor. If they found a light source they would move towards it, avoiding or moving obstacles on their way. These robots demonstrated that complex behaviour could arise from a simple design. Elmer and Elsie only had the equivalent of two nerve cells.
1961–1963	The Johns Hopkins University develops 'Beast'. Beast used a sonar to move around. When its batteries ran low it would find a power socket and plug itself in.
1969	Mowbot was the very first robot that would automatically mow the lawn.

1970	The Stanford Cart line follower was a mobile robot that was able to follow a white line, using a camera to see. It was radio linked to a large mainframe that made the calculations. At about the same time the Stanford Research Institute is building and doing research on Shakey the Robot, a robot named after its jerky motion. Shakey had a camera, a rangefinder, bump sensors and a radio link. Shakey was the first robot that could reason about its actions. This means that Shakey could be given very general commands, and that the robot would figure out the necessary steps to accomplish the given task. The Soviet Union explores the surface of the Moon with Lunokhod 1, a lunar rover.
1976	In its Viking program the NASA sends two unmanned spacecraft to Mars.
1980	The interest of the public in robots rises, resulting in robots that could be purchased for home use. These robots served entertainment or educational purposes. Examples include the RB5X, which still exists today and the HERO series. The Stanford Cart is now able to navigate its way through obstacle courses and make maps of its environment.
Early 1980s	The team of Ernst Dickmanns at Bundeswehr University Munich builds the first robot cars, driving up to 55 mph on empty streets.
1987	Hughes Research Laboratories demonstrates the first cross-country map and sensor-based autonomous operation of a robotic vehicle.
1989	Mark Tilden invents BEAM robotics.
1990s	Joseph Engelberger, father of the industrial robotic arm, works with colleagues to design the first commercially available autonomous mobile hospital robots, sold by Helpmate. The US Department of Defense funds the MDARS-I project, based on the Cybermotion indoor security robot.
1991	Edo. Franzi, André Guignard and Francesco Mondada developed Khepera, an autonomous small mobile robot intended for research activities. The project was supported by the LAMI-EPFL lab.
1993–1994	Dante I and Dante II were developed by Carnegie Mellon University. Both were walking robots used to explore live volcanoes.
1994	With guests on board, the twin robot vehicles VaMP and VITA-2 of Daimler-Benz and Ernst Dickmanns of UniBwM drive more than one thousand kilometers on a Paris three-lane highway in standard heavy traffic at speeds up to 130 km/h. They demonstrate autonomous driving in free lanes, convoy driving, and lane changes left and right with autonomous passing of other cars.
1995	Semi-autonomous ALVINN steered a car coast-to-coast under computer control for all but about 50 of the 2850 miles. Throttle and brakes, however, were controlled by a human driver.
1995	In the same year, one of Ernst Dickmanns' robot cars (with robot-controlled throttle and brakes) drove more than 1000 miles from Munich to Copenhagen and back, in traffic, at up to 120 mph, occasionally executing maneuvers to pass other cars (only in a few critical situations a safety driver took over). Active vision was used to deal with rapidly changing street scenes.
1995	The Pioneer programmable mobile robot becomes commercially available at an affordable price, enabling a widespread increase in robotics research and university study over the next decade as mobile robotics becomes a standard part of the university curriculum.
1996–1997	NASA sends the Mars Pathfinder with its rover Sojourner to Mars. The rover explores the surface, commanded from earth. Sojourner was equipped with a hazard avoidance system. This enabled Sojourner to autonomously find its way through unknown martian terrain.

1999	Sony introduces Aibo, a robotic dog capable of seeing, walking and interacting with its environment. The PackBot remote-controlled military mobile robot is introduced.
2001	Start of the Swarm-bots project. Swarm bots resemble insect colonies. Typically they consist of a large number of individual simple robots, that can interact with each other and together perform complex tasks.
2002	Appears Roomba, a domestic autonomous mobile robot that cleans the floor.
2003	Axxon Robotics purchases Intellibot, manufacturer of a line of commercial robots that scrub, vacuum, and sweep floors in hospitals, office buildings and other commercial buildings. Floor care robots from Intellibot Robotics LLC operate completely autonomously, mapping their environment and using an array of sensors for navigation an obstacle avoidance.
2004	Robosapien, a biomorphic toy robot designed by Mark Tilden is commercially available. In 'The Centibots Project' 100 autonomous robots work together to make a map of an unknown environment and search for objects within the environment. In the first DARPA Grand Challenge competition, fully autonomous vehicles compete against each other on a desert course.
2005	Boston Dynamics creates a quadruped robot intended to carry heavy loads across terrain too rough for vehicles.
2006	Sony stops making Aibo and HelpMate halts production, but a lower-cost PatrolBot customizable autonomous service robot system becomes available as mobile robots continue the struggle to become commercially viable. The US Department of Defense drops the MDARS-I project, but funds MDARS-E, an autonomous field robot. TALON-Sword, the first commercially available robot with grenade launcher and other integrated weapons options, is released. Honda's Asimo learns to run and climb stairs.
2007	In the DARPA Urban Grand Challenge, six vehicles autonomously comple a complex course involving manned vehicles and obstacles. Kiva Systems clever robots proliferate in distribution operations; these smart shelving units sort themselves according to the popularity of their contents. The Tug becomes a popular means for hospitals to move large cabinets of stock from place to place, while the Speci-Minder with Motivity begins carrying blood and other patient samples from nurses' stations to various labs. Seekur, the first widely available, non-military outdoor service robot, pulls a 3-ton vehicle across a parking lot, drives autonomously indoors and begins learning how to navigate itself outside. Meanwhile, PatrolBot learns to follow people and detect doors that are ajar.
2008	Boston Dynamics released video footage of a new generation BigDog able to walk on icy terrain and recover its balance when kicked from the side.
2010	The Multi Autonomous Ground-robotic International Challenge has teams of autonomous vehicles map a large dynamic urban environment, identify and track humans and avoid hostile objects.

Automated Guided Vehicle

An **automated guided vehicle** or **automatic guided vehicle** (AGV) is a mobile robot that follows markers or wires in the floor, or uses vision, magnets, or lasers for navigation. They are most often used in industrial applications to move materials around a manufacturing facility or a warehouse. Application of the automatic guided vehicle has broadened during the late 20th century.

Introduction

Automated guided vehicles (AGVs) increase efficiency and reduce costs by helping to automate a manufacturing facility or warehouse. The first AGV was invented by Barrett Electronics in 1953. The AGV *can tow objects* behind them in trailers to which they can autonomously attach. The trailers can be used to move raw materials or finished product. The AGV can also store objects on a bed. The objects can be placed on a set of motorized rollers (conveyor) and then pushed off by reversing them. AGVs are employed in nearly every industry, including, pulp, paper, metals, newspaper, and general manufacturing. Transporting materials such as food, linen or medicine in hospitals is also done.

An AGV can also be called a laser guided vehicle (LGV). In Germany the technology is also called *Fahrerlose Transportsysteme* (FTS) and in Sweden *förarlösa truckar*. Lower cost versions of AGVs are often called Automated Guided Carts (AGCs) and are usually guided by magnetic tape. AGCs are available in a variety of models and can be used to move products on an assembly line, transport goods throughout a plant or warehouse, and deliver loads.

The first AGV was brought to market in the 1950s, by Barrett Electronics of Northbrook, Illinois, and at the time it was simply a tow truck that followed a wire in the floor instead of a rail. In 1976, Egemin Automation (Holland, MI) started working on the development of an automatic driverless control system for use in several industrial and commercial applications. Out of this technology came a new type of AGV, which follows invisible UV markers on the floor instead of being towed by a chain. The first such system was deployed at the Willis Tower in Chicago, Illinois to deliver mail throughout its offices.

Over the years the technology has become more sophisticated and today automated vehicles are mainly Laser navigated *e.g.* LGV (Laser Guided Vehicle). In an automated process, LGVs are programmed to communicate with other robots to ensure product is moved smoothly through the warehouse, whether it is being stored for future use or sent directly to shipping areas. Today, the AGV plays an important role in the design of new factories and warehouses, safely moving goods to their rightful destination.

Navigation

Wired

A slot is cut in to the floor and a wire is placed approximately 1 inch below the surface. This slot is cut along the path the AGV is to follow. This wire is used to transmit a radio signal. A sensor is installed on the bottom of the AGV close to the ground. The sensor detects the relative position of the radio signal being transmitted from the wire. This information is used to regulate the steering circuit, making the AGV follow the wire.

Guide tape

AGVs (some known as automated guided carts or AGCs) use tape for the guide path. The tapes can be one of two styles : magnetic or colored. The AGC is fitted with the appropriate guide sensor to follow the path of the tape. One major advantage of tape over wired guidance is that it can be easily removed and relocated if the course needs to change. Colored tape is initially less expensive, but lacks the advantage of being embedded in high traffic areas where the tape may become damaged or dirty. A flexible magnetic bar can also be embedded in the floor like wire but works under the same provision as magnetic tape and so remains unpowered or passive. Another advantage of magnetic guide tape is the dual polarity. small pieces of magnetic tape may be placed to change states of the AGC based on polarity and sequence of the tags.

Laser Target Navigation

The navigation is done by mounting reflective tape on walls, poles or fixed machines. The AGV carries a laser transmitter and receiver on a rotating turret. The laser is transmitted and received by the same sensor. The angle and (sometimes) distance to any reflectors that in line of sight and in range are automatically calculated. This information is compared to the map of the reflector layout stored in the AGV's memory. This allows the navigation system to triangulate the current position of the AGV. The current position is compared to the path programmed in to the reflector layout map. The steering is adjusted accordingly to keep the AGV on track. It can then navigate to a desired target using the constantly updating position.

- Modulated Lasers The use of modulated laser light gives greater range and accuracy over pulsed laser systems. By emitting a continuous fan of modulated laser light a system can obtain an uninterrupted reflection as soon as the scanner achieves line of sight with a reflector. The reflection ceases at the trailing edge of the reflector which ensures an accurate and consistent measurement from every reflector on every scan. By using a modulated laser a system can achieve an angular resolution of ~ 0.1 mrad (0.006°) at 8 scanner revolutions per second.
- Pulsed Lasers A typical pulsed laser scanner emits pulsed laser light at a rate of 14,400 Hz which gives a maximum possible resolution of ~ 3.5 mrad (0.2°) at 8 scanner revolutions per second. To achieve a workable navigation, the readings must be interpolated based on the intensity of the reflected laser light, to identify the centre of the reflector.

Inertial (Gyroscopic) Navigation

Another form of an AGV guidance is inertial navigation. With inertial guidance, a computer control system directs and assigns tasks to the vehicles. Transponders are embedded in the floor of the work place. The AGV uses these transponders to verify that the vehicle is on course. A gyroscope is able to detect the slightest change in the direction of the vehicle and corrects it in order to keep the AGV on its path. The margin of error for the inertial method is ±1 inch.

Inertial can operate in nearly any environment including tight aisles or extreme temperatures. Inertial navigation can include use of magnets embedded in the floor of the facility that the vehicle can read and follow.

Unit-load AGV using natural-features navigation to carry steel to quality assurance lab

Natural Features (Natural Targeting) Navigation

Navigation without retrofitting of the workspace is called Natural Features or Natural Targeting Navigation. One method uses one or more range-finding sensors, such as a laser range-finder, as well as gyroscopes or inertial measurement units with Monte-Carlo/Markov localization techniques to understand where it is as it dynamically plans the shortest permitted path to its goal. The advantage of such systems is that they are highly flexible for on-demand delivery to any location. They can handle failure without bringing down the entire manufacturing operation, since AGVs can plan paths around the failed device. They also are quick to install, with less down-time for the factory.

Steering Control

To help an AGV navigate it can use three different steer control systems. The differential speed control is the most common. In this method there are two independent drive wheels. Each drive is driven at different speeds in order to turn or the same speed to allow the AGV to go forwards or backwards. The AGV turns in a similar fashion to a tank. This method of steering is the simplest as it does not require additional steering motors and mechanism. More often than not, this is seen on an AGV that is used to transport and turn in tight spaces or when the

AGV is working near machines. This setup for the wheels is not used in towing applications because the AGV would cause the trailer to jackknife when it turned.

The second type of steering used is steered wheel control AGV. This type of steering can be similar to a cars steering. But this is not very manoeuvrable. It is more common to use a three wheeled vehicle similar to a conventional three wheeled forklift. The drive wheel is the turning wheel. It is more precise in following the programmed path than the differential speed controlled method. This type of AGV has smoother turning. Steered wheel control AGV can be used in all applications; unlike the differential controlled. Steered wheel control is used for towing and can also at times have an operator control it.

The third type is a combination of differential and steered. Two independent steer/drive motors are placed on diagonal corners of the AGV and swivelling castors are placed on the other corners. It can turn like a car (rotating in an arc) in any direction. It can crab in any direction and it can drive in differential mode in any direction.

Vision Guidance

Vision-Guided AGVs can be installed with no modifications to the environment or infrastructure. They operate by using cameras to record features along the route, allowing the AGV to replay the route by using the recorded features to navigate. Vision-Guided AGVs use Evidence Grid technology, an application of probabilistic volumetric sensing, and was invented and initially developed by Dr. Moravec at Carnegie Mellon University. The Evidence Grid technology uses probabilities of occupancy for each point in space to compensate for the uncertainty in the performance of sensors and in the environment. The primary navigation sensors are specially designed stereo cameras. The vision-guided AGV uses 360-degree images and build a 3D map, which allows the vision-guided AGVs to follow a trained route without human assistance or the addition of special features, landmarks or positioning systems.

Geoguidance

A geoguided AGV recognizes its environment to establish its location. Without any infrastructure, the forklift equipped with geoguidance technology detects and identifies columns, racks and walls within the warehouse. Using these fixed references, it can position itself, in real time and determine its route. There are no limitations on distances to cover or number of pick-up or drop-off locations. Routes are infinitely modifiable.

Path Decision

AGVs have to make decisions on path selection. This is done through different methods : frequency select mode (wired navigation only), and path select mode (wireless navigation only) or via a magnetic tape on the floor not only to guide the AGV but also to issue steering commands and speed commands.

Frequency Select Mode

Frequency select mode bases its decision on the frequencies being emitted from the floor. When an AGV approaches a point on the wire which splits the AGV detects the two frequencies and through a table stored in its memory decides on the best path. The different frequencies are required only at the decision point for the AGV. The frequencies can change back to one set signal after this point. This method is not easily expandable and requires extra cutting meaning more money.

Path select mode

An AGV using the path select mode chooses a path based on preprogrammed paths. It uses the measurements taken from the sensors and compares them to values given to them by programmers. When an AGV approaches a decision point it only has to decide whether to follow path 1, 2, 3, *etc*. This decision is rather simple since it already knows its path from its programming. This method can increase the cost of an AGV because it is required to have a team of programmers to program the AGV with the correct paths and change the paths when necessary. This method is easy to change and set up.

Magnetic tape mode

The magnetic tape is laid on the surface of the floor or buried in a 10mm channel; not only does it provide the path for the AGV to follow but also strips of the tape in different combos of polarity, sequence, and distance laid alongside the track tell the AGV to change lane, speed up, slow down, and stop.

Traffic control

Flexible manufacturing systems containing more than one AGV may require it to have traffic control so the AGV's will not run into one another. Traffic control can be carried out locally or by software running on a fixed computer elsewhere in the facility. Local methods include zone control, forward sensing control, and combination control. Each method has its advantages and disadvantages.

Zone control

Zone control is the favorite of most environments because it is simple to install and easy to expand. Zone control uses a wireless transmitter to transmit a signal in a fixed area. Each AGV contains a sensing device to receive this signal and transmit back to the transmitter. If the area is clear the signal is set at "clear" allowing any AGV to enter and pass through the area. When an AGV is in the area the "stop" signal is sent and all AGV attempting to enter the area stop and wait for their turn. Once the AGV in the zone has moved out beyond the zone the "clear" signal is sent to one of the waiting AGVs. Another way to set up zone control traffic management is to equip each individual robot with its own small transmitter/receiver. The individual AGV then sends its own "do not enter" message to all the AGVs getting to close to its zone in the area. A problem with this method is if one zone goes down all the AGV's are at risk to collide with any other AGV. Zone control is a cost efficient way to control the AGV in an area.

Forward Sensing Control

Forward sensing control uses collision avoidance sensors to avoid collisions with other AGV in the area. These sensors include : sonic, which work like radar; optical, which uses an infrared sensor; and bumper, physical contact sensor. Most AGVs are equipped with a bumper sensor of some sort as a fail safe. Sonic sensors send a "chirp" or high frequency signal out and then wait for a reply from the outline of the reply the AGV can determine if an object is ahead of it and take the necessary actions to avoid collision. The optical uses an infrared transmitter/receiver and sends an infrared signal which then gets reflected back; working on a similar concept as the sonic sensor. The problems with these are they can only protect the AGV from so many sides. They are relatively hard to install and work with as well.

Combination Control

Combination control sensing is using collision avoidance sensors as well as the zone control sensors. The combination of the two helps to prevent collisions in any situation. For normal operation the zone control is used with the collision avoidance as a fail safe. For example, if the zone control system is down, the collision avoidance system would prevent the AGV from colliding.

System Management

Industries with AGVs need to have some sort of control over the AGVs. There are three main ways to control the AGV : locator panel, CRT color graphics display, and central logging and report.

A locator panel is a simple panel used to see which area the AGV is in. If the AGV is in one area for too long, it could mean it is stuck or broken down. CRT color graphics display shows real time where each vehicle is. It also gives a status of the AGV, its battery voltage, unique identifier, and can show blocked spots. Central logging used to keep track of the history of all the AGVs in the system. Central logging stores all the data and history from these vehicles which can be printed out for technical support or logged to check for up time.

AGV is a system often used in FMS to keep up, transport, and connect smaller subsystems into one large production unit. AGVs employ a lot of technology to ensure they do not hit one another and make sure they get to their destination. Loading and transportation of materials from one area to another is the main task of the AGV. AGV require a lot of money to get started with, but they do their jobs with high efficiency. In places such as Japan automation has increased and is now considered to be twice as efficient as factories in America. For a huge initial cost the total cost over time decreases.

Vehicle Types

AGVS Towing Vehicles were the first type introduced and are still a very popular type today. Towing vehicles can pull a multitude of trailer types and have capacities ranging from 8,000 pounds to 60,000 pounds. *AGVS Unit Load Vehicles* are equipped with decks, which permit unit load transportation and often automatic load transfer. The decks can either be lift and lower type, powered or non-powered roller, chain or belt decks or custom decks with multiple compartments.

- AGVS Pallet Trucks are designed to transport palletized loads to and from floor level; eliminating the need for fixed load stands.
- AGVS Fork Truck has the ability to service loads both at floor level and on stands. In some cases these vehicles can also stack loads in rack.

Forklift AGV with Stabilizer Pad

- AGVS Hybrid Vehicles are adapted from a standard CAT-style man-aboard truck so that they can run fully automated or be driven by a fork truck driver. These can be used for trailer loading as well as moving materials around warehouses. Most often, they are equipped with forks, but can be customized to accommodate most load types.

Hybrid AGV picking load

- Light Load AGVS are vehicles which have capacities in the neighborhood of 500 pounds or less and are used to transport small parts, baskets, or other light loads though a light manufacturing environment. They are designed to operate in areas with limited space.
- AGVS Assembly Line Vehicles are an adaptation of the light load AGVS for applications involving serial assembly processes.

Common AGV Applications

Automated Guided Vehicles can be used in a wide variety of applications to transport many different types of material including pallets, rolls, racks, carts, and containers. AGVs excel in applications with the following characteristics :

- Repetitive movement of materials over a distance
- Regular delivery of stable loads
- Medium throughput/volume
- When on-time delivery is critical and late deliveries are causing inefficiency
- Operations with at least two shifts
- Processes where tracking material is important

Raw Materials Handling

AGVs are commonly used to transport raw materials such as paper, steel, rubber, metal, and plastic. This includes transporting materials from receiving to the warehouse, and delivering materials directly to production lines.

Work-in-process Movement

Work-in-Process movement is one of the first applications where automated guided vehicles were used, and includes the repetitive movement of materials throughout the manufacturing process. AGVs can be used to move material from the warehouse to production/processing lines or from one process to another.

Pallet Handling

Pallet handling is an extremely popular application for AGVs as repetitive movement of pallets is very common in manufacturing and distribution facilities. AGVs can move pallets from the palletizer to stretch wrapping to the warehouse/ storage or to the outbound shipping docks.

Finished Product Handling

Moving finished goods from manufacturing to storage or shipping is the final movement of materials before they are delivered to customers. These movements often require the gentlest material handling because the products are complete and subject to damage from rough handling. Because AGVs operate with precisely controlled navigation and acceleration and deceleration this minimizes the potential for damage making them an excellent choice for this type of application

Trailer Loading

Automatic loading of trailers is a relatively new application for automated guided vehicles and becoming increasingly popular. AGVs are used to transport and load pallets of finished goods directly into standard, over-the-road trailers without any special dock equipment. AGVs can pick up pallets from conveyors, racking, or staging lanes and deliver them into the trailer in the specified loading pattern. Some Automatic Trailer Loading AGVs utilize Natural Targeting to view the walls of the trailer for navigation. These types of ATL AGVs can be either completely driverless or hybrid CAT-based vehicles.

Roll Handling

AGVs are used to transport rolls in many types of plants including paper mills, converters, printers, newspapers, steel producers, and plastics manufacturers. AGVs can store and stack rolls on the floor, in racking, and can even automatically load printing presses with rolls of paper.

AGVs are used to move sea containers in some maritime container terminals. The main benefits are reduced labour costs and a more reliable performance. This use of AGVs was pioneered by ECT in the Netherlands at the Delta terminal in the Port of Rotterdam.

Primary Application Industries

Efficient, cost effective movement of materials is an important, and common element in improving operations in many manufacturing plants and warehouses. Because automatic guided vehicles (AGVs) can delivery efficient, cost effective movement of materials, AGVs can be applied to various industries in standard or customized designs to best suit an industry's requirements. Industry's currently utilizing AGVs include (but are not limited to) :

[2] A forktruck vehicle delivering a pallet of finished goods

Pharmaceutical

AGVs are a preferred method of moving materials in the pharmaceutical industry. Because an AGV system tracks all movement provided by the AGVs, it supports process validation and cGMP (current Good Manufacturing Practice).

Chemical

AGVs deliver raw materials, move materials to curing storage warehouses, and provide transportation to other processing cells and stations. Common industries include rubber, plastics, and specialty chemicals.

Manufacturing

AGVs are often used in general manufacturing of products. AGVs can typically be found delivering raw materials, transporting work-in process, moving finished goods, removing scrap materials, and supplying packaging materials.

Automotive

AGV installations are found in Stamping Plants, Power Train (Engine and Transmission) Plants, and Assembly Plants delivering raw materials, transporting work-in process, and moving finished goods. AGVs are also used to supply specialized tooling which must be changed.

Paper and Print

AGVs can move paper rolls, pallets, and waste bins to provide all routine material movement in the production and warehousing (storage/retrieval) of paper, newspaper, printing, corrugating, converting, and plastic film.

Food and Beverage

AGVs can be applied to move materials in food processing (such as the loading of food or trays into sterilizers) and at the "end of line," linking the palletizer, stretch wrapper, and the warehouse. AGVs can load standard, over-the-road trailers with finished goods, and unload trailers to supply raw materials or packaging materials to the plant. AGVs can also store and retrieve pallets in the warehouse.

Hospital

AGVs are becoming increasingly popular in the healthcare industry for efficient transport, and are programmed to be fully integrated to automatically operate doors, elevators/lifts, cart washers, trash dumpers, *etc.* AGVs typically move linens, trash, regulated medical waste, patient meals, soiled food trays, and surgical case carts.

Warehousing

AGVs used in Warehouses and Distribution Centers logically move loads around the warehouses and prepare them for shipping/loading or receiving or move them from an induction conveyor to logical storage locations within the warehouse. Often, this type of use is accompanied by customized warehouse management software.

Battery Charging

AGVs utilize a number of battery charging options. Each option is dependent on the users preference. The most commonly used battery charging technologies are *Battery Swap, Automatic/Opportunity Charging,* and *Automatic Battery Swap.*

Battery Swap

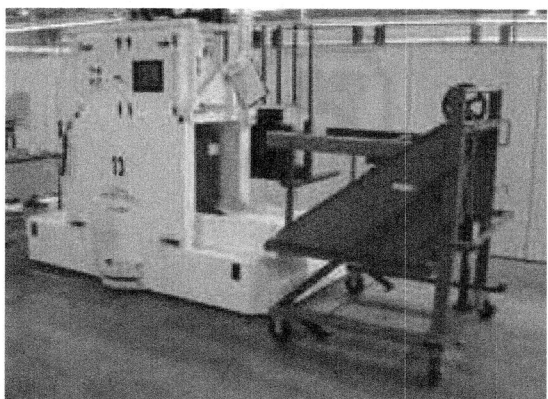

"*Battery swap technology*" requires an operator to manually remove the discharged battery from the AGV and place a fully charged battery in its place after approximately 8 – 12 hours (about one shift) of AGVs operation. 5 – 10 minutes is required to perform this with each AGV in the fleet.

Automatic and Opportunity Charging

"*Automatic and opportunity battery charging*" allows for continuous operation. On average an AGV charges for 12 minutes every hour for automatic charging and no manual intervention is required. If opportunity is being utilized the AGV will receive a charge whenever the opportunity arises. When a battery pack gets to a predetermined level the AGV will finish the current job that it has been assigned before it goes to the charging station.

Automatic Battery Swap

"*Automatic battery swap*" is an alternative to manual battery swap. It requires an additional piece of automation machinery, an automatic battery changer, to the overall AGV system. AGVs will pull up to the battery swap station and have their batteries automatically replaced with fully charged batteries. The automatic battery changer then places the removed batteries into a charging slot for automatic recharging. The automatic battery changer keeps track of the batteries in the system and pulls them only when they are fully charged.

While a battery swap system reduces the manpower required to swap batteries, recent developments in battery charging technology allow batteries to be charged more quickly and efficiently potentially eliminating the need to swap batteries.

Service Robot

Service robots assist human beings, typically by performing a job that is dirty, dull, distant, dangerous or repetitive, including household chores. They typically are autonomous and/or operated by a built-in control system, with manual override options. The term "service robot" is less well-defined. The International Federation of Robotics (IFR) has proposed a tentative definition, "A service robot is a robot which operates semi- or fully autonomously to perform services useful to the well-being of humans and equipment, excluding manufacturing operations."

Types

The possible applications of robots to assist in human chores is widespread. At present there are a number of main categories that these robots fall into.

Industrial

Industrial service robots can be used to carry out simple tasks, such as examining welding, as well as more complex, harsh-environment tasks, such as aiding in the dismantling of nuclear power stations. If the robot is an automatically controlled, reprogrammable, multipurpose manipulator programmable in three or more axes, which may be either fixed in place or mobile for use in industrial automation applications. It is called "Industrial Robot".

Restaurant and Bar

Many bars are starting to become automated through the use of robots, even producing complex cocktails. There are also robots used for Waiting.

Domestic

The Roomba vacuum cleaner is one of the most popular domestic service robots.

Domestic robots perform tasks that humans regularly perform around their homes such as cleaning floors, mowing the lawn and pool maintenance. They can also provide assistance to the disabled and infirm as well as becoming robot butlers.

Scientific

Robotic systems perform many functions such as repetitive tasks performed in research. These range from the multiple repetitive tasks made by gene samplers and sequencers, to systems which can almost replace the scientist in designing and running experiments, analysing data and even forming hypotheses. The ADAM at the University of Aberystwyth in Wales can "[make] logical assumptions based on information programmed into it about yeast metabolism and the way proteins and genes work in other species. It then set about proving that its predictions were correct."

Autonomous scientific robots perform tasks which humans would find difficult or impossible, from the deep sea to outer space. The Woods Hole Sentry can descend to 4,500 metres and allows a higher payload as it does not need a support ship or the oxygen and other facilities demanded by human piloted vessels. Robots in space include the Mars rovers which could carry out sampling and photography in the harsh environment of the atmosphere on Mars.

Self-reconfiguring Modular Robot

Modular self-reconfiguring robotic systems or **self-reconfigurable modular robots** are autonomous kinematic machines with variable morphology. Beyond conventional actuation, sensing and control typically found in fixed-morphology robots, self-reconfiguring robots are also able to deliberately change their own shape by rearranging the connectivity of their parts, in order to adapt to new circumstances, perform new tasks, or recover from damage.

For example, a robot made of such components could assume a worm-like shape to move through a narrow pipe, reassemble into something with spider-like legs to cross uneven terrain, then form a third arbitrary object (like a ball or wheel that can spin itself) to move quickly over a fairly flat terrain; it can also be used for making "fixed" objects, such as walls, shelters, or buildings.

In some cases this involves each module having 2 or more connectors for connecting several together. They can contain electronics, sensors, computer processors, memory, and power supplies; they can also contain actuators that are used for manipulating their location in the environment and in relation with each other. A feature found in some cases is the ability of the modules to automatically connect and disconnect themselves to and from each other, and to form into many objects or perform many tasks moving or manipulating the environment.

By saying "self-reconfiguring" or "self-reconfigurable" it means that the mechanism or device is capable of utilizing its own system of control such as with actuators or stochastic means to change its overall structural shape. Having the quality of being "modular" in "self-reconfiguring modular robotics" is to say that the same module or set of modules can be added to or removed from the system, as opposed to being generically "modularized" in the broader sense. The underlying intent is to have an indefinite number of identical modules, or a finite and relatively small set of identical modules, in a mesh or matrix structure of self-reconfigurable modules.

Self-reconfiguration is also different from the concept of self-replication, and self-replication is not necessarily a quality that a self-reconfigurable module or collection of such modules can or must possess. A matrix of N-number of modules does not need to be able to increase the quantity of modules to greater than N to be considered self-reconfigurable. It is sufficient for self-reconfigurable modules to be a device that is produced at a conventional factory, where dedicated machines stamp or mold components, and factory workers on an assembly line assemble the components to build each module.

There are two basic types of methods of segment articulation that self-reconfigurable mechanisms can utilize to reshape their structures, chain reconfiguration and lattice reconfiguration.

Structure and Control

Modular robots are usually composed of multiple building blocks of a relatively small repertoire, with uniform docking interfaces that allow transfer of mechanical forces and moments, electrical power and communication throughout the robot.

The modular building blocks usually consist of some primary structural actuated unit, and potentially additional specialized units such as grippers, feet, wheels, cameras, payload and energy storage and generation.

A taxonomy of Architectures

- Modular self-reconfiguring robotic systems can be generally classified into several architectural groups by the geometric arrangement of their unit (lattice vs. chain). Several systems exhibit hybrid properties, and modular robots have also been classified into the two categories of Mobile Configuration Change (MCC) and Whole Body Locomotion (WBL). Lattice architecture have their units connecting their docking interfaces at points into virtual cells of some regular grid. This network of docking points can be compared to atoms in a crystal and the grid to the lattice of that crystal. Therefore the kinematical features of lattice robots can be characterized by their corresponding crystallographic displacement groups. Usually few units are sufficient to accomplish a reconfiguration step. Lattice architectures allows a simpler mechanical design and a simpler computational representation and reconfiguration planning that can be more easily scaled to complex systems.

- Chain architecture do not use a virtual network of docking points for their units. The units are able to reach any point in the space and are therefore more versatile, but a chain of many units may be necessary to reach a point making it usually more difficult to accomplish a reconfiguration step. Such systems are also more computationally difficult to represent and analyze.

- Hybrid architecture takes advantages of both previous architectures. Control and mechanism are designed for lattice reconfiguration but also allow to reach any point in the space.

Modular robotic systems can also be classified according to the way by which units are reconfigured (moved) into place.

- Deterministic reconfiguration relies on units moving or being directly manipulated into their target location during reconfiguration. The exact location of each unit is known at all times. Reconfiguration times can be guaranteed, but sophisticated feedback control is necessary to assure precise manipulation. Macro-scale systems are usually deterministic.
- Stochastic reconfiguration relies on units moving around using statistical processes (like Brownian motion). The exact location of each unit only known when it is connected to the main structure, but it may take unknown paths to move between locations. Reconfiguration times can be guaranteed only statistically. Stochastic architectures are more favorable at micro scales.

Modular robotic systems are also generally classified depending on the design of the modules.

- Homogeneous modular robot systems have many modules of the same design forming a structure suitable to perform the required task. An advantage over other systems is that they are simple to scale in size (and possibly function), by adding more units. A commonly described disadvantage is limits to functionality - these systems often require more modules to achieve a given function, than heterogeneous systems.
- Heterogeneous modular robot systems have different modules, each of which do specialized functions, forming a structure suitable to perform a task. An advantage is compactness, and the versatility to design and add units to perform any task. A commonly described disadvantage is an increase in complexity of design, manufacturing, and simulation methods.

Other modular robotic systems exist which are not self-reconfigurable, and thus do not formally belong to this family of robots though they may have similar appearance. For example, self-assembling systems may be composed of multiple modules but cannot dynamically control their target shape. Similarly, tensegrity robotics may be composed of multiple interchangeable modules but cannot self-reconfigure.

Motivation and Inspiration

There are two key motivations for designing modular self-reconfiguring robotic systems.

- Functional advantage : Self reconfiguring robotic systems are potentially more robust and more adaptive than conventional systems. The reconfiguration ability allows a robot or a group of robots to disassemble and reassemble machines to form new morphologies that are better suitable for new tasks, such as changing from a legged robot to a snake robot and then to a rolling robot. Since robot parts are interchangeable (within a robot and between different robots), machines can also replace faulty parts autonomously, leading to self-repair.

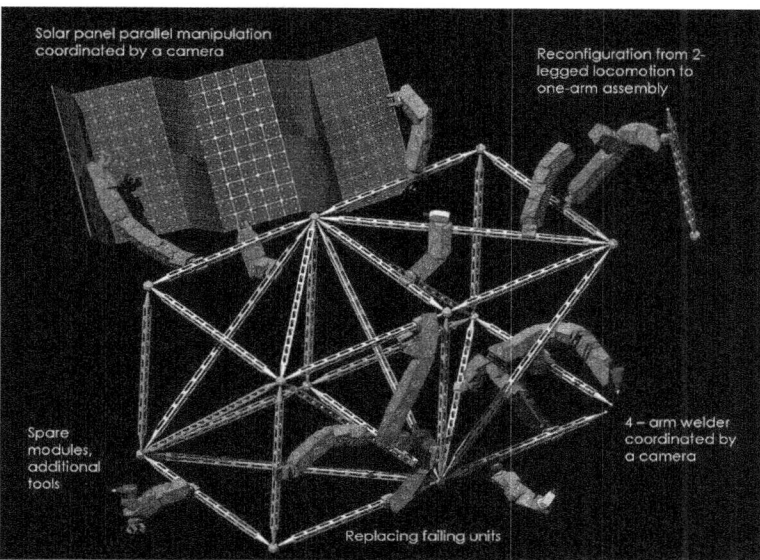

Autonomous modular robotics in space

- Economic advantage : Self reconfiguring robotic systems can potentially lower overall robot cost by making a range of complex machines out of a single (or relatively few) types of mass-produced modules.

Both these advantages have not yet been fully realized. A modular robot is likely to be inferior in performance to any single custom robot tailored for a specific task. However, the advantage of modular robotics is only apparent when considering multiple tasks that would normally require a set of different robots.

The added degrees of freedom make modular robots more versatile in their potential capabilities, but also incur a performance tradeoff and increased mechanical and computational complexities.

The quest for self-reconfiguring robotic structures is to some extent inspired by envisioned applications such as long-term space missions, that require long-term self-sustaining robotic ecology that can handle unforeseen situations and may require self repair. A second source of inspiration are biological systems that are self-constructed out of a relatively small repertoire of lower-level building blocks (cells or amino acids, depending on scale of interest). This architecture underlies biological systems' ability to physically adapt, grow, heal, and even self replicate – capabilities that would be desirable in many engineered systems.

Application Areas

Given these advantages, where would a modular self-reconfigurable system be used? While the system has the promise of being capable of doing a wide variety of things, finding the "killer application" has been somewhat elusive. Here are several examples :

Space Exploration

One application that highlights the advantages of self-reconfigurable systems is long-term space missions. These require long-term self-sustaining robotic ecology that can handle unforeseen situations and may require self repair. Self-reconfigurable systems have the ability to handle tasks that are not known a prioritise especially compared to fixed configuration systems. In addition, space missions are highly volume and mass constrained. Sending a robot system that can reconfigure to achieve many tasks is better than sending many robots that each can do one task.

Telepario

Another example of an application has been coined "telepario" by CMU professors Todd Mowry and Seth Goldstein. What the researchers propose to make are moving, physical, three-dimensional replicas of people or objects, so lifelike that human senses would accept them as real. This would eliminate the need for cumbersome virtual reality gear and overcome the viewing angle limitations of modern 3D approaches. The replicas would mimic the shape and appearance of a person or object being imaged in real time, and as the originals moved, so would their replicas. One aspect of this application is that the main development thrust is geometric representation rather than applying forces to the environment as in a typical robotic manipulation task. This project is widely known as claytronics or Programmable matter (noting that programmable matter is a much more general term, encompassing functional programmable materials, as well).

Bucket of Stuff

A third long term vision for these systems has been called "bucket of stuff". In this vision, consumers of the future have a container of self-reconfigurable modules say in their garage, basement, or attic. When the need arises, the consumer calls forth the robots to achieve a task such as "clean the gutters" or "change the oil in the car" and the robot assumes the shape needed and does the task. One source of inspiration for the development of these systems comes from the application. A second source is biological systems that are self-constructed out of a relatively small repertoire of lower-level building blocks (cells or amino acids, depending on scale of interest). This architecture underlies biological systems' ability to physically adapt, grow, heal, and even self replicate – capabilities that would be desirable in many engineered systems.

History and State of the Art

The roots of the concept of modular self-reconfigurable robots can be traced back to the "quick change" end effector and automatic tool changers in computer numerical controlled machining centers in the 1970s. Here, special modules each with a common connection mechanism could be automatically swapped out on the end of a robotic arm. However, taking the basic concept of the common connection mechanism and applying it to the whole robot was introduced by Toshio Fukuda with the CEBOT (short for cellular robot) in the late 1980s.

The early 1990s saw further development from Greg Chirikjian, Mark Yim, Joseph Michael, and Satoshi Murata. Chirikjian, Michael, and Murata developed lattice reconfiguration systems and Yim developed a chain based system. While these researchers started with from a mechanical engineering emphasis, designing and building modules then developing code to program them, the work of Daniela Rus and Wei-min Shen developed hardware but had a greater impact on the programming aspects. They started a trend towards provable or verifiable distributed algorithms for the control of large numbers of modules.

One of the more interesting hardware platforms recently has been the MTRAN II and III systems developed by Satoshi Murata *et al.* This system is a hybrid chain and lattice system. It has the advantage of being able to achieve tasks more easily like chain systems, yet reconfigure like a lattice system.

More recently new efforts in stochastic self-assembly have been pursued by Hod Lipson and Eric Klavins. A large effort at CMU headed by Seth Goldstein and Todd Mowry has started looking at issues in developing millions of modules.

Many tasks have been shown to be achievable, especially with chain reconfiguration modules. This demonstrates the versatility of these systems however, the other two advantages, robustness and low cost have not been demonstrated. In general the prototype systems developed in the labs have been fragile and expensive as would be expected during any initial development.

There is a growing number of research groups actively involved in modular robotics research. To date, about 30 systems have been designed and constructed, some of which are shown below.

Some Current Systems

A chain self-reconfiguration system. Each module is about 50 mm on a side, and has 1 rotational DOF. It is part of the PolyBot modular robot family that has demonstrated many modes of locomotion including walking : biped, 14 legged, slinky-like, snake-like : concertina in a gopher hole, inchworm gaits, rectilinear undulation and sidewinding gaits, rolling like a tread at up to 1.4 m/s, riding a tricycle, climbing : stairs, poles pipes, ramps *etc.*

M-TRAN III

A hybrid type self-reconfigurable system. Each module is two cube size (65 mm side), and has 2 rotational DOF and 6 flat surfaces for connection. It is the 3rd M-TRAN prototypes. Compared with the former (M-TRAN II), speed and reliability of connection is largely improved. As a chain type system, locomotion by CPG (Central Pattern Generator) controller in vaious shapes has been demonstrated by M-TRAN II. As a lattice type system, it can change its configuration, *e.g.*, between a 4 legged walker to a caterpillar like robot.

AMOEBA-I, a three-module reconfigurable mobile robot was developed in Shenyang Institute of Automation (SIA), Chinese Academy of Sciences (CAS) by Liu J G et al..AMOEBA-I has nine kinds of non-isomorphic configurations and high mobility under unstructured environments.Four generations of its platform have been developed and a series of researches have been carried out on their reconfiguration mechanism, non-isomorphic configurations, tipover stability, and reconfiguration planning. Experiments have demonstrated that such kind structure permits good mobility and high flexibility to uneven terrain. Being hyper-redundant, modularized and reconfigurable, AMOEBA-I has many possible applications such as Urban Search and Rescue (USAR) and space exploration.

Stochastic-3D

High spatial resolution for arbitrary three-dimensional shape formation with modular robots can be accomplished using lattice system with large quantities of very small, prospectively microscopic modules. At small scales, and with large quantities of modules, deterministic control over reconfiguration of individual modules will become unfeasible, while stochastic mechanisms will naturally prevail. Microscopic size of modules will make the use of electromagnetic actuation and interconnection prohibitive, as well, as the use of on-board power storage.

Three large scale prototypes were built in attempt to demonstrate dynamically programmable three-dimensional stochastic reconfiguration in a neutral-buoyancy environment. The first prototype used electromagnets for module reconfiguration and interconnection. The modules were 100 mm cubes and weighed 0.81 kg. The second prototype used stochastic fluidic reconfiguration and interconnection mechanism. Its 130 mm cubic modules weighed 1.78 kg each and made reconfiguration experiments excessively slow. The current third implementation inherits the fluidic reconfiguration principle. The lattice grid size is 80 mm, and the reconfiguration experiments are under way.

Molecubes

This hybrid self-reconfiguring system was built by the Cornell Computational Synthesis Lab to physically demonstrate artificial kinematic self-reproduction. Each module is a 0.65 kg cube with 100 mm long edges and one rotational degree of freedom. The axis of rotation is aligned with the cube's longest diagonal. Physical self-reproduction of a three- and a four-module robots was demonstrated. It was also shown that, disregarding the gravity constraints, an infinite number of self-reproducing chain meta-structures can be built from Molecubes.

The Programmable Parts

The programmable parts are stirred randomly on an air-hockey table by randomly actuated air jets. When they collide and stick, they can communicate and decide whether to stay stuck, or if and when to detach. Local interaction rules can be devised and optimized to guide the robots to make any desired global shape.

SuperBot

The SuperBot modules fall into the hybrid architecture. The modules have three degrees of freedom each. The design is based on two previous systems : Conro (by the same research group) and MTRAN. Each module can connect to another module through one of its six dock connectors. They can communicate and share power through their dock connectors. Several locomotion gaits have been developed for different arrangements of modules. For high-level communication the modules use hormone-based control, a distributed, scalable protocol that does not require the modules to have unique ID's.

Miche

The Miche system is a modular lattice system capable of arbitrary shape formation. Each module is an autonomous robot module capable of connecting to and communicating with its immediate neighbors. When assembled into a structure, the modules form a system that can be virtually sculpted using a computer interface and a distributed process. The group of modules collectively decide who is on the final shape and who is not using algorithms that minimize the information transmission and storage. Finally, the modules not in the structure let go and fall off under the control of an external force, in this case gravity.

The Distributed Flight Array The Distributed Flight Array is a modular robot consisting of hexagonal-shaped single-rotor units that can take on just about any shape or form. Although each unit is capable of generating enough thrust to lift itself off the ground, on its own it is incapable of flight much like a helicopter cannot fly without its tail rotor. However, when joined together, these units evolve into a sophisticated multi-rotor system capable of coordinated flight and much more.

Roombots

Roombots have a hybrid architecture. Each module has three degree of freedom, two of them using the diametrical axis within a regular cube, and a third (center) axis of rotation connecting the two spherical parts. All three axes are continuously rotatory. The outer Roombots DOF is using the same axis-orientation as Molecubes, the third, central Roombots axis enables the module to rotate its two outer DOF against each other. This novel feature enables a single Roombots module to locomote on flat terrain, but also to climb a wall, or to cross a concave, perpendicular edge. Convex edges require the assembly of at least two modules into a Roombots "Metamodule". Each module has ten available connector slots, currently two of them are equipped with an active connection mechanism based on mechanical latches. Roombots are designed for two tasks : to eventually shape objects of daily life, *e.g.* furniture, and to locomote, *e.g.* as a quadruped or a tripod robot made from multiple modules.

Sambot

Being inspired form social insects, multicellar organism and mophogenetic robots. The aim of the Sambot is to develop swarm robotics and conduct research on the swarm intelligenc, self-assembly and co-evolution of the body and brain for autonomous morphogeneous. Differing from swarm robot, self-reconfigurable robot and morphgenetic robot, the research focuses on self-assembly swarm modular robots that interact and dock as an autonomous mobile module with others to achieve swarm intelligence and furtherly discuss the autonomous construction in space station and exploratary tools and artificial complex structures. Each Sambot robot can run as an autonomos individual in wheel and besides, using combination of the sensors and docking mechanism, the robot can interact and dock with the environments and other robots. By the advantage of motion and connection, Sambot swarms can aggregate into a symbiotic or whole organism and generate locomotion as the bionic articular robots. In this case, some self-assembling, self-organizing, self-reconfigurating, and self-repairing function and research are available in design and appliction view. Inside the modular robot whose size is 80(W)X80(L)X102 mm, MCU (ARM and AVR), communication (Zigbee), sensors, power, IMU, positioning modules are embeded.

Moteins

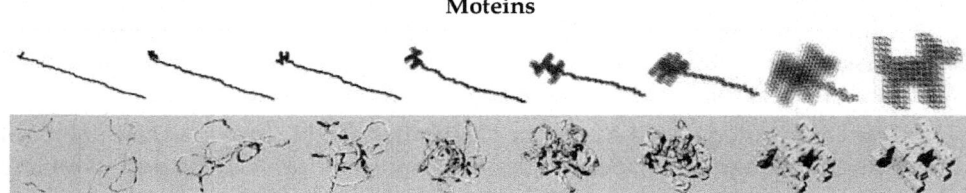

Motein

It is mathematically proven that physical strings of simple shapes can be folded into any continuous area or volumetric shape. Moteins employ such shape-universal folding strategies, with one or two degrees of freedom and simple actuators with only two or three states.

Quantitative Accomplishment

- The robot with most active modules has 56 units <polybot centipede, PARC>
- The smallest actuated modular unit has a size of 12 mm <smart pebble, MIT>
- The largest actuated modular unit (by volume) has the size of 8 m^3 <(GHFC)giant helium filled catoms, CMU>
- The strongest actuation modules are able to lift 5 identical horizontally cantilevered units.<PolyBot g1v5, PARC>
- The fastest modular robot can move at 23 unit-sizes/second.<CKbot, dynamic rolling, ISER'06>

- The largest simulated system contained many 100,000's of units.
- Challenges, solutions, and opportunities

Since the early demonstrations of early modular self-reconfiguring systems, the size, robustness and performance has been continuously improving. In parallel, planning and control algorithms have been progressing to handle thousands of units. There are, however, several key steps that are necessary for these systems to realize their promise of *adaptability, robustness and low cost*. These steps can be broken down into challenges in the hardware design, in planning and control algorithms and in application. These challenges are often intertwined.

Hardware Design Challenges

The extent to which the promise of self-reconfiguring robotic systems can be realized depends critically on the numbers of modules in the system. To date, only systems with up to about 50 units have been demonstrated, with this number stagnating over almost a decade. There are a number of fundamental limiting factors that govern this number :

- Limits on strength, precision, and field robustness (both mechanical and electrical) of bonding/docking interfaces between modules
- Limits on motor power, motion precision and energetic efficiency of units, (*i.e.* specific power, specific torque)
- Hardware/software design. Hardware that is designed to make the software problem easier. Self-reconfiguring systems have more tightly coupled hardware and software than any other existing system.

Planning and Control Challenges

Though algorithms have been developed for handling thousands of units in ideal conditions, challenges to scalability remain both in low-level control and high-level planning to overcome realistic constraints :

- Algorithms for parallel-motion for large scale manipulation and locomotion
- Algorithms for robustly handling a variety of failure modes, from misalignments, dead-units (not responding, not releasing) to units that behave erratically.
- Algorithms that determine the optimal configuration for a given task
- Algorithms for optimal (time, energy) reconfiguration plan
- Efficient and scalable (asynchronous) communication among multiple units

Application Challenges

Though the advantages of Modular self-reconfiguring robotic systems is largely recognized, it has been difficult to identify specific application domains where benefits can be demonstrated in the short term. Some suggested applications are

- Space exploration and Space colonization applications, *e.g.* Lunar colonization
- Construction of large architectural systems
- Deep sea exploration/mining
- Search and rescue in unstructured environments
- Rapid construction of arbitrary tools under space/weight constraints
- Disaster relief shelters for displaced peoples
- Shelters for impoverished areas which require little on-the-ground expertise to assemble

Grand Challenges

Several robotic fields have identified "Grand Challenges" that act as a catalyst for development and serve as a short-term goal in absence of immediate "killer apps". The Grand Challenge is not in itself a research agenda or milestone, but a means to stimulate and evaluate coordinated progress across multiple technical frontiers. Several Grand Challenges have been proposed for the modular self-reconfiguring robotics field :

- Demonstration of a system with >1000 units. Physical demonstration of such a system will inevitably require rethinking key hardware and algorithmic issues, as well as handling noise and error.
- Robosphere. A self-sustaining robotic ecology, isolated for a long period of time (1 year) that needs to sustain operation and accomplish unforeseen tasks without any human presence.
- Self replication A system with many units capable of self replication by collecting scattered building blocks will require solving many of the hardware and algorithmic challenges.
- Ultimate Construction A system capable of making objects out of the components of, say, a wall.
- Biofilter analogy If the system is ever made small enough to be injected into a mammal, one task may be to monitor molecules in the blood stream and allow some to pass and others not to, somewhat like the Blood–brain barrier. As a challenge, an analogy may be made where system must be able to :
    o be inserted into a hole one module's diameter.
    o travel some specified distance in a channel that is say roughly 40 x 40 module diameters in area.
    o form a barrier fully conforming to the channel (whose shape is non-regular, and unknown beforehand).

o   allow some objects to pass and others not to (not based on size).
o   Since sensing is not the emphasis of this work, the actual detection of the passable objects should be made trivial.

Inductive Transducers

A unique potential solution that can be exploited is the use of inductors as transducers. This could be useful for dealing with docking and bonding problems. At the same time it could also be beneficial for its capabilities of docking detection (alignment and finding distance), power transmission, and (data signal) communication. A proof-of-concept video can be seen here. The rather limited exploration down this avenue is probably a consequence of the historical lack of need in any applications for such an approach.

Modular Robotics Google Group

Modular Robotics Google Group is an open public forum dedicated to announcements of events in the field of Modular Robotics. This medium is used to disseminate calls to workshops, special issues and other academic activities of interest to modular robotics researchers. The founders of this Google group intend it to facilitate the exchange of information and ideas within the community of modular robotics researchers around the world and thus promote acceleration of advancements in modular robotics. Anybody who is interested in objectives and progress of Modular Robotics can join this Google group and learn about the new developments in this field.

Educational Robot

Robots are used as educational assistants to teachers. From the 1980s, robots such as turtles were used in schools and programmed using the Logo language.

There are robot kits like Lego Mindstorms, BIOLOID, OLLO from ROBOTIS, or BotBrain Educational Robots can help children to learn about mathematics, physics, programming, and electronics. Robotics have also been introduced into the lives of elementary and high school students in the form of robot competitions with the company FIRST (For Inspiration and Recognition of Science and Technology). The organization is the foundation for the FIRST Robotics Competition, FIRST LEGO League, Junior FIRST LEGO League, and FIRST Tech Challenge competitions.

There have also been devices shaped like robots such as the teaching computer, Leachim, and 2-XL, a robot shaped game/ teaching toy based on an 8-track tape player, both invented Michael J. Freeman.

Collaborative Robots

A *collaborative robot* or *cobot* is a robot that can safely and effectively interact with human workers while performing simple industrial tasks. However, end-effectors and other environmental conditions may create hazards, and as such risk assessments should be done before using any industrial motion-control application.

The collaborative robots most widely used in industries today are manufactured by Universal Robots in Denmark.

Rethink Robotics—founded by Rodney Brooks, previously with iRobot—introduced Baxter in September 2012; as an industrial robot designed to safely interact with neighboring human workers, and be programmable for performing simple tasks. Baxters stop if they detect a human in the way of their robotic arms and have prominent off switches. Intended for sale to small businesses, they are promoted as the robotic analogue of the personal computer. As of May 2014, 190 companies in the US have bought Baxters and they are being used commercially in the UK.

Military Robots

Some experts and academics have questioned the use of robots for military combat, especially when such robots are given some degree of autonomous functions. There are also concerns about technology which might allow some armed robots to be controlled mainly by other robots. The US Navy has funded a report which indicates that, as military robots become more complex, there should be greater attention to implications of their ability to make autonomous decisions. One researcher states that autonomous robots might be more humane, as they could make decisions more effectively. However, other experts question this.

One robot in particular, the EATR, has generated public concerns over its fuel source, as it can continually refuel itself using organic substances. Although the engine for the EATR is designed to run on biomass and vegetation specifically selected by its sensors, which it can find on battlefields or other local environments, the project has stated that chicken fat can also be used.

Manuel De Landa has noted that "smart missiles" and autonomous bombs equipped with artificial perception can be considered robots, as they make some of their decisions autonomously. He believes this represents an important and dangerous trend in which humans are handing over important decisions to machines.

ROBOTS IN SOCIETY

Roughly half of all the robots in the world are in Asia, 32% in Europe, and 16% in North America, 1% in Australasia and 1% in Africa. 40% of all the robots in the world are in Japan, making Japan the country with the highest number of robots.

ROBOETHICS

The term **roboethics** was coined by roboticist Gianmarco Veruggio in 2002, who also served as chair of an Atelier funded by the European Robotics Research Network to outline areas where research may be needed. The road map effec-

tively divided ethics of artificial intelligence into two sub-fields to accommodate researchers' differing interests :
- Machine ethics is concerned with the behavior of artificial moral agents (AMAs)
- Roboethics is concerned with the behavior of humans, how humans design, construct, use and treat robots and other artificially intelligent beings

Main Positions on Roboethics

Since the First International Symposium on Roboethics (Sanremo, Italy, 2004), three main ethical positions emerged from the robotics community :
- Not interested in ethics (This is the attitude of those who consider that their actions are strictly technical, and do not think they have a social or a moral responsibility in their work)
- Interested in short-term ethical questions (This is the attitude of those who express their ethical concern in terms of "good" or "bad," and who refer to some cultural values and social conventions)
- Interested in long-term ethical concerns (This is the attitude of those who express their ethical concern in terms of global, long-term questions)

Disciplines Involved in Roboethics

The design of Roboethics requires the combined commitment of experts of several disciplines, who, working in transnational projects, committees, commissions, have to adjust laws and regulations to the problems resulting from the scientific and technological achievements in Robotics and AI.

In all likelihood, it is to be expected that the birth of new curricula studiorum and specialties, necessary to manage a subject so complex, just as it happened with Forensic Medicine. In particular, the main fields involved in Roboethics are : robotics, computer science, artificial intelligence, philosophy, ethics, theology, biology, physiology, cognitive science, neurosciences, law, sociology, psychology, and industrial design.

Principles

As roboethics is a human-centered ethics, it should comply with the principles state in the most important and widely accepted Charters of Human Rights :
- Human dignity and human rights.
- Equality, justice and equity.
- Benefit and harm.
- Respect for cultural diversity and pluralism.
- Non-discrimination and non-stigmatization.

- Autonomy and individual responsibility.
- Informed consent.
- Privacy.
- Confidentiality.
- Solidarity and cooperation.
- Social responsibility.
- Sharing of benefits.
- Responsibility towards the biosphere.

General Ethical Issues in Science and Technology

Roboethics shares with the other fields of science and technology most of the ethical problems derived from the Second and Third Industrial Revolutions :
- Dual-use technology.
- Environmental impact of technology.
- Effects of technology on the global distribution of wealth.
- Digital divide, socio-technological gap.
- Fair access to technological resources.
- Dehumanization of humans in the relationship with the machines.
- Technology addiction.
- Anthropomorphization of the machines.

History

Since antiquity, the discussion of ethics in relation to the treatment of non-human and even non-living things and their potential "spirituality" have been discussed. With the development machinery and eventually robots, this philosophy was also applied to robotics. The first publication directly addressing roboethics was developed by Isaac Asimov as his Three Laws of Robotics in 1942, in the context of his science fiction works, although the term "roboethics" was created by Gianmarco Veruggio in 2002.

Relationship to Unemployment

A recent example of human replacement involves Taiwanese technology company Foxconn who, in July 2011, announced a three-year plan to replace workers with more robots. At present the company uses ten thousand robots but will increase them to a million robots over a three-year period.

Service robots of different varieties including medical robots, underwater robots, surveillance robots, demolition robots and other types of robots that carry out a multitude of jobs are gaining in numbers. Service robots are everyday tools for mankind. They can clean floors, mow lawns and guard homes and will also assist old and handicapped people, do some surgeries, inspect pipes and sites that are hazardous to people, fight fires and defuse bombs.

CONTEMPORARY USES

At present, there are two main types of robots, based on their use : general-purpose autonomous robots and dedicated robots.

Robots can be classified by their specificity of purpose. A robot might be designed to perform one particular task extremely well, or a range of tasks less well. Of course, all robots by their nature can be re-programmed to behave differently, but some are limited by their physical form. For example, a factory robot arm can perform jobs such as cutting, welding, gluing, or acting as a fairground ride, while a pick-and-place robot can only populate printed circuit boards.

General-purpose Autonomous Robots

General-purpose autonomous robots can perform a variety of functions independently. General-purpose autonomous robots typically can navigate independently in known spaces, handle their own re-charging needs, interface with electronic doors and elevators and perform other basic tasks. Like computers, general-purpose robots can link with networks, software and accessories that increase their usefulness. They may recognize people or objects, talk, provide companionship, monitor environmental quality, respond to alarms, pick up supplies and perform other useful tasks. General-purpose robots may perform a variety of functions simultaneously or they may take on different roles at different times of day. Some such robots try to mimic human beings and may even resemble people in appearance; this type of robot is called a humanoid robot. Humanoid robots are still in a very limited stage, as no humanoid robot can, as of yet, actually navigate around a room that it has never been in. Thus, humanoid robots are really quite limited, despite their intelligent behaviors in their well-known environments.

Factory Robots

Car Production

Over the last three decades, automobile factories have become dominated by robots. A typical factory contains hundreds of industrial robots working on fully automated production lines, with one robot for every ten human workers. On an automated production line, a vehicle chassis on a conveyor is welded, glued, painted and finally assembled at a sequence of robot stations.

Packaging

Industrial robots are also used extensively for palletizing and packaging of manufactured goods, for example for rapidly taking drink cartons from the end of a conveyor belt and placing them into boxes, or for loading and unloading machining centers.

Electronics

Mass-produced printed circuit boards (PCBs) are almost exclusively manufactured by pick-and-place robots, typically with SCARA manipulators, which remove tiny electronic components from strips or trays, and place them on to PCBs with great accuracy. Such robots can place hundreds of thousands of components per hour, far out-performing a human in speed, accuracy, and reliability.

Automated Guided Vehicles (AGVs)

Mobile robots, following markers or wires in the floor, or using vision or lasers, are used to transport goods around large facilities, such as warehouses, container ports, or hospitals.

Early AGV-Style Robots

Limited to tasks that could be accurately defined and had to be performed the same way every time. Very little feedback or intelligence was required, and the robots needed only the most basic exteroceptors (sensors). The limitations of these AGVs are that their paths are not easily altered and they cannot alter their paths if obstacles block them. If one AGV breaks down, it may stop the entire operation.

Interim AGV-Technologies

Developed to deploy triangulation from beacons or bar code grids for scanning on the floor or ceiling. In most factories, triangulation systems tend to require moderate to high maintenance, such as daily cleaning of all beacons or bar codes. Also, if a tall pallet or large vehicle blocks beacons or a bar code is marred, AGVs may become lost. Often such AGVs are designed to be used in human-free environments.

Intelligent AGVs (i-AGVs)

Such as SmartLoader, SpeciMinder, ADAM, Tug Eskorta, and MT 400 with Motivity are designed for people-friendly workspaces. They navigate by recognizing natural features. 3D scanners or other means of sensing the environment in two or three dimensions help to eliminate cumulative errors in dead-reckoning calculations of the AGV's current position. Some AGVs can create maps of their environment using scanning lasers with simultaneous localization and mapping (SLAM) and use those maps to navigate in real time with other path planning and obstacle avoidance algorithms. They are able to operate in complex environments and perform non-repetitive and non-sequential tasks such as transporting photomasks in a semiconductor lab, specimens in hospitals and goods in warehouses. For dynamic areas, such as warehouses full of pallets, AGVs require additional strategies using three-dimensional sensors such as time-of-flight or stereovision cameras.

Dirty, Dangerous, Dull or Inaccessible Tasks

There are many jobs which humans would rather leave to robots. The job may be boring, such as domestic cleaning, or dangerous, such as exploring inside a volcano. Other jobs are physically inaccessible, such as exploring another planet, cleaning the inside of a long pipe, or performing laparoscopic surgery.

Space Probes

Almost every unmanned space probe ever launched was a robot. Some were launched in the 1960s with very limited abilities, but their ability to fly and land (in the case of Luna 9) is an indication of their status as a robot. This includes the Voyager probes and the Galileo probes, and others.

Telerobots

Teleoperated robots, or telerobots, are devices remotely operated from a distance by a human operator rather than following a predetermined sequence of movements, but which has semi-autonomous behaviour. They are used when a human cannot be present on site to perform a job because it is dangerous, far away, or inaccessible. The robot may be in another room or another country, or may be on a very different scale to the operator. For instance, a laparoscopic surgery robot allows the surgeon to work inside a human patient on a relatively small scale compared to open surgery, significantly shortening recovery time. They can also be used to avoid exposing workers to the hazardous and tight spaces such as in duct cleaning. When disabling a bomb, the operator sends a small robot to disable it. Several authors have been using a device called the Longpen to sign books remotely. Teleoperated robot aircraft, like the Predator Unmanned Aerial Vehicle, are increasingly being used by the military. These pilotless drones can search terrain and fire on targets. Hundreds of robots such as iRobot's Packbot and the Foster-Miller TALON are being used in Iraq and Afghanistan by the U.S. military to defuse roadside bombs or improvised explosive devices (IEDs) in an activity known as explosive ordnance disposal (EOD).

Automated Fruit Harvesting Machines

Used to pick fruit on orchards at a cost lower than that of human pickers.

Domestic Robots

Domestic robots are simple robots dedicated to a single task work in home use. They are used in simple but unwanted jobs, such as vacuum cleaning and floor washing, and lawn mowing.

Military Robots

Military robots include the SWORDS robot which is currently used in ground-based combat. It can use a variety of weapons and there is some discussion of giving it some degree of autonomy in battleground situations.

Unmanned combat air vehicles (UCAVs), which are an upgraded form of UAVs, can do a wide variety of missions, including combat. UCAVs are being designed such as the BAE Systems Mantis which would have the ability to fly themselves, to pick their own course and target, and to make most decisions on their own. The BAE Taranis is a UCAV built by Great Britain which can fly across continents without a pilot and has new means to avoid detection. Flight trials are expected to begin in 2011.

The AAAI has studied this topic in depth and its president has commissioned a study to look at this issue.

Some have suggested a need to build "Friendly AI", meaning that the advances which are already occurring with AI should also include an effort to make AI intrinsically friendly and humane. Several such measures reportedly already exist, with robot-heavy countries such as Japan and South Korea having begun to pass regulations requiring robots to be equipped with safety systems, and possibly sets of 'laws' akin to Asimov's Three Laws of Robotics. An official report was issued in 2009 by the Japanese government's Robot Industry Policy Committee. Chinese officials and researchers have issued a report suggesting a set of ethical rules, and a set of new legal guidelines referred to as "Robot Legal Studies." Some concern has been expressed over a possible occurrence of robots telling apparent falsehoods.

Mining Robots

Mining robots are designed to solve a number of problems currently facing the mining industry, including skills shortages, improving productivity from declining ore grades, and achieving environmental targets. Due to the hazardous nature of mining, in particular underground mining, the prevalence of autonomous, semi-autonomous, and tele-operated robots has greatly increased in recent times. A number of vehicle manufacturers provide autonomous trains, trucks and loaders that will load material, transport it on the mine site to its destination, and unload without requiring human intervention. One of the world's largest mining corporations, Rio Tinto, has recently expanded its autonomous vehicle fleet to the world's largest, consisting of 150 autonomous Komatsu trucks, operating in Western Australia.

Drilling, longwall and rockbreaking machines are now also available as autonomous robots. The Atlas Copco Rig Control System can autonomously execute a drilling plan on a drilling rig, moving the rig into position using GPS, set up the drill rig and drill down to specified depths. Similarly, the Transmin Rocklogic system can automatically plan a path to position a rockbreaker at a selected destination. These systems greatly enhance the safety and efficiency of mining operations.

Robot Kits

Robotic kits like Lego Mindstorms, BIOLOID, OLLO from ROBOTIS, or Bot-Brain Educational Robots can help children to learn about mathematics, physics, programming, and electronics.

Robot Competitions

Robotics have also been introduced into the lives of elementary and high school students with the company FIRST (For Inspiration and Recognition of Science and Technology). The organization is the foundation for the FIRST Robotics Competition, FIRST LEGO League, Junior FIRST LEGO League, and FIRST Tech Challenge competitions.

Healthcare

Robots in healthcare have two main functions. Those which assist an individual, such as a sufferer of a disease like Multiple Sclerosis, and those which aid in the overall systems such as pharmacies and hospitals.

Home Automation for the Elderly and Disabled

Robots have developed over time from simple basic robotic assistants, such as the Handy 1, through to semi-autonomous robots, such as FRIEND which can assist the elderly and disabled with common tasks.

The population is aging in many countries, especially Japan, meaning that there are increasing numbers of elderly people to care for, but relatively fewer young people to care for them. Humans make the best carers, but where they are unavailable, robots are gradually being introduced.

FRIEND is a semi-autonomous robot designed to support disabled and elderly people in their daily life activities, like preparing and serving a meal. FRIEND make it possible for patients who are paraplegic, have muscle diseases or serious paralysis (due to strokes *etc.*), to perform tasks without help from other people like therapists or nursing staff.

Pharmacies

Script Pro manufactures a robot designed to help pharmacies fill prescriptions that consist of oral solids or medications in pill form. The pharmacist or pharmacy technician enters the prescription information into its information system. The system, upon determining whether or not the drug is in the robot, will send the information to the robot for filling. The robot has 3 different size vials to fill determined by the size of the pill. The robot technician, user, or pharmacist determines the needed size of the vial based on the tablet when the robot is stocked. Once the vial is filled it is brought up to a conveyor belt that delivers it to a holder that spins the vial and attaches the patient label. Afterwards it is set on another conveyor that delivers the patient's medication vial to a slot labeled with the patient's name on an LED read out. The pharmacist or technician then checks the contents of the vial to ensure it's the correct drug for the correct patient and then seals the vials and sends it out front to be picked up. The robot is a very time efficient device that the pharmacy depends on to fill prescriptions.

McKesson's Robot RX is another healthcare robotics product that helps pharmacies dispense thousands of medications daily with little or no errors. The robot can be ten feet wide and thirty feet long and can hold hundreds of different kinds of medications and thousands of doses. The pharmacy saves many resources like staff members that are otherwise unavailable in a resource scarce industry. It uses an electromechanical head coupled with a pneumatic system to capture each dose and deliver it to its either stocked or dispensed location. The head moves along a single axis while it rotates 180 degrees to pull the medications. During this process it uses barcode technology to verify its pulling the correct drug. It then delivers the drug to a patient specific bin on a conveyor belt. Once the bin is filled with all of the drugs that a particular patient needs and that the robot stocks, the bin is then released and returned out on the conveyor belt to a technician waiting to load it into a cart for delivery to the floor.

Research Robots

While most robots today are installed in factories or homes, performing labour or life saving jobs, many new types of robot are being developed in laboratories around the world. Much of the research in robotics focuses not on specific industrial tasks, but on investigations into new types of robot, alternative ways to think about or design robots, and new ways to manufacture them. It is expected that these new types of robot will be able to solve real world problems when they are finally realized.

Nanorobots

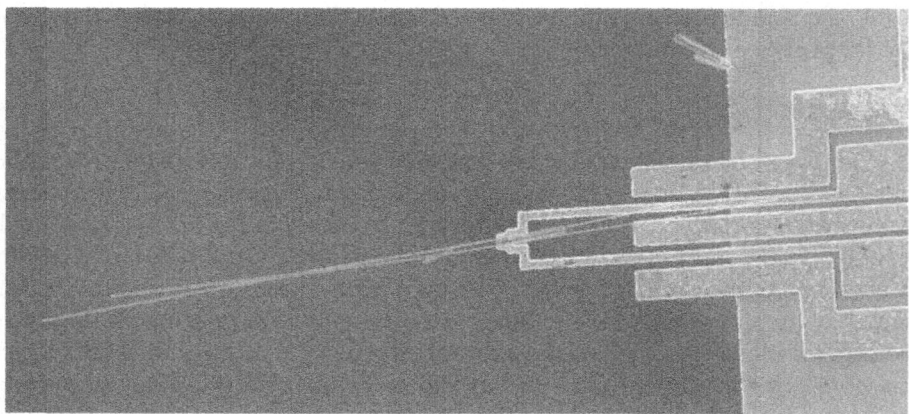

A microfabricated electrostatic gripper holding some silicon nanowires.

Nanorobotics is the emerging technology field of creating machines or robots whose components are at or close to the microscopic scale of a nanometer (10^{-9} meters). Also known as "nanobots" or "nanites", they would be constructed from molecular machines. So far, researchers have mostly produced only parts of these complex systems, such as bearings, sensors, and synthetic molecular motors, but functioning robots have also been made such as the entrants to the Nanobot Robocup contest. Researchers also hope to be able to create entire robots as small as viruses or bacteria, which could perform tasks on a tiny scale. Possible applications include micro surgery (on the level of individual cells), utility fog, manufacturing, weaponry and cleaning. Some people have suggested that if there were nanobots which could reproduce, the earth would turn into "grey goo", while others argue that this hypothetical outcome is nonsense.

Reconfigurable Robots

A few researchers have investigated the possibility of creating robots which can alter their physical form to suit a particular task, like the fictional T-1000. Real robots are nowhere near that sophisticated however, and mostly consist of a small number of cube shaped units, which can move relative to their neighbours. Algorithms have been designed in case any such robots become a reality.

Soft Robots

Robots with silicone bodies and flexible actuators (air muscles, electroactive polymers, and ferrofluids), controlled using fuzzy logic and neural networks, look and feel different from robots with rigid skeletons, and can have different behaviors.

Swarm Robots

A swarm of robots from the open-source micro-robotic project

Inspired by colonies of insects such as ants and bees, researchers are modeling the behavior of swarms of thousands of tiny robots which together perform a useful task, such as finding something hidden, cleaning, or spying. Each robot is quite simple, but the emergent behavior of the swarm is more complex. The whole set of robots can be considered as one single distributed system, in the same way an ant colony can be considered a superorganism, exhibiting swarm intelligence. The largest swarms so far created include the iRobot swarm, the SRI/MobileRobots CentiBots project and the Open-source Micro-robotic Project swarm, which are being used to research collective behaviors. Swarms are also more resistant to failure. Whereas one large robot may fail and ruin a mission, a swarm can continue even if several robots fail. This could make them attractive for space exploration missions, where failure is normally extremely costly.

Haptic Interface Robots

Robotics also has application in the design of virtual reality interfaces. Specialized robots are in widespread use in the haptic research community. These robots, called "haptic interfaces", allow touch-enabled user interaction with real and virtual environments. Robotic forces allow simulating the mechanical properties of "virtual" objects, which users can experience through their sense of touch.

Future Development

Technological Trends

Various techniques have emerged to develop the science of robotics and robots. One method is evolutionary robotics, in which a number of differing robots are submitted to tests. Those which perform best are used as a model to create a subsequent "generation" of robots. Another method is developmental robotics, which tracks changes and development within a single robot in the areas of problem-solving and other functions.

Technological Development

Overall Trends

Japan hopes to have full-scale commercialization of service robots by 2025. Much technological research in Japan is led by Japanese government agencies, particularly the Trade Ministry.

As robots become more advanced, eventually there may be a standard computer operating system designed mainly for robots. Robot Operating System is an open-source set of programs being developed at Stanford University, the Massachusetts Institute of Technology and the Technical University of Munich, Germany, among others. ROS provides ways to program a robot's navigation and limbs regardless of the specific hardware involved. It also provides high-level commands for items like image recognition and even opening doors. When ROS boots up on a robot's computer, it would obtain data on attributes such as the length and movement of robots' limbs. It would relay this data to higher-level algorithms. Microsoft is also developing a "Windows for robots" system with its Robotics Developer Studio, which has been available since 2007.

New Functions and Abilities

The Caterpillar Company is making a dump truck which can drive itself without any human operator.

Many future applications of robotics seem obvious to people, even though they are well beyond the capabilities of robots available at the time of the prediction. As early as 1982 people were confident that someday robots would : 1. clean parts by removing molding flash 2. spray paint automobiles with absolutely no human presence 3. pack things in boxes — for example, orient and nest chocolate candies in candy boxes 4. make electrical cable harness 5. load trucks with boxes — a packing problem 6. handle soft goods, such as garments and shoes 7. shear sheep 8. prosthesis 9. cook fast food and work in other service industries 10. household robot.

Generally such predictions are overly optimistic in timescale.

Reading Robot

A literate or 'reading robot' named Marge has intelligence that comes from software. She can read newspapers, find and correct misspelled words, learn about banks like Barclays, and understand that some restaurants are better places to eat than others.

STRUCTURE OF INDUSTRIAL ROBOTS OR MANIPULATORS :

Robots used in industries mostly perform the tasks of picking and placing different things. These functions are similar to that of a human arm, hold something and then place and fit that at the require spot. Such robotic arms are also called as manipulators. They have different types of Base Bodies.

Structure of Industrial Robots or Manipulators

Robotic arms are used in assembly lines to pick the components of products to be assembled and place and fit them at the right place. Such as in a production line of canned food, lids are placed, cans are picked by robotic arms and placed in the container packets or like in the assembly line of cars, the frames of cars move on the conveying system one by one and robotic arms or manipulators working around the frame fit different components on to it and finally completed cars come out of the assembly line.

The functions that can be performed by a manipulator, the reach of its end effectors, the orientation in which it can work and the overall working space for a manipulator is determined by the structure of the body of the manipulator. Also the types of joints used to connect different members of the manipulator along with the basic Body of the robot determine the overall degrees of freedom of the manipulator in its motion.

The body of the manipulator moves the end effectors to the target point where the object is grasped by the gripper fitted to the wrist. Wrist orients the object in the required direction. The wrist is fitted on the end point of the manipulator. The base Body, wrist and gripper constitute the basic structure of a manipulator.

Types of Base Bodies of Manipulators

Based on the working space, axes of movements and degrees of freedom of the manipulator as whole the Base Bodies can be classified into different types.

Linear : This type of manipulator arm as the name suggests can move along only a single direction. Linear manipulator base body comprises of one prismatic joint, that is, a slider which can move only along an axis. Such robots can only perform simple tasks like picking and placing objects from one spot to other. This can be the most basic type of manipulator possible. The range of this manipulator arm is limited by the length of the prismatic joint. The motion of the linear base body can be defined by a single variable and can be obtained by one linear actuator.

Robotics

Robots are not just machines, they are many steps ahead a typical machine. Robots like machines can perform different tough jobs easily but the advancement is that they can do it by their own. Once programmed robots can perform required tasks repeatedly in exactly the same way.

Types of Base Bodies of Manipulators

As the number of members and the degrees of freedom of the joints connecting those increases the robotic arm can work in more complex space and orientation.

Planer : The planer base body of a manipulator can be seen as one linear manipulator attached to another linear manipulator with axes of the two manipulators being perpendicular to each other in the same plane. Thus the end of the manipulator can move along two mutually perpendicular axes and can cover all the points in the plane defined by the range of motion along the two axes. The motion of the planer base body can be defined by two variables and motion can be achieved by two linear actuators. Such manipulator has two degrees of freedom.

Examples of tasks planer manipulators can do; curve tracing or printing on sheets, metal sheet cutting for specified design, welding on a planer metal piece or drilling at different spots in a plane.

Cartesian : Such robot body allows the manipulator to move along the three Cartesian axes only, which are the three mutually orthogonal axes. The working space of such robots will be a cuboid whose dimensions are defined by the range of motion of the manipulator along each axis. Cartesian robot base consists of three prismatic joints arranged perpendicular to each other. It requires three variables to define its position, has equal number of actuators and has the same number of degrees of freedom.

One such example can be the hoisting robot used in a workshop, shipping yard or construction site which moves on rails and the hoisting hook can be moved up and down through the cables.

The types of robot base bodies discussed so far only comprise of prismatic joints and the motion along any of the axes is only linear, there is no rotary motion. Though the end effectors of a Cartesian manipulator can reach every point defined in the space but still it is not used for every purpose. It will be slow for moving between different points in the space and cannot effectively orient its end effectors properly for certain jobs. Thus revolute and cylindrical joints are included in the base body to obtain more efficient structure of manipulators.

AUTOMATIC TYPE ROBOT

Manipulation Robotic Systems, an integral part of many industrial manipulators, are mainly divided on the basis of the type of control system they have. Automatic robots, a type of manipulation robotic system, are considered to be one of the earliest robotic systems.

Automatic Type Robots are an integral part of several industrial robotics systems, and are supposed to be the earliest type of robotic system present in the market today. Automatic robots are divided into four main categories, mainly based on their characteristics and application.

Manipulation robotic system can be can be classified into three main types :
- o   Autonomous controlled
- o   Remotely controlled
- o   Manually controlled

An autonomous robotic system is mainly used as industrial robots whereas the remotely controlled robots are used in areas or environments which are inaccessible or harmful to humans. The manually controlled system is used for handling goods or for transportation purposes.

Classification of Autonomous Robotic system

Out of the three types of manipulation robotic systems, the autonomous system can be further classified into
- Programmable
- Non Programmable
- Adaptive
- Intelligent

Programmable and Non Programmable Automatic Robots

Out of these, non-programmable robots are of the most basic type. In fact a non-programmable robot is not even considered a robot, but a manipulator devoid of any reprogrammable controlling device. One example of such robots is the mechanical arm used in industries. Non-programmable robots are generally attached to programmable equipment used in manufacturing industries for mass production.

A programmable robot, as the name suggests, is a first generation robot with an actuator with the facility of each of its joints being reprogrammable according to the kind of application. The function and application of the robots can be changed just by reprogramming the robot, however once programmed, they perform a specific function in a fixed sequence and fixed pattern. All the industrial robots are of programmable type which would perform a monotonous motion both in the presence or absence of any part in its grip. The main drawback of this type of robot is that, once programmed, it can be used to hold an object of a specific type and shape and that too placed in a particular position. As this type of robot cannot change its position when required, it is always a bit difficult to use in a changing application system.

Adaptive Robots

Adaptive robots are also industrial robots, but of a kind more sophisticated than programmable robots. Unlike programmable robots, adaptive robots can adapt to a certain extent and, after evaluating a particular situation, perform the action required. In order to enable them to perform these tasks, adaptive robots are equipped with sensors and control system. The sensors sense the change in environmental conditions, and the control system, by assessing the signals from sensors, provides the required motion. Adaptive robots are generally used in situations wherein it is difficult to program a robot to perform actions in a particular pattern due to obstacles or other moving parts. Adaptive robots are used in functions such as welding, spray painting, *etc.*

Intelligent Robots

Intelligent robots, as the name suggests, are the most intelligent of all types, with several sensors and microprocessors for storing, analyzing, and processing the data. Intelligent robots can perform any kind of work because of their ability to analyze any situation and provide the necessary movement according to that. For this, the system is provided with several manipulators, having their own controllers.

THE BRAINS AND BODY OF A ROBOT

Robots are an essential part of technology today. However, most people overlook the similarities robots have with humans, in terms of a control center and a structure. Two key components of most robots are the brains, in the form of microcontroller chips, and the body.

Introduction

From an anatomical point of view, most robots generally have a brain, a body, a power source, sensors, and action and feedback mechanisms. Specifically, I would like to discuss the brains and body of a robot.

Brains

Robot brains come in a wide variety of forms. In fact, some robots are built without brains and are controlled by people through remote control. Robots can also be built with a brain that is spread out in different parts of the robot. For instance, basic chips can be used to operate individual parts of the robot, such as an arm or leg, with these individual parts working independently of the other parts. Furthermore, robots can be built with brains that are located far away from its body, such as in a computer.

All things considered, the number one choice for robot brains is the microcontroller chip. Like the microprocessor chips found in computers, microcontrollers are similar, except that microcontroller chips are somewhat like a tiny computer themselves. A small amount of memory and storage space is built directly into the microcontroller chip. When comparing the microprocessor chips found in computers and microcontroller chips, the chips found in computers dedicate their channels to high speed memory connectors, whereas microcontroller chips have a much larger variety of input and output ports. These ports can connect to buttons, sensors, and other devices.

Although it may not seem evident, we are surrounded by microcontrollers. These useful chips are found in vehicles, household appliances such as dishwashers, VCRs, TVs, radios, and other home and work appliances. Fortunately, the high demand for these microcontroller chips has made these chips inexpensive and abundant.

Body

Although the body of a robot may seem unimportant in comparison to the other parts of a robot, many who attempt to build a robot fail to incorporate a framework into their design. Many who build homemade robots end up with robots that are too susceptible due to circuit boards that are exposed and wires sticking out. This usually leads to a robot that either collapses or moves unevenly. A proper frame for a robot not only ensures that the robot stays in one piece, but also protects the robot from injury. An excellent example of the use of a reliable framework is in a production line robot. These robots have special designs that enable them to efficiently operate in environments that most humans would find potentially dangerous. Production line robot design involves a certain number of axes that determines the versatility of the robot. However, it is the kinematics of the assembly line robot that determines the arrangement of the robot. Some parts of the robot are bolted together while others are welded together, allowing for both rigid and dynamic parts in the same robot. It is the kinematics of the framework of a production line robot that defines its purpose, whether for welding car parts together or lifting heavy loads.

Another underestimated aspect of a robot is the visual appeal. Those who desire to take up robot building as a hobby should learn that showmanship is critical in how others will view your robot. Despite the fact that a robot may be technically impressive, the finishing touches to the appearance of the robot is what draws attention. The M&M robot is an example of an attractive and innovative robot body design.

HOW DOES A ROBOT'S VOICE RECOGNITION SYSTEM WORKS?

Ever wondered how a robot is able to perform tasks after taking commands from the user? It is the Robot's Voice Recognition system that identifies the words and performs the desired action. Want to know how?

Introduction

Voice recognition is the process by which a robot identifies what is ordered to it and performs an action based on the information received. It is to note that not all robots have this functionality; however the system can also be integrated at a later stage.

The voice recognition system works on the basis of the frequency and amplitudes of the spoken words. A signal is generated and sent to the main operating unit of the robot, after dissecting the received words into various frequencies and amplitudes.

How Does the System Works?

When an order is given to a robot in the form of words the voice recognition system breaks down each and every word into constituent frequencies of each alphabet. The voice recognition system has a pre-programmed database, which has a unique frequency for each and every letter. This database helps the robot to identify the word and perform the right action. However, a good amount of training needs to be given by the user for facilitating the robot to initially form a table comprising of major frequencies of words and alphabets.

The table once formed, acts as a quick reference for the robot to identify the frequently used words. When a word is spoken, the robot identifies the sound, determines the exact frequency and looks up in the table to perceive the right word. If the robot is not able to find the right frequency, it finds the frequency of the alphabet closest to the one needed and thus recognizes the whole word.

Frequent Training Required

In order to increase the accuracy, the robot should be trained repetitively to identify the right frequency. Moreover, more the training provided, less is the variation. This means that in case of any type of voice modulation or variation, the system will not try to match the perceived signal to many frequencies but will neglect the frequency and won't perform any action. However, if very few frequencies are matched for a particular variation, it may misinterpret a word or choose a word similar in sound.

It is for this reason that most of the robots with voice recognition system are highly trained by the users. More the robot is trained, quicker is the process of the voice recognition system to identify the word and send a signal to the controlling unit, performing the desired action.

POSITION AND ORIENTATION OF THE OBJECTS IN ROBOTIC AUTOMATION

To study the mechanics of robotic manipulators comprehensively and then apply it for the mechanical synthesis of the manipulators, we will first look at the basic topics of mechanics involved in the mechanics of manipulators.

Industrial Robots result in increased levels of automation in industrial processes. Robot use started in repetitive, simple tasks, and now they are also being used also in precision works. This transition happened because of the sophisticated synthesis of robots and inclusion of electronics and computer control. An industrial robot is designed such that it can be programmed to be used in different applications.

The mechanics of the industrial robots and manipulators is not a new field of engineering in itself. It is actually a combination of different branches of engineering. The mechanics, static and dynamic analysis and design, comes from mechanical engineering, and the analysis of motion and path planning of the manipulators is done with the help of mathematical tools.

The Position and Orientation of Objects

For the design and analysis of the manipulators, the basic thing to keep track of and to plan is the position as well as the orientation of objects. The objects for which we are concerned are the components of the manipulators, like their links, joints, and tools, and also the objects in the space around the manipulator as well as all the objects with which the manipulator interacts.

For the effective design of the manipulator, the position and orientation of the objects concerned need to be represented suitably such that it can be mathematically processed to make the manipulator move in the desired manner.

Attaching Frame to the Object

To define the position and orientation of the object and to keep track of its movement a coordinate frame is rigidly attached to the object. The motion of the object, that is, the change in position and orientation of the object is now given by the frame attached to the object.

Reference Coordinate Frame

A Reference Coordinate Frame is defined with respect to which the position and orientations of all the other frames attached to the objects are described.

Transformation of Frames

We can define any frame as the reference frame. For convenience we keep changing the reference frames. So the position and orientation of the frames attached to the objects need to be transformed from one reference frame to the other. There are mathematical methods for the transformation of the frames.

Mechanics of Robotic Manipulators

Robotics is perhaps the outcome of the desire to synthesize such machines which can perform tasks like humans do and even beyond, that is, performing task which humans cannot do, and that too in a way even better than what humans do. This series introduces you to Mechanics of Robotic Manipulators.

Chapter 2

INDUSTRIAL ROBOT

Articulated industrial robot operating in a foundry.

An **industrial robot** is defined by ISO 8373 as an *automatically controlled, reprogrammable, multipurpose manipulator programmable in three or more axes*. The field of robotics may be more practically defined as the study, design and use of robot systems for manufacturing (a top-level definition relying on the prior definition of *robot*).

Typical applications of robots include welding, painting, assembly, pick and place (such as packaging, palletizing and SMT), product inspection, and testing; all accomplished with high endurance, speed, and precision.

TYPES AND FEATURES

The most commonly used robot configurations are articulated robots, SCARA robots, delta robots and cartesian coordinate robots, (gantry robots or x-y-z robots). In the context of general robotics, most types of robots would fall into the category of robotic arms (inherent in the use of the word *manipulator* in ISO standard 8373). Robots exhibit varying degrees of autonomy :

- Some robots are programmed to faithfully carry out specific actions over and over again (repetitive actions) without variation and with a high degree of accuracy. These actions are determined by programmed routines that specify the direction, acceleration, velocity, deceleration, and distance of a series of coordinated motions.
- Other robots are much more flexible as to the orientation of the object on which they are operating or even the task that has to be performed on the object itself, which the robot may even need to identify. For example, for more precise guidance, robots often contain machine vision sub-systems acting as their visual sensors, linked to powerful computers or controllers. Artificial intelligence, or what passes for it, is becoming an increasingly important factor in the modern industrial robot.

HISTORY OF INDUSTRIAL ROBOTICS

The earliest known industrial robot, conforming to the ISO definition was completed by "Bill" Griffith P. Taylor in 1937 and published in Meccano Magazine, March 1938. The crane-like device was built almost entirely using Meccano parts, and powered by a single electric motor. Five axes of movement were possible, including *grab* and *grab rotation*. Automation was achieved using punched paper tape to energise solenoids, which would facilitate the movement of the crane's control levers. The robot could stack wooden blocks in pre-programmed patterns. The number of motor revolutions required for each desired movement was first plotted on graph paper. This information was then transferred to the paper tape, which was also driven by the robot's single motor. Chris Shute built a complete replica of the robot in 1997.

George Devol applied for the first robotics patents in 1954 . The first company to produce a robot was Unimation, founded by Devol and Joseph F. Engelberger in 1956, and was based on Devol's original patents. Unimation robots were also called *programmable transfer machines* since their main use at first was to transfer objects from one point to another, less than a dozen feet or so apart. They used hydraulic actuators and were programmed in *joint coordinates*, i.e. the angles of the various joints were stored during a teaching phase and replayed in operation. They were accurate to within 1/10,000 of an inch. Unimation later licensed their technology to Kawasaki Heavy Industries and GKN, manufacturing Unimates in Japan and England respectively. For some time Unimation's only competitor was Cincinnati Milacron Inc. of Ohio. This changed radically in the late 1970s when several big Japanese conglomerates began producing similar industrial robots.

In 1969 Victor Scheinman at Stanford University invented the Stanford arm, an all-electric, 6-axis articulated robot designed to permit an arm solution. This allowed it accurately to follow arbitrary paths in space and widened the potential use of the robot to more sophisticated applications such as assembly and welding. Scheinman then designed a second arm for the MIT AI Lab, called the "MIT arm." Scheinman, after receiving a fellowship from Unimation to develop his designs, sold those designs to Unimation who further developed them with support from General Motors and later marketed it as the Programmable Universal Machine for Assembly (PUMA).

Industrial robotics took off quite quickly in Europe, with both ABB Robotics and KUKA Robotics bringing robots to the market in 1973. ABB Robotics (formerly ASEA) introduced IRB 6, among the world's first *commercially available* all electric micro-processor controlled robot. The first two IRB 6 robots were sold to Magnusson in Sweden for grinding and polishing pipe bends and were installed in production in January 1974. Also in 1973 KUKA Robotics built its first robot, known as FAMULUS, also one of the first articulated robots to have six electro-mechanically driven axes.

Interest in robotics increased in the late 1970s and many US companies entered the field, including large firms like General Electric, and General Motors (which formed joint venture FANUC Robotics with FANUC LTD of Japan). U.S. startup companies included Automatix and Adept Technology, Inc. At the height of the robot boom in 1984, Unimation was acquired by Westinghouse Electric Corporation for 107 million U.S. dollars. Westinghouse sold Unimation to Stäubli Faverges SCA of France in 1988, which is still making articulated robots for general industrial and cleanroom applications and even bought the robotic division of Bosch in late 2004.

Only a few non-Japanese companies ultimately managed to survive in this market, the major ones being : Adept Technology, Stäubli-Unimation, the Swedish-Swiss company ABB Asea Brown Boveri, the German company KUKA Robotics and the Italian company Comau.

TECHNICAL DESCRIPTION

Number of axes – two axes are required to reach any point in a plane; three axes are required to reach any point in space. To fully control the orientation of the end of the arm (*i.e.* the *wrist*) three more axes (yaw, pitch, and roll) are required. Some designs trade limitations in motion possibilities for cost, speed, and accuracy.

Defining Parameters

- Number of axes – two axes are required to reach any point in a plane; three axes are required to reach any point in space. To fully control the orientation of the end of the arm (*i.e.* the wrist) three more axes (yaw, pitch, and roll) are required. Some designs trade limitations in motion possibilities for cost, speed, and accuracy.

- Degrees of freedom – this is usually the same as the number of axes.
- Working envelope – the region of space a robot can reach.
- Kinematics – the actual arrangement of rigid members and joints in the robot, which determines the robot's possible motions. Classes of robot kinematics include articulated, cartesian, parallel and SCARA.
- Carrying capacity or payload – how much weight a robot can lift.
- Speed – how fast the robot can position the end of its arm. This may be defined in terms of the angular or linear speed of each axis or as a compound speed *i.e.* the speed of the end of the arm when all axes are moving.
- Acceleration – how quickly an axis can accelerate. Since this is a limiting factor a robot may not be able to reach its specified maximum speed for movements over a short distance or a complex path requiring frequent changes of direction.
- Accuracy – how closely a robot can reach a commanded position. When the absolute position of the robot is measured and compared to the commanded position the error is a measure of accuracy. Accuracy can be improved with external sensing for example a vision system or Infra-Red.
- Repeatability – how well the robot will return to a programmed position. This is not the same as accuracy. It may be that when told to go to a certain X-Y-Z position that it gets only to within 1 mm of that position. This would be its accuracy which may be improved by calibration. But if that position is taught into controller memory and each time it is sent there it returns to within 0.1 mm of the taught position then the repeatability will be within 0.1 mm.

Accuracy and repeatability are different measures. Repeatability is usually the most important criterion for a robot and is similar to the concept of 'precision' in measurement. ISO 9283 sets out a method whereby both accuracy and repeatability can be measured. Typically a robot is sent to a taught position a number of times and the error is measured at each return to the position after visiting 4 other positions. Repeatability is then quantified using the standard deviation of those samples in all three dimensions. A typical robot can, of course make a positional error exceeding that and that could be a problem for the process. Moreover the repeatability is different in different parts of the working envelope and also changes with speed and payload. ISO 9283 specifies that accuracy and repeatability should be measured at maximum speed and at maximum payload. But this results in pessimistic values whereas the robot could be much more accurate and repeatable at light loads and speeds. Repeatability in an industrial process is also subject to the accuracy of the end effector, for example a gripper, and even to the design of the 'fingers' that match the gripper to the object being grasped. For example, if a robot picks a screw by its head, the screw could be at a random angle. A subsequent attempt to insert the screw into a hole could easily fail. These and similar scenarios can be improved with 'lead-ins' *e.g.* by making the entrance to the hole tapered.

- Motion control – for some applications, such as simple pick-and-place assembly, the robot need merely return repeatably to a limited number of pre-taught positions. For more sophisticated applications, such as welding and finishing (spray painting), motion must be continuously controlled to follow a path in space, with controlled orientation and velocity.
- Power source – some robots use electric motors, others use hydraulic actuators. The former are faster, the latter are stronger and advantageous in applications such as spray painting, where a spark could set off an explosion; however, low internal air-pressurisation of the arm can prevent ingress of flammable vapours as well as other contaminants.
- Drive – some robots connect electric motors to the joints via gears; others connect the motor to the joint directly (direct drive). Using gears results in measurable 'backlash' which is free movement in an axis. Smaller robot arms frequently employ high speed, low torque DC motors, which generally require high gearing ratios; this has the disadvantage of backlash. In such cases the harmonic drive is often used.
- Compliance - this is a measure of the amount in angle or distance that a robot axis will move when a force is applied to it. Because of compliance when a robot goes to a position carrying its maximum payload it will be at a position slightly lower than when it is carrying no payload. Compliance can also be responsible for overshoot when carrying high payloads in which case acceleration would need to be reduced.

Robot Programming and Interfaces

The setup or programming of motions and sequences for an industrial robot is typically taught by linking the robot controller to a laptop, desktop computer or (internal or Internet) network.

A robot and a collection of machines or peripherals is referred to as a workcell, or cell. A typical cell might contain a parts feeder, a molding machine and a robot. The various machines are 'integrated' and controlled by a single computer or PLC. How the robot interacts with other machines in the cell must be programmed, both with regard to their positions in the cell and synchronizing with them.

Software : The computer is installed with corresponding interface software. The use of a computer greatly simplifies the programming process. Specialized robot software is run either in the robot controller or in the computer or both depending on the system design.

There are two basic entities that need to be taught (or programmed) : positional data and procedure. For example in a task to move a screw from a feeder to a hole the positions of the feeder and the hole must first be taught or programmed. Secondly the procedure to get the screw from the feeder to the hole must be programmed along with any I/O involved, for example a signal to indicate when the screw is in the feeder ready to be picked up. The purpose of the robot software is to facilitate both these programming tasks.

Teaching the robot positions may be achieved a number of ways :

Positional commands The robot can be directed to the required position using a GUI or text based commands in which the required X-Y-Z position may be specified and edited.

Teach pendant : Robot positions can be taught via a teach pendant. This is a handheld control and programming unit. The common features of such units are the ability to manually send the robot to a desired position, or "inch" or "jog" to adjust a position. They also have a means to change the speed since a low speed is usually required for careful positioning, or while test-running through a new or modified routine. A large emergency stop button is usually included as well. Typically once the robot has been programmed there is no more use for the teach pendant.

Lead-by-the-nose is a technique offered by many robot manufacturers. In this method, one user holds the robot's manipulator, while another person enters a command which de-energizes the robot causing it to go limp. The user then moves the robot by hand to the required positions and/or along a required path while the software logs these positions into memory. The program can later run the robot to these positions or along the taught path. This technique is popular for tasks such as paint spraying.

Offline programming is where the entire cell, the robot and all the machines or instruments in the workspace are mapped graphically. The robot can then be moved on screen and the process simulated. A robotics simulator is used to create embedded applications for a robot, without depending on the physical operation of the robot arm and end effector. The advantages of robotics simulation is that it saves time in the design of robotics applications. It can also increase the level of safety associated with robotic equipment since various "what if" scenarios can be tried and tested before the system is activated.[8] Robot simulation software provides a platform to teach, test, run, and debug programs that have been written in a variety of programming languages.

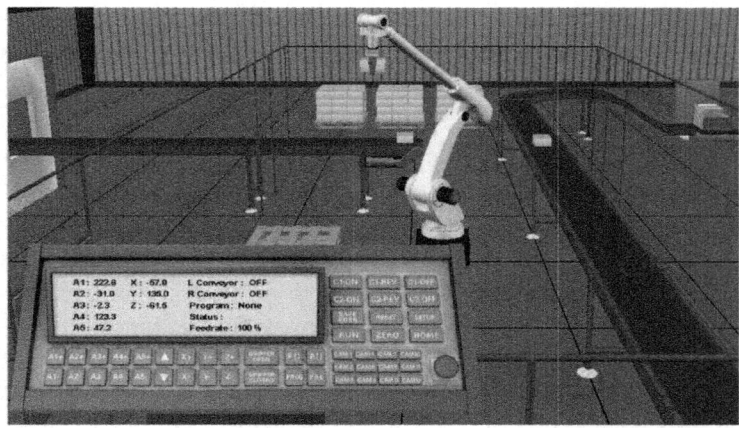

RoboLogix Robotics Simulator.

Robot simulation tools allow for robotics programs to be conveniently written and debugged off-line with the final version of the program tested on an actual robot. The ability to preview the behavior of a robotic system in a virtual world allows for a variety of mechanisms, devices, configurations and controllers to be tried and tested before being applied to a "real world" system. Robotics simulators have the ability to provide real-time computing of the simulated motion of an industrial robot using both geometric modeling and kinematics modeling.

Others In addition, machine operators often use user interface devices, typically touchscreen units, which serve as the operator control panel. The operator can switch from program to program, make adjustments within a program and also operate a host of peripheral devices that may be integrated within the same robotic system. These include end effectors, feeders that supply components to the robot, conveyor belts, emergency stop controls, machine vision systems, safety interlock systems, bar code printers and an almost infinite array of other industrial devices which are accessed and controlled via the operator control panel.

The teach pendant or PC is usually disconnected after programming and the robot then runs on the program that has been installed in its controller. However a computer is often used to 'supervise' the robot and any peripherals, or to provide additional storage for access to numerous complex paths and routines.

End-of-arm Tooling

The most essential robot peripheral is the end effector, or end-of-arm-tooling (EOT). Common examples of end effectors include welding devices (such as MIG-welding guns, spot-welders, *etc.*), spray guns and also grinding and deburring devices (such as pneumatic disk or belt grinders, burrs, *etc.*), and grippers (devices that can grasp an object, usually electromechanical or pneumatic). Another common means of picking up an object is by vacuum. End effectors are frequently highly complex, made to match the handled product and often capable of picking up an array of products at one time. They may utilize various sensors to aid the robot system in locating, handling, and positioning products.

Controlling Movement

For a given robot the only parameters necessary to completely locate the end effector (gripper, welding torch, *etc.*) of the robot are the angles of each of the joints or displacements of the linear axes (or combinations of the two for robot formats such as SCARA). However there are many different ways to define the points. The most common and most convenient way of defining a point is to specify a Cartesian coordinate for it, *i.e.* the position of the 'end effector' in mm in the X, Y and Z directions relative to the robot's origin. In addition, depending on the types of joints a particular robot may have, the orientation of the end effector in yaw, pitch, and roll and the location of the tool point relative to the robot's faceplate must also be specified. For a jointed arm these coordinates must be converted to joint angles by the robot controller and such conversions are known as Cartesian

Transformations which may need to be performed iteratively or recursively for a multiple axis robot. The mathematics of the relationship between joint angles and actual spatial coordinates is called kinematics.

Positioning by Cartesian coordinates may be done by entering the coordinates into the system or by using a teach pendant which moves the robot in X-Y-Z directions. It is much easier for a human operator to visualize motions up/down, left/right, *etc.* than to move each joint one at a time. When the desired position is reached it is then defined in some way particular to the robot software in use, *e.g.* P1 - P5 below.

Typical Programming

Most articulated robots perform by storing a series of positions in memory, and moving to them at various times in their programming sequence. For example, a robot which is moving items from one place to another might have a simple 'pick and place' program similar to the following:

Define Points P1–P5 :

1. Safely above workpiece
2. 10 cm Above bin A
3. At position to take part from bin A
4. 10 cm Above bin B
5. At position to take part from bin B.

Define Program :

1. Move to P1
2. Move to P2
3. Move to P3
4. Close gripper
5. Move to P2
6. Move to P4
7. Move to P5
8. Open gripper
9. Move to P4
10. Move to P1 and finish

Singularities

The American National Standard for Industrial Robots and Robot Systems — Safety Requirements defines a singularity as "a condition caused by the collinear alignment of two or more robot axes resulting in unpredictable robot motion and velocities." It is most common in robot arms that utilize a "triple-roll wrist". This is a wrist about which the three axes of the wrist, controlling yaw, pitch, and roll, all pass through a common point. An example of a wrist singularity is when the path through which the robot is traveling causes the first and third axes of the robot's wrist (*i.e.* robot's axes 4 and 6) to line up. The second wrist axis then attempts to spin 360° in zero time to maintain the orientation of the end effector. Another common term for this singularity is a "wrist flip". The result of a singularity can be quite dramatic and can have adverse effects on the robot arm, the end effector, and the process. Some industrial robot manufacturers have attempted to side-step the situation by slightly altering the robot's path to prevent this condition. Another method is to slow the robot's travel speed, thus reducing the speed required for the wrist to make the transition. The ANSI/RIA has mandated that robot manufacturers shall make the user aware of singularities if they occur while the system is being manually manipulated.

A second type of singularity in wrist-partitioned vertically articulated six-axis robots occurs when the wrist center lies on a cylinder that is centered about axis 1 and with radius equal to the distance between axes 1 and 4. This is called a shoulder singularity. Some robot manufacturers also mention alignment singularities, where axes 1 and 6 become coincident. This is simply a sub-case of shoulder singularities. When the robot passes close to a shoulder singularity, joint 1 spins very fast.

The third and last type of singularity in wrist-partitioned vertically articulated six-axis robots occurs when the wrist's center lies in the same plane as axes 2 and 3.

Singularities are closely related to the phenomena of Gimbal Lock, which has a similar root cause of axes becoming lined up.

A video illustrating these three types of singular configurations is available here.

RECENT AND FUTURE DEVELOPMENTS

As of 2012, the robotic arm business is approaching a mature state, where they can provide enough speed, accuracy and ease of use for most of the applications. Vision guidance (aka machine vision) is bringing much flexibility to robotic cells.

Hand in hand with increasing off-line programmed applications, robot calibration is becoming more and more important in order to guarantee a good positioning accuracy.

Other developments include downsizing industrial arms for light industrial use such as production of small products, sealing and dispensing, quality control, and handling samples in the laboratory. Such robots are usually classified as "bench top" robots. Robots are used in pharmaceutical research in a technique called high-throughput screening. Bench-top robots are also used in consumer applications (micro-robotic arms). Industrial arms may be used in combination with or even mounted on automated guided vehicles (AGVs) to make the automation chain more flexible between pick-up and drop-off.

Another recent development is the introduction of lower cost easily programmed robots, which are often low-force low-speed robots enabling them to work in unrestricted areas. They are programmed, for example, by a person grasping the robot and moving it through its intended cycle of operation.

MARKET STRUCTURE

According to the International Federation of Robotics (IFR) study *World Robotics 2012*, there were at least 1,153,000 operational industrial robots by the end of 2011. This number is estimated to reach 1,575,000 by the end of 2015.

For the year 2011 the IFR estimates the worldwide sales of industrial robots with US$8.5 billion. Including the cost of software, peripherals and systems engineering, the annual turnover for robot systems is estimated to be US$25.5 billion in 2011.

The Japanese government estimates the industry could surge from about $5.2 billion in 2006 to $26 billion in 2010 and nearly $70 billion by 2025. In 2005, there were over 370,000 operational industrial robots in Japan. A 2007 national technology roadmap by the Trade Ministry calls for 1 million industrial robots to be installed throughout the country by 2025.

Estimated worldwide annual supply of industrial robots (in units):

Year	supply
1998	69,000
1999	79,000
2000	99,000
2001	78,000
2002	69,000
2003	81,000
2004	97,000
2005	120,000
2006	112,000
2007	114,000
2008	113,000
2009	60,000
2010	118,000
2011	166,000
2013	168,000

DEALING WITH DECOMMISSIONED INDUSTRIAL ROBOTS

After a robot has outlived its normal utility, its disposal becomes a challenge for the enterprise using it. Resale, sending to a scrap yard, using it for land-fill, and recycling are some of the options available for decommissioned robots.

What are Robots?

By definition, robots are aids created to make work easier, faster, more accurate, and safer to do. They are mechanically driven and have some artificial intelligence that can be programmed to perform different commands. Robots have been developed for many reasons, but have largely found their major use in the manufacturing sector. Some of the work that can be performed by industrial robots is the lifting of heavy weights, painting, drilling, welding, and handling chemicals and hazardous materials. These robots are mainly fixed and have limited movement.

Maintenance departments and facilities that handle hazardous material also use robots to do work. Exploration missions to the moon and more recently to Mars used robotic equipment to survey and collect data for research centers on earth. Exploration robots are built to withstand extreme conditions and are mobile; they need to be mobile to conduct geological surveys and collect data and samples. Anti-terrorism agencies and the military use robots to neutralize dangerous things like bombs or mines. Personal robots are rare since they need to be programmed to do many different tasks.

Decommissioning of Robots

When industrial robots stop working or are replaced with newer versions, it is called "decommissioning." The problem of disposal, of course, starts after the robot has been decommissioned. Resale, sending to a scrap yard, using it for land-fill, and recycling are some of the options available for decommissioned robots. The fate that a decommissioned robot meets depends on the nature of work for which it was developed, the materials used for making it, and the law of the land.

Recycling

If no potential customer for the used equipment comes forward, retired industrial robots should ideally be sent to recycling plants for the proper disposal of the different materials used to build them. Most robots are built using plastic and metals, which should not pose any dangers, but within these robots are many electronic sensors, motion detectors, batteries, motors, and other part that may contain harmful materials. This does not apply to all robots, but particularly does for robots that are constructed as a single operational unit.

Robotic attachments like used pick and place robots, painting robots, and precision welding, positioning, and manufacturing robots, as well as all robots that receive commands from a central command, do not pose a high pollution risk as they are composed mostly of mechanical parts. Nevertheless, all used robots, if possible, should be sent to robot recycling units who specialize in dismantling them and extracting all the reusable parts and materials.

Storing

Robots that have been used in certain hazardous operations such as inside nuclear power plants may never get proper disposal due to the nature of the work they have to do. When they are retired or break down, replacement is done and the old robot is moved to a storage area within the secured perimeter.

Exploration robots are also never built for cycling since there fate is not predictable. Robots on the moon and mars are some example of these kinds of irretrievable robots.

Back on earth the problem is getting bigger with the risk of contamination of water reservoirs and underground water by harmful elements that are contained within the artificial intelligence and sensors that controls the robots mobility and precision. People who are environmentally conscious have created innovative ways by which they can contribute to reducing pollution.

Retired Robots and Art

Artists have begun using them in designs to decorate offices, businesses, and homes and parks. Some of these innovative artists create sculpture depicting different scenarios that modern day people face on a daily bases. Some interior decorators use retired robots which have been extracted of all harmful elements to decorate robot enthusiasts homes and rooms with them. Robotic arms used for welding can be used as lamp holders or as a coat rest.

The Problem of E-Waste

With the world population increasing on a daily basis and demand for produces and resources increasing, there is urgency now more than ever to act responsibly towards e-waste. With few recycling plants distributed around the world and more e-waste being produced on a daily basis humanity has to think twice before we find ourselves buried in the waste itself.

With proper management and financial support, e-waste like used robots can be recycled and extracted parts can be reused on other models of robots. This requires a lot of time and man power since they need to be dismantled piece by piece in order to perform the job correctly.

TYPES OF INDUSTRIAL ROBOTS

Workerbot

The **workerbot** is a trademark, which was developed by the pi4_robotics GmbH to describe an industrial robot, which was modeled with its possibilities of movement and its sensory abilities of a human. The industrial robot has two arms with seven degrees of freedom. In the arms of force sensors are integrated, enabling the robot to work while the forces occurring measure and similar to humans the gripping process or machining processes to adapt to the forces occurring accordingly. The robot is also equipped with cameras that it can detect its environment and react to it. This industrial robot has been developed within the EU funded project PISA (*Flexible Assembly* Systems through Workplace-Sharing and Time-Sharing Human-Machine Cooperation). *The project* consortium consists of the lead company pi4_robotics GmbH and the Fraunhofer IPK, the Universidad Politécnica de Madrid and the company EICAS Automazione S.p.A. The aim is to enable the use of highly flexible industrial robots manufacturing companies within the European Union cost production and to prevent the migration to low-wage countries. The workerbot is still the first operative humanoid factory worker worldwide that can be acquired by purchase. In context with the workerbot there is the first webshop for humanoid robots.

Articulated Robot

A six-axis articulated welding robot reaching into a fixture to weld.

An **articulated robot** is a robot with rotary joints (*e.g.* a legged robot or an industrial robot). Articulated robots can range from simple two-jointed structures to systems with 10 or more interacting joints. They are powered by a variety of means, including electric motors.

Some types of robots, such as robotic arms, can be articulated or non-articulated.

Definitions

Articulated Robot : An articulated robot is one which uses rotary joints to access its work space. Usually the joints are arranged in a "chain", so that one joint supports another further in the chain.

Continuous Path : A control scheme whereby the inputs or commands specify every point along a desired path of motion. The path is controlled by the coordinated motion of the manipulator joints.

Degrees Of Freedom (DOF) : The number of independent motions in which the end effector can move, defined by the number of axes of motion of the manipulator.

Gripper : A device for grasping or holding, attached to the free end of the last manipulator link; also called the robot's hand or end-effector.

Payload : The maximum payload is the amount of weight carried by the robot manipulator at reduced speed while maintaining rated precision. Nominal payload is measured at maximum speed while maintaining rated precision. These ratings are highly dependent on the size and shape of the payload.

Pick And Place Cycle : Pick and place Cycle is the time, in seconds, to execute the following motion sequence : Move down one inch, grasp a rated payload; move up one inch; move across twelve inches; move down one inch; ungrasp; move up one inch; and return to start location.

Reach : The maximum horizontal distance from the center of the robot base to the end of its wrist.

Accuracy : The difference between the point that a robot is trying to achieve and the actual resultant position. Absolute accuracy is the difference between a point instructed by the robot control system and the point actually achieved by the manipulator arm, while repeatability is the cycle-to-cycle variation of the manipulator arm when aimed at the same point.

Repeatability : The ability of a system or mechanism to repeat the same motion or achieve the same points when presented with the same control signals. The cycle-to-cycle error of a system when trying to perform a specific task

Resolution : The smallest increment of motion or distance that can be detected or controlled by the control system of a mechanism. The resolution of any joint is a function of encoder pulses per revolution and drive ratio, and dependent on the distance between the tool center point and the joint axis.

Robot Program : A robot communication program for IBM and compatible personal computers. Provides terminal emulation and utility functions. This program can record all of the user memory, and some of the system memory to disk files.

Maximum Speed : The compounded maximum speed of the tip of a robot moving at full extension with all joints moving simultaneously in complimentary directions. This speed is the theoretical maximum and should under no circumstances be used to estimate cycle time for a particular application. A better measure of real world speed is the standard twelve inch pick and place cycle time. For critical applications, the best indicator of achievable cycle time is a physical simulation.

Servo Controlled : Controlled by a driving signal which is determined by the error between the mechanism's present position and the desired output position.

Via Point : A point through which the robot's tool should pass without stopping; via points are programmed in order to move beyond obstacles or to bring the arm into a lower inertia posture for part of the motion.

Work Envelope : A three-dimensional shape that defines the boundaries that the robot manipulator can reach; also known as reach envelope.

Cartesian Coordinate Robot

A **cartesian coordinate robot** (also called **linear robot**) is an industrial robot whose three principal axes of control are linear (*i.e.* they move in a straight line rather than rotate) and are at right angles to each other. The three sliding joints correspond to moving the wrist up-down,in-out,back-forth. Among other advantages, this mechanical arrangement simplifies the Robot control arm solution. Cartesian coordinate robots with the horizontal member supported at both ends are sometimes called **Gantry robots**. They are often quite large.

A popular application for this type of robot is a computer numerical control machine (CNC machine). The simplest application is used in milling and drawing machines where a pen or router translates across an x-y plane while a tool is raised and lowered onto a surface to create a precise design.

Paint Robot

Industrial paint robots have been used for decades in automotive paint applications from the first hydraulic versions - which are still in use today but are of inferior quality and safety - to the latest electronic offerings. The newest robots are accurate and deliver results with uniform film builds and exact thicknesses.

Originally industrial paint robots were large and expensive, but today the price of the robots have come down to the point that general industry can now afford to have the same level of automation that only the big automotive manufacturers could once afford.

The selection of today's paint robot is much greater varying in size and payload to allow many configuration for painting items of all sizes. The prices vary as well as the new robot market becomes more competitive and the used market continues to expand.

Painting robots are generally equipped with five or six axis, three for the base motions and up to three for applicator orientation. These robots can be used in any explosion hazard Class 1 Division 1 environment.

FANUC P-50iA Paint Robot

Robocrane

The **Robocrane** is a kind of manipulator resembling a Stewart platform but using an octahedral assembly of cables instead of struts. Like the Stewart platform, the Robocrane has six degrees of freedom (x, y, z, pitch, roll, & yaw).

It was developed by James S. Albus of the National Institute of Standards and Technology (NIST), using the Real-Time Control System which is a hierarchical control system. Given its unusual ability to "fly" tools around a work site, it has many possible applications, including stone carving, ship building, bridge construction, inspection, pipe or beam fitting and welding.

Dr. Albus invented and developed a new generation of robot cranes based on six cables and six winches configured as a Stewart platform. The NIST RoboCraneTM has the capacity to lift and precisely manipulate heavy loads over large volumes with fine control in all six degrees of freedom. Laboratory RoboCranes have demonstrated the ability to manipulate tools such as saws, grinders, and welding torches, and to lift and precisely position heavy objects such as steel beams and cast iron pipe. In 1992, the RoboCrane was selected by Construction Equipment magazine as one of the 100 most significant new products of the year for construction and related industries. It was also selected by Popular Science magazine for the "Best of What's New" award as one of the 100 top products, technologies, and scientific achievements of 1992.

A version of the RoboCrane has been commercially developed for the United States Air Force to enable rapid paint stripping, inspection, and repainting of very large military aircraft such as the C-5 Galaxy. RoboCrane is expected to save the United States Air Force $8 million annually at each of its maintenance facilities. This project was recognized in 2008 by a National Laboratories Award for technology transfer. Potential future applications of the RoboCrane include ship building, construction of high rise buildings, highways, bridges, tunnels, and port facilities; cargo handling, ship-to-ship cargo transfer on the high seas, radioactive and toxic waste clean-up; and underwater applications such as salvage, drilling, cable maintenance, and undersea waste site management.

SCARA

The **SCARA** acronym stands for **Selective Compliance Assembly Robot Arm** or **Selective Compliance Articulated Robot Arm**.

In 1981, Sankyo Seiki, Pentel and NEC presented a completely new concept for assembly robots. The robot was developed under the guidance of Hiroshi Makino, a professor at the University of Yamanashi. The robot was called Selective Compliance Assembly Robot Arm, SCARA. Its arm was rigid in the Z-axis and pliable in the XY-axes, which allowed it to adapt to holes in the XY-axes.

By virtue of the SCARA's parallel-axis joint layout, the arm is slightly compliant in the X-Y direction but rigid in the 'Z' direction, hence the term : Selective Compliant. This is advantageous for many types of assembly operations, *i.e.*, inserting a round pin in a round hole without binding.

The second attribute of the SCARA is the jointed two-link arm layout similar to our human arms, hence the often-used term, Articulated. This feature allows the arm to extend into confined areas and then retract or "fold up" out of the way. This is advantageous for transferring parts from one cell to another or for loading/ unloading process stations that are enclosed.

SCARA's are generally faster and cleaner than comparable Cartesian robot systems. Their single pedestal mount requires a small footprint and provides an easy, unhindered form of mounting. On the other hand, SCARA's can be more expensive than comparable Cartesian systems and the controlling software requires inverse kinematics for linear interpolated moves. This software typically comes with the SCARA though and is usually transparent to the end-user.

Serial Manipulator

Serial manipulators are the most common industrial robots. They are designed as a series of links connected by motor-actuated joints that extend from a base to an end-effector. Often they have an anthropomorphic arm structure described as having a "shoulder", an "elbow", and a "wrist".

Serial robots usually have six joints, because it requires at least six degrees of freedom to place a manipulated object in an arbitrary position and orientation in the workspace of the robot.

A popular application for serial robots in today's industry is the pick-and-place assembly robot, called a SCARA robot, which has four degrees of freedom.

Structure

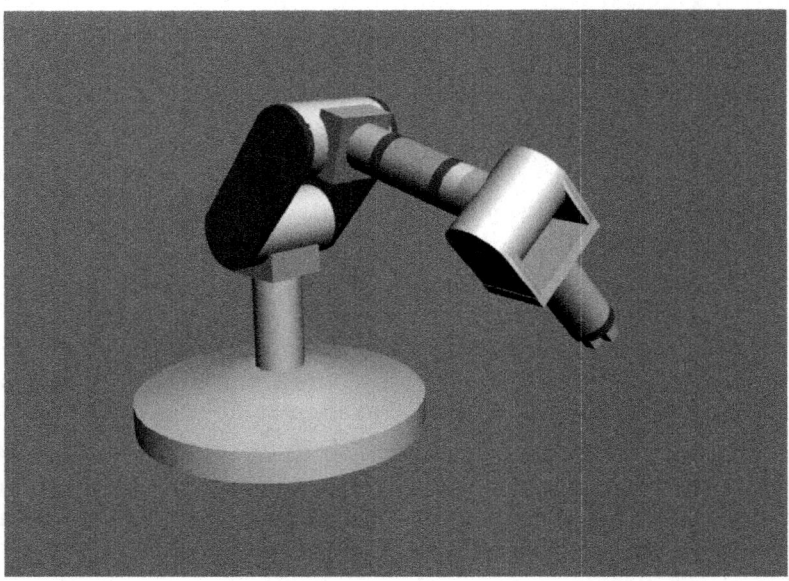

An example of a serial manipulator with six DOF in a kinematic chain.

In its most general form, a serial robot consists of a number of rigid links connected with joints. Simplicity considerations in manufacturing and control have led to robots with only revolute or prismatic joints and orthogonal, parallel and/or intersecting joint axes (instead of arbitrarily placed joint axes). Donald L. Pieper derived the first practically relevant result in this context, referred to as 321 kinematic structure : *The inverse kinematics of serial manipulators with six revolute joints, and with three consecutive joints intersecting, can be solved in closed-form, i.e. analytically* This result had a tremendous influence on the design of industrial robots.

The main advantage of a serial manipulator is a large workspace with respect to the size of the robot and the floor space it occupies. The main disadvantages of these robots are :

- the low stiffness inherent to an open kinematic structure,
- errors are accumulated and amplified from link to link,
- the fact that they have to carry and move the large weight of most of the actuators, and
- the relatively low effective load that they can manipulate.

Kinematics

The position and orientation of a robot's end effector are derived from the joint positions by means of a geometric model of the robot arm. For serial robots, the mapping from joint positions to end-effector pose is easy, the inverse mapping is more difficult. Therefore, most industrial robots have special designs that reduce the complexity of the inverse mapping.

Workspace

The reachable workspace of a robot's end-effector is the manifold of reachable frames. The dextrous workspace consists of the points of the reachable workspace where the robot can generate velocities that span the complete tangent space at that point, *i.e.*, it can translate the manipulated object with three degrees of freedom, and rotate the object with three degrees of rotation freedom. The relationships between joint space and Cartesian space coordinates of the object held by the robot are in general multiple-valued : the same pose can be reached by the serial arm in different ways, each with a different set of joint coordinates. Hence the reachable workspace of the robot is divided in configurations (also called assembly modes), in which the kinematic relationships are locally one-to-one.

Singularity

A singularity is a configuration of a serial manipulator in which the joint parameters no longer completely define the position and orientation of the end-effector. Singularities occur in configurations when joint axes align in a way that reduces the ability of the arm to position the end-effector. For example when a serial manipulator is fully extended it is in what is known as the boundary singularity.

At a singularity the end-effector loses one or more degrees of twist freedom (instantaneously, the end-effector cannot move in these directions). Serial robots with less than six independent joints are always singular in the sense that they can never span a six-dimensional twist space. This is often called an architectural singularity. A singularity is usually not an isolated point in the workspace of the robot, but a sub-manifold.

Redundant Manipulator

A redundant manipulator has more than six degrees of freedom which means that it has additional joint parameters that allow the configuration of the robot to change while it holds its end-effector in a fixed position and orientation.

A typical redundant manipulator has seven joints, for example three at the shoulder, one elbow joint and three at the wrist. This manipulator can move its elbow around a circle while it maintains a specific position and orientation of its end-effector.

A snake robot has many more than six degrees of freedom and is often called hyper-redundant.

Robot Welding

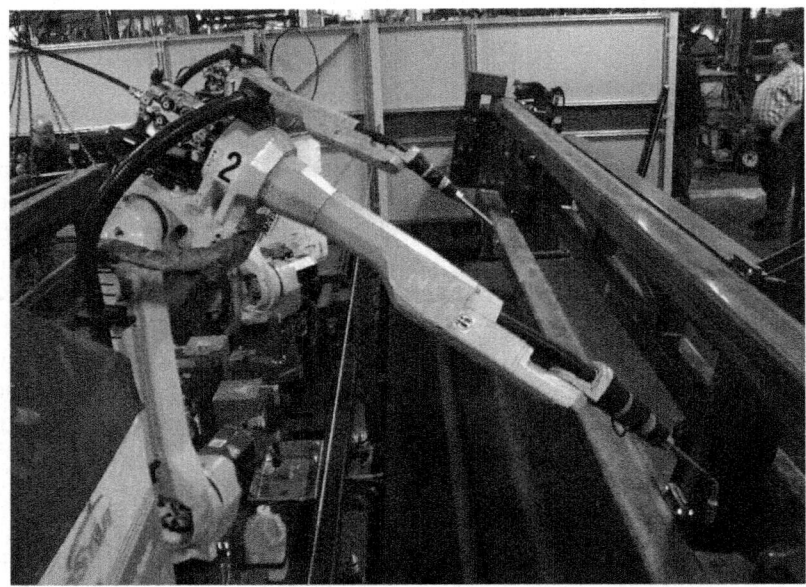

A set of six-axis robots used for welding.

Robot welding is the use of mechanized programmable tools (robots), which completely automate a welding process by both performing the weld and handling the part. Processes such as gas metal arc welding, while often automated, are not necessarily equivalent to robot welding, since a human operator sometimes prepares the materials to be welded. Robot welding is commonly used for resistance spot welding and arc welding in high production applications, such as the automotive industry.

Robot welding is a relatively new application of robotics, even though robots were first introduced into US industry during the 1960s. The use of robots in welding did not take off until the 1980s, when the automotive industry began using robots extensively for spot welding. Since then, both the number of robots used in industry and the number of their applications has grown greatly. In 2005, more than 120,000 robots were in use in North American industry, about half of them for welding. Growth is primarily limited by high equipment costs, and the resulting restriction to high-production applications.

Robot arc welding has begun growing quickly just recently, and already it commands about 20% of industrial robot applications. The major components of arc welding robots are the manipulator or the mechanical unit and the controller, which acts as the robot's "brain". The manipulator is what makes the robot move, and the design of these systems can be categorized into several common types, such as the SCARA robot and cartesian coordinate robot, which use different coordinate systems to direct the arms of the machine.

The robot may weld a pre-programmed position, be guided by machine vision, or by a combination of the two methods. However, the many benefits of robotic welding have proven to make it a technology that helps many original equipment manufacturers increase accuracy, repeat-ability, and throughput

The technology of signature image processing has been developed since the late 1990s for analyzing electrical data in real time collected from automated, robotic welding, thus enabling the optimization of welds.

5DX

The **5DX** is an automated X-ray inspection robot, which belongs to the set of automated test equipment robots and industrial robots utilizing machine vision. The 5DX is manufactured by Agilent Technologies. The 5DX is a non-destructive structural test X-ray machine, using laminography (tomography) to take 3D slices of an assembled printed circuit board. It is used in the assembled printed circuit board (PCB) electronics manufacturing industry to provide process feedback to a surface mount technology assembly line, as well as defect capture.

The 5DX is one of several tools used by many companies in the electronics manufacturing services sector to provide a means of structurally testing both the visible and hidden joints of components on assembled printed circuit boards for defects. These joints are referred to as integrated circuit to PCB interconnects.

Structural Defects

- Bridging (causing electrical shorts)
- Opens
- Missing or damaged components
- Non-wetting
- Billboards
- Tombstones
- Lifted leads
- Solder balls
- Voids (found in ball grid arrays and paste)
- Insufficient solder
- Excess solder
- Misalignment

5DX Technology

The 5DX uses a gantry robot to move the assembled printed circuit board underneath an X-ray source to be able to see the components' joints that require inspection. The positioning of board is guided with the use of CAD data, which represents the outer layers of a printed circuit board's electrical design.

A predefined laser surface map is used to bring the PCB into the plane of focus (depth of field) so that a slice of the component's joints can be taken. A slice will remove obstructions above or below the plane of focus so that only the regions of interest remain.

History of the 5DX

Agilent Technologies the OEM of the 5DX and x6000 has announced its plan to exit the X-ray business and focus resources in other areas of automated manufacturing test.

Product Revision History

- 3DX (end of life) originally developed by Four Pi Systems which was acquired by Hewlett Packard.
- 5DX Series I (end of life)
- 5DX Series 2/II/2L (end of life)
- 5DX Series 3
- 5DX Series 5000
- x6000 or x6k

ASEA IRB

The **ASEA IRB** is an industrial robot series for material handling, packing, transportation, polishing, welding, and grading. Built in 1975, the robot allowed movement in 5 axes with a lift capacity of 6 kg. It was the world's first fully electrically driven and microprocessor-controlled robot, using Intel's first chipset.

The ASEA IRB was constructed by Björn Weichbrodt, Ove Kullborg, Bengt Nilsson and Herbert Kaufmann and was manufactured by ASEA in Sweden/Västerås. The first model, IRB 6, was developed in 1972-1973 on assignment by the ASEA CEO Curt Nicolin and was shown for the first time at the end of August 1973. The example shown in the Swedish National Museum of Science and Technology is the first robot that was sold. It was bought by Magnussons in Genarp to wax and polish stainless steel tubes bent at 90° angles. This robot was donated to the museum during ASEA's 100-year anniversary in 1983.

The IRB 6 sold 1900 copies during the next 17 years . It became the Swedish symbol for a new Labour market, shared between man and robot.

Delta Robot

A **delta robot** is a type of parallel robot. It consists of three arms connected to universal joints at the base. The key design feature is the use of parallelograms in the arms, which maintains the orientation of the end effector. By contrast, a Stewart platform can change the orientation of its end effector.

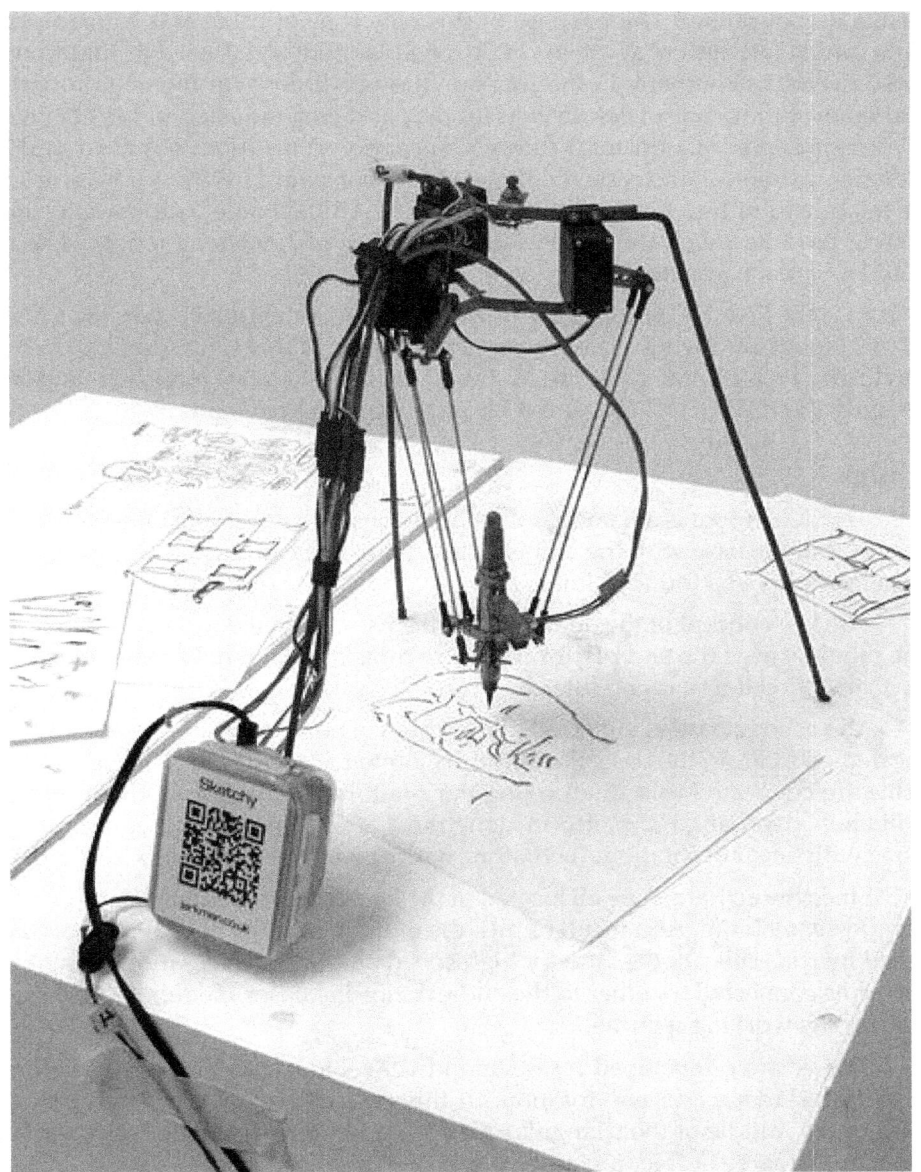

Sketchy, **a portrait-drawing delta robot**

Delta robots have popular usage in picking and packaging in factories because they can be quite fast, some executing up to 300 picks per minute.

History

The delta robot (a parallel arm robot) was invented in the early 1980s by a research team led by professor Reymond Clavel at the École Polytechnique Fédérale de Lausanne. The purpose of this new type of robot was to manipulate light and small objects at a very high speed, an industrial need at that time. In 1987, the Swiss company Demaurex purchased a license for the delta robot and started the production of delta robots for the packaging industry. In 1991 Reymond Clavel presented his doctoral thesis 'Conception d'un robot parallèle rapide à 4 degrés de liberté', and received the golden robot award in 1999 for his work and development of the delta robot. Also in 1999, ABB Flexible Automation started selling its delta robot, the FlexPicker. By the end of 1999 delta robots were also sold by Sigpack Systems.

In 2009, FANUC released the newest version of the Delta robot, the FANUC M-1iA Robot, and would later released variations of this Delta robot for heavier payloads. FANUC released the M-3iA in 2010 for heavier payloads, and most recently the FANUC M-2iA Robot for medium-sized payloads in 2012.

Design

The delta robot is a parallel robot, *i.e.* it consists of multiple kinematic chains connecting the base with the end-effector. The robot can also be seen as a spatial generalisation of a four-bar linkage.

The key concept of the delta robot is the use of parallelograms which restrict the movement of the end platform to pure translation, *i.e.* only movement in the X, Y or Z direction with no rotation.

The robot's base is mounted above the workspace and all the actuators are located on it. From the base, three middle jointed arms extend. The ends of these arms are connected to a small triangular platform. Actuation of the input links will move the triangular platform along the X, Y or Z direction. Actuation can be done with linear or rotational actuators, with or without reductions (direct drive).

Since the actuators are all located in the base, the arms can be made of a light composite material. As a result of this, the moving parts of the delta robot have a small inertia. This allows for very high speed and high accelerations. Having all the arms connected together to the end-effector increases the robot stiffness, but reduces its working volume.

The version developed by Reymond Clavel has four degrees of freedom : three translations and one rotation. In this case a fourth leg extends from the base to the middle of the triangular platform giving to the end effector a fourth, rotational degree of freedom around the vertical axis.

Currently other versions of the delta robot have been developed :

- Delta with 6 degrees of freedom : developed by the company Fanuc, on which a serial kinematic with 3 rotational degrees of freedom is placed on the end effector
- Delta with 4 degrees of freedom : developed by the company Adept, which has 4 parallelogram directly connected to the end-platform instead of having a forth leg coming in the middle of the end-effector

- Delta direct drive : a 3 degrees of freedom delta having the motor directly connected to the arms. Accelerations can be very high, from 30 up to 100 g.
- Pocket Delta : developed by the company Asyril, a version of the delta Robot adapted for high-precision applications.
- Delta Cube : developed by the LSRO, a delta robot built in a monolithic design, having flexure-hinges joints. This robot is adapted for ultra-high-precision applications.
- Several "linear delta" arrangements have been developed where the motors drive linear actuators rather than rotating an arm. Such linear delta arrangements can have much larger working volumes than rotational delta arrangements.

The majority of delta robots use rotary actuators. Vertical linear actuators have recently been used (using a linear delta design) to produce a novel design of 3D printer. These offer advantages over conventional leadscrew-based 3D printers of quicker access to a larger build volume for a comparable investment in hardware.

Applications

Industries that take advantage of the high speed of delta robots are the packaging industry, medical and pharmaceutical industry. For its stiffness it is also used for surgery. Other applications include high precision assembly operations in a clean room for electronic components. The structure of a delta robot can also be used to create haptic controllers. More recently, the technology has been adapted to 3D printers. These printers can be built for about a thousand dollars and compete well with the traditional Cartesian printers from the RepRap project.

Programmable Universal Machine for Assembly

The **PUMA** (*Programmable Universal Machine for Assembly*, or *Programmable Universal Manipulation Arm*) is an industrial robot arm developed by Victor Scheinman at pioneering robot company Unimation. Initially developed for General Motors, the PUMA was based on earlier designs Scheinman invented while at Stanford University.

Unimation produced PUMAs for years until being purchased by Westinghouse, and later by Swiss company Stäubli . Nokia Robotics manufactured about 1500 PUMA robots during the 1980s, the Puma-650 being their most popular model with customers. Some own Nokia Robotics products were also designed, like Nokia NS-16 Industrial Robot or NRS-15 . Nokia sold their Robotics division in 1990.

In 2002, General Motors Controls, Robotics and Welding (CRW) organization donated the original prototype PUMA robot to the Smithsonian Institution's National Museum of American History. It joins a distinguished collection of historically important robots that includes an early Unimate and the Odetics Odex 1.

RNA Automation

RNA Automation, a member of Rhein-Nadel Automation, was established in Birmingham UK in 1986, and has progressed into becoming the major supplier of parts handling equipment in the UK. The company operates in the area of specialised automation engineering, providing automatic parts handling equipment for high volume production in the cosmetics, pharmaceutical, electronics, food and metal working industries, with seven manufacturing facilities across Europe and North America and a network of sales and service outlets across the globe.

Company History

Founded in 1972, Rhein-Nadel Automation is the market leader throughout Europe and operates worldwide in parts handling technology. Rhein-Nadel Automation is a member of the Rheinnadel Group, which has been based since 1898 in the town of Aachen, Germany. Rhein-Nadel Automation have four manufacturing sites in Germany and individual manufacturing plants in the UK, Spain and Switzerland, together with a world-wide, decentralised distribution and service network with 28 subsidiaries.

Time Line

1968 The department of the needle factory, so far only responsible for the internal construction of operating material and machines, trades under the name Rheinnadel Maschinenbau and begins working for external customers.

1972 Rheinnadel Maschinenbau now also uses the firm name Rhein-Nadel Automation GmbH and concentrates exclusively on the area of Feeding Technology.

1980-1989 Rhein-Nadel Automation expands. The permanent establishment in Ergolding and the subsidiaries in Switzerland and Great Britain are founded.

1990-1999 Rhein-Nadel Automation continues to grow. The Spanish company Vibrant S.A. is purchased.

2000 With RNA Automated Systems Inc., the first subsidiary outside of Europe takes up its commercial activities in Canada.

2004 The Rheinnadel Group concentrates its activities on the area of Automation. RNA broadens its range with the components and continues to extend its development capacities.

RNA automation specialise in automated feeder and specialist handling systems offering vibratory bowl feeders, Linear Feeders, Centrifugal Feeders and Step Feeder systems. RNA also supply a range of specialist handling equipment including Vision Guided Robots, Tablet Inspection, Vision Inspection, Tray Loading and Bottle Handling.

Technology

RNA supplies a range of vibratory centrifugal feeders, linear and conveyor feed systems, and hopper elevator systems in different formates. RNA bowl tops and drive unitsis designed to work in harmony alongside each other and handle components from all areas of industry, providing the widest possible range of shapes and sizes made of any kinds of materials can be catered for.

In association with Hoppmann Corporation, RNA offers a full range of tooled centrifugal feeders. The centrifugal feeder systems can be interfaced and supplied as a complete packaging line with further production transportation via downstream conveyor systems to subsequent packaging operations such as capping, labelling, flow wrapping and cartoning machines.

The RNA Step Feeder is a compact low noise alternative to a bowl feeding system and in many circumstances components can be tooled without the aid or air. The RNA step feeder is manufactured in such a way that the static hopper, which has a very low filling height, can be loaded manually. The components are collected from the hopper by elevating plates, pre-sorted and fed to the top without any further feeding technique until they reach the desired transfer height.

Directional Carpet Linear feeders are developed to orientate components where a step feeder or bowl feeder is not suitable They enable gentle handling of oriented components from bulk to escapement devices creating product accumulation prior to a machine or handling unit. Large Heavy & bulky components contaminated in oil or swarf can be oriented and conveyed to a machine using a carpet feeder.

Specialist Handling

RNA are the sole agents in the UK & Ireland for the SVIA range of Vision Guided Robotic Systems. All systems have a Robotic arm for handling and manipulating the product, a camera system and share the same pc based control system and in most cases integrated with a standard ABB robot controller.

RNA alongside camera specialists Machine Vision Technology has developed a tablet inspection system to inspect and sort tablets up to and over 1000 parts per minute. The specification of a tablet inspection system is as follows :

A 600mm diameter vibratory bowl feeder in Stainless Steel and a variable speed controller. A speed of 800 - 1000 tablets per minute Outputs onto a conveyor with a reject sort facility. With up to four double speed progressive scan cameras and lighting mounted along the conveyor. A high speed PC with a 17 inch LCD display mounted in a SS enclosure. The system is designed to pharmaceutical standards The software complies with FDA requirements, all documentation is to CFR 21 pt 11, with the fully validated version option.

Unimate

Unimate was the first industrial robot, which worked on a General Motors assembly line at the Inland Fisher Guide Plant in Ewing Township, New Jersey, in 1961.

It was created by George Devol in the 1950s using his original patent filed in 1954 and granted in 1961. The patent begins :

The present invention relates to the automatic operation of machinery, particularly the handling apparatus, and to automatic control apparatus suited for such machinery.

Devol, together with Joseph Engelberger, his apprentice, started the world's first robot manufacturing company, Unimation.

The machine undertook the job of transporting die castings from an assembly line and welding these parts on auto bodies, a dangerous task for workers, who might be poisoned by exhaust gas or lose a limb if they were not careful.

The original Unimate consisted of a large computer-like box, joined to another box and was connected to an arm, with systematic tasks stored in a drum memory.

The Unimate also appeared on The Tonight Show hosted by Johnny Carson on which it knocked a golf ball into a cup, poured a beer, waved the orchestra conductor's baton and grasped an accordion and waved it around.

In 2003 the Unimate was inducted into the Robot Hall of Fame.

UWA Telerobot

The **UWA telerobot** is a teleoperable robot belonging to the school of mechanical and civil engineering at the University of Western Australia.

Development

The UWA telerobot is a historic landmark for the Internet and The University of Western Australia. It was the first telerobot device made available for general use on the Internet in 1994. The UWA telerobot was originally developed as part of a PhD thesis by Kenneth Taylor and was the subject of a later PhD by Barney Dalton.

The first robot on the Internet, a plastic toy robot with only 2 degrees of freedom, was placed online by a team under Ken Goldberg at the University of Southern California only three weeks before the UWA team released their website. The USC robot only lasted for seven months. The UWA robot is still online today, although the original robot was replaced in 1996 and the robot is no longer available for unrestricted public access, though interested parties can request permission.

Implementation

The current UWA telerobot is an ABB IRB1400 model 6 DOF serial chain robot fitted with a pneumatic gripper attachment. The robot runs on a standard ABB S4 Robot Controller linked to a Linux server and which in turn communicates with a second server running ABB's RobComm software and a National Instruments Labview application that was custom written for the task by James Trevelyan with assistance from Perth-based Icon Technologies and students. The robot forms part of the UWA telelabs project.

The Telerobot has undergone many changes to its control structure over time. Originally controlled via static html web pages using CGI, work by Dalton saw the introduction of an augmented reality Java-based interface that met with limited success. Control is currently by way of a downloadable LabVIEW client application that incorporates real-time video streaming, with access control provided by the Telelabs system.

Current Status

The robot continues to be the basis for research and group projects undertaken by Mechatronics Engineering students and staff at UWA, Primarily involving the addition of new features or capabilities to the system. The robot is also used as a teaching aid for a course in mechanisms and multibody systems run by Karol Miller.

Chapter 3

ROBOTICS

Robotics is the branch of mechanical engineering, electrical engineering and computer science that deals with the design, construction, operation, and application of robots, as well as computer systems for their control, sensory feedback, and information processing. These technologies deal with automated machines that can take the place of humans in dangerous environments or manufacturing processes, or resemble humans in appearance, behavior, and/or cognition. Many of today's robots are inspired by nature contributing to the field of bio-inspired robotics.

The concept of creating machines that can operate autonomously dates back to classical times, but research into the functionality and potential uses of robots did not grow substantially until the 20th century. Throughout history, robotics has been often seen to mimic human behavior, and often manage tasks in a similar fashion. Today, robotics is a rapidly growing field, as technological advances continue, research, design, and building new robots serve various practical purposes, whether domestically, commercially, or militarily. Many robots do jobs that are hazardous to people such as defusing bombs, mines and exploring shipwrecks.

ETYMOLOGY

The word *robotics* was derived from the word *robot*, which was introduced to the public by Czech writer Karel Čapek in his play *R.U.R. (Rossum's Universal Robots)*, which was published in 1920. The word *robot* comes from the Slavic word *robota*, which means labour. The play begins in a factory that makes artificial people called *robots*, creatures who can be mistaken for humans – similar to the modern ideas of androids. Karel Čapek himself did not coin the word. He wrote a short letter in reference to an etymology in the *Oxford English Dictionary* in which he named his brother Josef Čapek as its actual originator.

According to the *Oxford English Dictionary*, the word *robotics* was first used in print by Isaac Asimov, in his science fiction short story "Liar!", published in May 1941 in *Astounding Science Fiction*. Asimov was unaware that he was coining the term; since the science and technology of electrical devices is *electronics*, he assumed *robotics* already referred to the science and technology of robots. In some of Asimov's other works, he states that the first use of the word *robotics* was in his short story *Runaround* (Astounding Science Fiction, March 1942). However, the original publication of "Liar!" predates that of "Runaround" by ten months, so the former is generally cited as the word's origin.

HISTORY OF ROBOTICS

In 1927 the *Maschinenmensch* ("machine-human") gynoid humanoid robot (also called "Parody", "Futura", "Robotrix", or the "Maria impersonator") was the first depiction of a robot ever to appear on film was played by German actress Brigitte Helm in Fritz Lang's film Metropolis.

In 1942 the science fiction writer Isaac Asimov formulated his Three Laws of Robotics.

In 1948 Norbert Wiener formulated the principles of cybernetics, the basis of practical robotics.

Fully autonomous robots only appeared in the second half of the 20th century. The first digitally operated and programmable robot, the Unimate, was installed in 1961 to lift hot pieces of metal from a die casting machine and stack them. Commercial and industrial robots are widespread today and used to perform jobs more cheaply, or more accurately and reliably, than humans. They are also employed in jobs which are too dirty, dangerous, or dull to be suitable for humans. Robots are widely used in manufacturing, assembly, packing and packaging, transport, earth and space exploration, surgery, weaponry, laboratory research, safety, and the mass production of consumer and industrial goods.

Date	Significance	Robot Name	Inventor
Third century B.C. and earlier	One of the earliest descriptions of automata appears in the *Lie Zi* text, on a much earlier encounter between King Mu of Zhou and a mechanical engineer known as Yan Shi, an 'artificer'. The latter allegedly presented the king with a life-size, human-shaped figure of his mechanical handiwork.		Yan Shi
First century A.D. and earlier	Descriptions of more than 100 machines and automata, including a fire engine, a wind organ, a coin-operated machine, and a steam-powered engine, in *Pneumatica* and *Automata* by Heron of Alexandria		Ctesibius, Philo of Byzantium, Heron of Alexandria, and others
c. 420 B.C.E	A wooden, steam propelled bird, which was able to fly		Archytas of Tarentum
1206	Created early humanoid automata, programmable automaton band	Robot band, hand-washing automaton, automated moving peacocks	Al-Jazari

1495	Designs for a humanoid robot	Mechanical knight	Leonardo da Vinci
1738	Mechanical duck that was able to eat, flap its wings, and excrete	Digesting Duck	Jacques de Vaucanson
1898	Nikola Tesla demonstrates first radio-controlled vessel.	Teleautomaton	Nikola Tesla
1921	First fictional automatons called "robots" appear in the play *R.U.R.*	Rossum's Universal Robots	Karel Čapek
1930s	Humanoid robot exhibited at the 1939 and 1940 World's Fairs	Elektro	Westinghouse Electric Corporation
1948	Simple robots exhibiting biological behaviors	Elsie and Elmer	William Grey Walter
1956	First commercial robot, from the Unimation company founded by George Devol and Joseph Engelberger, based on Devol's patents	Unimate	George Devol
1961	First installed industrial robot.	Unimate	George Devol
1973	First industrial robot with six electromechanically driven axes	Famulus	KUKA Robot Group
1974	The world's first microcomputer controlled electric industrial robot, IRB 6 from ASEA, was delivered to a small mechanical engineering company in southern Sweden. The design of this robot had been patented already 1972.	IRB 6	ABB Robot Group
1975	Programmable universal manipulation arm, a Unimation product	PUMA	Victor Scheinman

ROBOTIC ASPECTS

There are many types of robots; they are used in many different environments and for many different uses, although being very diverse in application and form they all share three basic similarities when it comes to their construction.

Robotic Construction

First : Robots all have some kind of mechanical construction, a frame, form or shape that usually is the solution/result for a set task or problem. For example if you want a robot to travel across heavy dirt or mud, you might think to use tracker treads, So the form your robot might be a box with tracker treads. The treads being the mechanical construction for traveling across the problem of heavy mud or dirt. This mechanical aspect usually deals with a real world application of an object or of itself, example lifting, moving, carrying, flying, swimming, running, walking...*etc.* The mechanical aspect is mostly the creators solution to completing the assign task and dealing with the physics of the environment around it, example: gravity, friction, resistance...*etc.* Form follows function.

Electrical Aspect

Second : Robots have an electrical aspect to them in them, in the form of wires, sensors, circuits, batteries ...*etc.* Example : the tracker tread robot that was mention earlier, it will need some kind of power to actually move the tracker treads. That power comes in the form of electricity, which will have to travel through a wire and originate from a battery, a basic electrical circuit. Even gas powered machines that get their power mainly form gas still require an electrical current to start the gas using process which is why most gas powered machines like cars, have batteries. The electrical aspect of robots is used for movement : as in the control of motors which are used mostly were motion is needed. Sensing : electrical signals are used to determine things like heat, sound, position, and energy status. Operation : robots need some level of electrical energy supplied to their motors and/or sensors in order to be turned on, and do basic operations.

Third : All robots contain some level of computer programming (code), A program is how a robot decides when or how to do something. For example : what if you wanted the tractor tread robot (from our previous examples) to move across a muddy road, even though it has the correct mechanical construction, and it receives the correct amount of power from its battery, it doesn't go anywhere. Why? What actually tells the robot to move? A program. Even if you had a remote control and you pushed a button telling it to move forward it will still need a program relating the button you pushed to the action of moving forward. Programs are the core essence of a robot, it could have excellent mechanical/electrical construction, but if its program is poorly constructed its performance will be very poor or it may not perform at all. There are three different types of robotic programs, RC,

AI and hybrid. RC stands for Remote Control, a robot with this type of program has a preexisting set of commands that it will only do if and when it receives a signal from a control source, most of the time the control source is a human being with a remote control. AI stand for artificial Intelligence, robots with this kind of programing interact with their environment on their own without a control source. Robots with AI create solutions to objects/problems they encounter by using their preexisting programing to decide, understand, learn and/or create. Hybrid is a form of program that incorporates both AI and RC functions, For example : your robot may work completely on its own, encounter a problem, come up with two solutions like an AI system, and then rely completely on you to decide what to do like a RC system. Robots have three aspect of construction mechanical, electrical and programing.

COMPONENTS

Power Source

At present mostly (lead-acid) batteries are used as a power source. Many different types of batteries can be used as a power source for robots. They range from lead acid batteries which are safe and have relatively long shelf lives but are rather heavy to silver cadmium batteries that are much smaller in volume and are currently much more expensive. Designing a battery powered robot needs to take into account factors such as safety, cycle lifetime and weight. Generators, often some type of internal combustion engine, can also be used. However, such designs are often mechanically complex and need fuel, require heat dissipation and are relatively heavy. A tether connecting the robot to a power supply would remove the power supply from the robot entirely. This has the advantage of saving weight and space by moving all power generation and storage components elsewhere. However, this design does come with the drawback of constantly having a cable connected to the robot, which can be difficult to manage. Potential power sources could be :

- pneumatic (compressed gases)
- hydraulics (liquids)
- flywheel energy storage
- organic garbage (through anaerobic digestion)
- faeces (human, animal); may be interesting in a military context as faeces of small combat groups may be reused for the energy requirements of the robot assistant

Actuation

Actuators are like the "muscles" of a robot, the parts which convert stored energy into movement. By far the most popular actuators are electric motors that spin a wheel or gear, and linear actuators that control industrial robots in factories.

But there are some recent advances in alternative types of actuators, powered by electricity, chemicals, or compressed air.

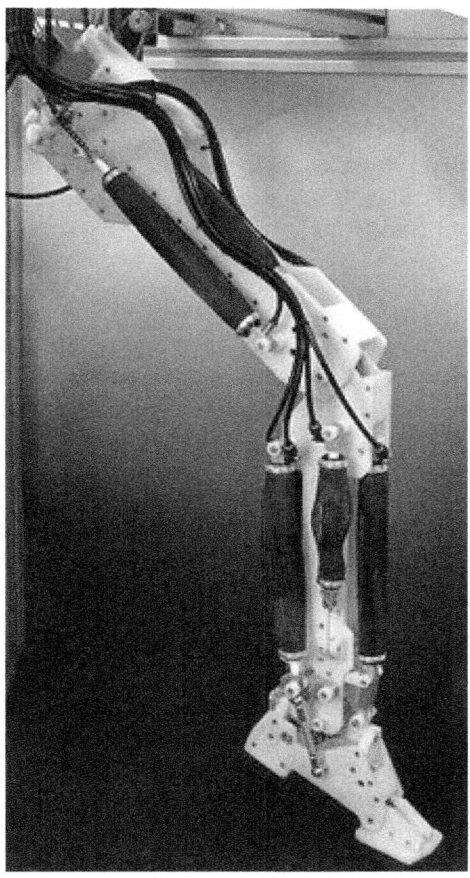

A robotic leg powered by air muscles

Electric Motors

The vast majority of robots use electric motors, often brushed and brushless DC motors in portable robots or AC motors in industrial robots and CNC machines. These motors are often preferred in systems with lighter loads, and where the predominant form of motion is rotational.

Linear Actuators

Various types of linear actuators move in and out instead of by spinning, and often have quicker direction changes, particularly when very large forces are needed such as with industrial robotics. They are typically powered by compressed air (pneumatic actuator) or an oil (hydraulic actuator).

Series Elastic Actuators

A spring can be designed as part of the motor actuator, to allow improved force control. It has been used in various robots, particularly walking humanoid robots.

Air Muscles

Pneumatic artificial muscles, also known as air muscles, are special tubes that contract (typically up to 40%) when air is forced inside them. They have been used for some robot applications.

Muscle Wire

Muscle wire, also known as Shape Memory Alloy, Nitinol or Flexinol Wire, is a material that contracts slightly (typically under 5%) when electricity runs through it. They have been used for some small robot applications.

Electroactive Polymers

EAPs or EPAMs are a new plastic material that can contract substantially (up to 380% activation strain) from electricity, and have been used in facial muscles and arms of humanoid robots, and to allow new robots to float, fly, swim or walk.

Piezo Motors

Recent alternatives to DC motors are piezo motors or ultrasonic motors. These work on a fundamentally different principle, whereby tiny piezoceramic elements, vibrating many thousands of times per second, cause linear or rotary motion. There are different mechanisms of operation; one type uses the vibration of the piezo elements to walk the motor in a circle or a straight line. Another type uses the piezo elements to cause a nut to vibrate and drive a screw. The advantages of these motors are nanometer resolution, speed, and available force for their size. These motors are already available commercially, and being used on some robots.

Elastic Nanotubes

Elastic nanotubes are a promising artificial muscle technology in early-stage experimental development. The absence of defects in carbon nanotubes enables these filaments to deform elastically by several percent, with energy storage levels of perhaps 10 J/cm^3 for metal nanotubes. Human biceps could be replaced with an 8 mm diameter wire of this material. Such compact "muscle" might allow future robots to outrun and outjump humans.

Sensing

Sensors allow robots to receive information about a certain measurement of the environment, or internal components. This is essential for robots to perform their tasks, and act upon any changes in the environment to calculate the appropriate response. They are used for various forms of measurements, to give the robots warnings about safety or malfunctions, and to provide real time information of the task it is performing.

Touch

Current robotic and prosthetic hands receive far less tactile information than the human hand. Recent research has developed a tactile sensor array that mimics the mechanical properties and touch receptors of human fingertips. The sensor array is constructed as a rigid core surrounded by conductive fluid contained by an elastomeric skin. Electrodes are mounted on the surface of the rigid core and are connected to an impedance-measuring device within the core. When the artificial skin touches an object the fluid path around the electrodes is deformed, producing impedance changes that map the forces received from the object. The researchers expect that an important function of such artificial fingertips will be adjusting robotic grip on held objects.

Scientists from several European countries and Israel developed a prosthetic hand in 2009, called SmartHand, which functions like a real one — allowing patients to write with it, type on a keyboard, play piano and perform other fine movements. The prosthesis has sensors which enable the patient to sense real feeling in its fingertips.

Vision

Computer vision is the science and technology of machines that see. As a scientific discipline, computer vision is concerned with the theory behind artificial systems that extract information from images. The image data can take many forms, such as video sequences and views from cameras.

In most practical computer vision applications, the computers are pre-programmed to solve a particular task, but methods based on learning are now becoming increasingly common.

Computer vision systems rely on image sensors which detect electromagnetic radiation which is typically in the form of either visible light or infra-red light. The sensors are designed using solid-state physics. The process by which light propagates and reflects off surfaces is explained using optics. Sophisticated image sensors even require quantum mechanics to provide a complete understanding of the image formation process. Robots can also be equipped with multiple vision sensors to be better able to compute the sense of depth in the environment. Like human eyes, robots' "eyes" must also be able to focus on a particular area of interest, and also adjust to variations in light intensities.

There is a subfield within computer vision where artificial systems are designed to mimic the processing and behavior of biological system, at different levels of complexity. Also, some of the learning-based methods developed within computer vision have their background in biology.

Other

Other common forms of sensing in robotics use LIDAR, RADAR and SONAR.

Manipulation

Puma, one of the first industrial robots

Robots need to manipulate objects; pick up, modify, destroy, or otherwise have an effect. Thus the "hands" of a robot are often referred to as *end effectors*, while the "arm" is referred to as a *manipulator*. Most robot arms have replaceable effectors, each allowing them to perform some small range of tasks. Some have a fixed manipulator which cannot be replaced, while a few have one very general purpose manipulator, for example a humanoid hand.

Mechanical Grippers

One of the most common effectors is the gripper. In its simplest manifestation it consists of just two fingers which can open and close to pick up and let go of a range of small objects. Fingers can for example be made of a chain with a metal wire run through it. Hands that resemble and work more like a human hand include the Shadow Hand, the Robonaut hand, ... Hands that are of a mid-level complexity include the Delft hand. Mechanical grippers can come in various types, including friction and encompassing jaws. Friction jaws use all the force of the gripper to hold the object in place using friction. Encompassing jaws cradle the object in place, using less friction.

Vacuum Grippers

Vacuum grippers are very simple astrictive devices, but can hold very large loads provided the prehension surface is smooth enough to ensure suction.

Pick and place robots for electronic components and for large objects like car windscreens, often use very simple vacuum grippers.

General Purpose Effectors

Some advanced robots are beginning to use fully humanoid hands, like the Shadow Hand, MANUS, and the Schunk hand. These are highly dexterous manipulators, with as many as 20 degrees of freedom and hundreds of tactile sensors.

LOCOMOTION

Rolling Robots

For simplicity most mobile robots have four wheels or a number of continuous tracks. Some researchers have tried to create more complex wheeled robots with only one or two wheels. These can have certain advantages such as greater efficiency and reduced parts, as well as allowing a robot to navigate in confined places that a four wheeled robot would not be able to.

Two-wheeled Balancing Robots

Balancing robots generally use a gyroscope to detect how much a robot is falling and then drive the wheels proportionally in the same direction, to counterbalance the fall at hundreds of times per second, based on the dynamics of an inverted pendulum. Many different balancing robots have been designed. While the Segway is not commonly thought of as a robot, it can be thought of as a component of a robot, when used as such Segway refer to them as RMP (Robotic Mobility Platform). An example of this use has been as NASA's Robonaut that has been mounted on a Segway.

One-wheeled Balancing Robots

A one-wheeled balancing robot is an extension of a two-wheeled balancing robot so that it can move in any 2D direction using a round ball as its only wheel. Several one-wheeled balancing robots have been designed recently, such as Carnegie Mellon University's "Ballbot" that is the approximate height and width of a person, and Tohoku Gakuin University's "BallIP". Because of the long, thin shape and ability to maneuver in tight spaces, they have the potential to function better than other robots in environments with people.

Spherical Orb Robots

Several attempts have been made in robots that are completely inside a spherical ball, either by spinning a weight inside the ball, or by rotating the outer shells of the sphere. These have also been referred to as an orb bot or a ball bot.

Six-wheeled Robots

Using six wheels instead of four wheels can give better traction or grip in outdoor terrain such as on rocky dirt or grass.

Tracked Robots

Tank tracks provide even more traction than a six-wheeled robot. Tracked wheels behave as if they were made of hundreds of wheels, therefore are very common for outdoor and military robots, where the robot must drive on very rough terrain. However, they are difficult to use indoors such as on carpets and smooth floors. Examples include NASA's Urban Robot "Urbie".

Walking Applied To Robots

Walking is a difficult and dynamic problem to solve. Several robots have been made which can walk reliably on two legs, however none have yet been made which are as robust as a human. There has been much study on human inspired walking, such as AMBER lab which was established in 2008 by the Mechanical Engineering Department at Texas A&M University. Many other robots have been built that walk on more than two legs, due to these robots being significantly easier to construct. Walking robots can be used for uneven terrains, which would provide better mobility and energy efficiency than other locomotion methods. Hybrids too have been proposed in movies such as I, Robot, where they walk on 2 legs and switch to 4 (arms+legs) when going to a sprint. Typically, robots on 2 legs can walk well on flat floors and can occasionally walk up stairs. None can walk over rocky, uneven terrain. Some of the methods which have been tried are :

ZMP Technique

The Zero Moment Point (ZMP) is the algorithm used by robots such as Honda's ASIMO. The robot's onboard computer tries to keep the total inertial forces (the combination of Earth's gravity and the acceleration and deceleration of walking), exactly opposed by the floor reaction force (the force of the floor pushing back on the robot's foot). In this way, the two forces cancel out, leaving no moment (force causing the robot to rotate and fall over). However, this is not exactly how a human walks, and the difference is obvious to human observers, some of whom have pointed out that ASIMO walks as if it needs the lavatory. ASIMO's walking algorithm is not static, and some dynamic balancing is used . However, it still requires a smooth surface to walk on.

Hopping

Several robots, built in the 1980s by Marc Raibert at the MIT Leg Laboratory, successfully demonstrated very dynamic walking. Initially, a robot with only one leg, and a very small foot, could stay upright simply by hopping. The movement is the same as that of a person on a pogo stick. As the robot falls to one side, it would jump slightly in that direction, in order to catch itself. Soon, the algorithm was generalised to two and four legs. A bipedal robot was demonstrated running and even performing somersaults. A quadruped was also demonstrated which could trot, run, pace, and bound.

Dynamic Balancing (Controlled Falling)

A more advanced way for a robot to walk is by using a dynamic balancing algorithm, which is potentially more robust than the Zero Moment Point technique, as it constantly monitors the robot's motion, and places the feet in order to maintain stability. This technique was recently demonstrated by Anybots' Dexter Robot, which is so stable, it can even jump. Another example is the TU Delft Flame.

Passive Dynamics

Perhaps the most promising approach utilizes passive dynamics where the momentum of swinging limbs is used for greater efficiency. It has been shown that totally unpowered humanoid mechanisms can walk down a gentle slope, using only gravity to propel themselves. Using this technique, a robot need only supply a small amount of motor power to walk along a flat surface or a little more to walk up a hill. This technique promises to make walking robots at least ten times more efficient than ZMP walkers, like ASIMO.

Other Methods Of Locomotion

Flying

Two robot snakes. Left one has 64 motors (with 2 degrees of freedom per segment), the right one 10.

A modern passenger airliner is essentially a flying robot, with two humans to manage it. The autopilot can control the plane for each stage of the journey, including takeoff, normal flight, and even landing. Other flying robots are uninhabited, and are known as unmanned aerial vehicles (UAVs). They can be smaller and lighter without a human pilot on board, and fly into dangerous territory for military surveillance missions. Some can even fire on targets under command. UAVs are also being developed which can fire on targets automatically, without the need for a command from a human. Other flying robots include cruise missiles, the Entomopter, and the Epson micro helicopter robot. Robots such as the Air Penguin, Air Ray, and Air Jelly have lighter-than-air bodies, propelled by paddles, and guided by sonar.

Snaking

Several snake robots have been successfully developed. Mimicking the way real snakes move, these robots can navigate very confined spaces, meaning they may one day be used to search for people trapped in collapsed buildings. The Japanese ACM-R5 snake robot can even navigate both on land and in water.

Skating

A small number of skating robots have been developed, one of which is a multi-mode walking and skating device. It has four legs, with unpowered wheels, which can either step or roll. Another robot, Plen, can use a miniature skateboard or roller-skates, and skate across a desktop.

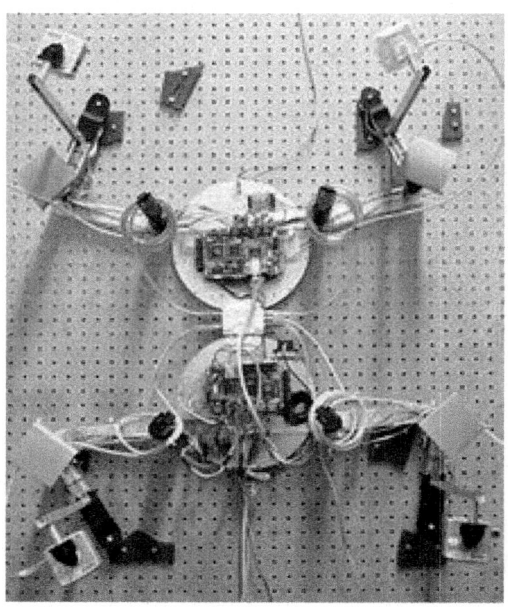

Capuchin Climbing Robot.

Climbing

Several different approaches have been used to develop robots that have the ability to climb vertical surfaces. One approach mimics the movements of a human climber on a wall with protrusions; adjusting the center of mass and moving each limb in turn to gain leverage. An example of this is Capuchin, built by Dr. Ruixiang Zhang at Stanford University, California. Another approach uses the specialized toe pad method of wall-climbing geckoes, which can run on smooth surfaces such as vertical glass. Examples of this approach include Wallbot and Stickybot. China's "Technology Daily" November 15, 2008 reported New Concept Aircraft (ZHUHAI) Co., Ltd. Dr. Li Hiu Yeung and his research group have recently successfully developed the bionic gecko robot "Speedy Freelander". According to Dr. Li introduction, this gecko robot can rapidly climbing up and down in a variety of building walls, ground and vertical wall fissure or walking upside down on the ceiling, it is able to adapt on smooth glass, rough or sticky dust walls as well as the various surface of metallic materials and also can automatically identify obstacles, circumvent the bypass and flexible and realistic movements. Its flexibility and speed are comparable to the natural gecko. A third approach is to mimic the motion of a snake climbing a pole.

Swimming (Piscine)

It is calculated that when swimming some fish can achieve a propulsive efficiency greater than 90%. Furthermore, they can accelerate and maneuver far better than any man-made boat or submarine, and produce less noise and water disturbance. Therefore, many researchers studying underwater robots would like to copy this type of locomotion. Notable examples are the Essex University Computer Science Robotic Fish, and the Robot Tuna built by the Institute of Field Robotics, to analyze and mathematically model thunniform motion. The Aqua Penguin, designed and built by Festo of Germany, copies the streamlined shape and propulsion by front "flippers" of penguins. Festo have also built the Aqua Ray and Aqua Jelly, which emulate the locomotion of manta ray, and jellyfish, respectively.

Sailing

Sailboat robots have also been developed in order to make measurements at the surface of the ocean. A typical sailboat robot is *Vaimos* built by IFREMER and ENSTA-Bretagne. Since the propulsion of sailboat robots uses the wind, the energy of the batteries is only used for the computer, for the communication and for the actuators (to tune the rudder and the sail). If the robot is equipped with solar panels, the robot could theoretically navigate forever. The two main competitions of sailboat robots are WRSC, which takes place every year in Europe, and Sailbot.

Environmental Interaction and Navigation

Though a significant percentage of robots in commission today are either human controlled, or operate in a static environment, there is an increasing interest in robots that can operate autonomously in a dynamic environment. These robots require some combination of navigation hardware and software in order to traverse their environment. In particular unforeseen events (*e.g.* people and other obstacles that are not stationary) can cause problems or collisions. Some highly advanced robots such as ASIMO, and Meinü robot have particularly good robot navigation hardware and software. Also, self-controlled cars, Ernst Dickmanns' driverless car, and the entries in the DARPA Grand Challenge, are capable of sensing the environment well and subsequently making navigational decisions based on this information. Most of these robots employ a GPS navigation device with waypoints, along with radar, sometimes combined with other sensory data such as LIDAR, video cameras, and inertial guidance systems for better navigation between waypoints.

Human-robot Interaction

If robots are to work effectively in homes and other non-industrial environments, the way they are instructed to perform their jobs, and especially how they will be told to stop will be of critical importance. The people who interact with them may have little or no training in robotics, and so any interface will need to be extremely intuitive. Science fiction authors also typically assume that robots will eventually be capable of communicating with humans through speech, gestures, and facial expressions, rather than a command-line interface. Although speech would be the most natural way for the human to communicate, it is unnatural for the robot. It will probably be a long time before robots interact as naturally as the fictional C-3PO.

Speech Recognition

Interpreting the continuous flow of sounds coming from a human, in real time, is a difficult task for a computer, mostly because of the great variability of speech. The same word, spoken by the same person may sound different depending on local acoustics, volume, the previous word, whether or not the speaker has a cold, *etc.*. It becomes even harder when the speaker has a different accent. Nevertheless, great strides have been made in the field since Davis, Biddulph, and Balashek designed the first "voice input system" which recognized "ten digits spoken by a single user with 100% accuracy" in 1952. Currently, the best systems can recognize continuous, natural speech, up to 160 words per minute, with an accuracy of 95%.

Robotic Voice

Other hurdles exist when allowing the robot to use voice for interacting with humans. For social reasons, synthetic voice proves suboptimal as a communication medium, making it necessary to develop the emotional component of robotic voice through various techniques.

Gestures

One can imagine, in the future, explaining to a robot chef how to make a pastry, or asking directions from a robot police officer. In both of these cases, making hand gestures would aid the verbal descriptions. In the first case, the robot would be recognizing gestures made by the human, and perhaps repeating them for confirmation. In the second case, the robot police officer would gesture to indicate "down the road, then turn right". It is likely that gestures will make up a part of the interaction between humans and robots. A great many systems have been developed to recognize human hand gestures.

Facial Expression

Facial expressions can provide rapid feedback on the progress of a dialog between two humans, and soon may be able to do the same for humans and robots. Robotic faces have been constructed by Hanson Robotics using their elastic polymer called Frubber, allowing a large number of facial expressions due to the elasticity of the rubber facial coating and embedded subsurface motors (servos). The coating and servos are built on a metal skull. A robot should know how to approach a human, judging by their facial expression and body language. Whether the person is happy, frightened, or crazy-looking affects the type of interaction expected of the robot. Likewise, robots like Kismet and the more recent addition, Nexi can produce a range of facial expressions, allowing it to have meaningful social exchanges with humans.

Artificial Emotions

Artificial emotions can also be generated, composed of a sequence of facial expressions and/or gestures. As can be seen from the movie Final Fantasy : The Spirits Within, the programming of these artificial emotions is complex and requires a large amount of human observation. To simplify this programming in the movie, presets were created together with a special software program. This decreased the amount of time needed to make the film. These presets could possibly be transferred for use in real-life robots.

Personality

Many of the robots of science fiction have a personality, something which may or may not be desirable in the commercial robots of the future. Nevertheless, researchers are trying to create robots which appear to have a personality : *i.e.* they use sounds, facial expressions, and body language to try to convey an internal state, which may be joy, sadness, or fear. One commercial example is Pleo, a toy robot dinosaur, which can exhibit several apparent emotions.

Control

The mechanical structure of a robot must be controlled to perform tasks. The control of a robot involves three distinct phases – perception, processing, and action (robotic paradigms). Sensors give information about the environment or the robot itself (*e.g.* the position of its joints or its end effector). This information is then processed to be stored or transmitted, and to calculate the appropriate signals to the actuators (motors) which move the mechanical.

RuBot II can resolve manually Rubik cubes

The processing phase can range in complexity. At a reactive level, it may translate raw sensor information directly into actuator commands. Sensor fusion may first be used to estimate parameters of interest (*e.g.* the position of the robot's gripper) from noisy sensor data. An immediate task (such as moving the gripper in a certain direction) is inferred from these estimates. Techniques from control theory convert the task into commands that drive the actuators.

At longer time scales or with more sophisticated tasks, the robot may need to build and reason with a "cognitive" model. Cognitive models try to represent the robot, the world, and how they interact. Pattern recognition and computer vision can be used to track objects. Mapping techniques can be used to build maps of the world. Finally, motion planning and other artificial intelligence techniques may be used to figure out how to act. For example, a planner may figure out how to achieve a task without hitting obstacles, falling over, *etc*.

Autonomy Levels

Control systems may also have varying levels of autonomy.

1. Direct interaction is used for haptic or tele-operated devices, and the human has nearly complete control over the robot's motion.
2. Operator-assist modes have the operator commanding medium-to-high-level tasks, with the robot automatically figuring out how to achieve them.
3. An autonomous robot may go for extended periods of time without human interaction. Higher levels of autonomy do not necessarily require more complex cognitive capabilities. For example, robots in assembly plants are completely autonomous, but operate in a fixed pattern.

Another classification takes into account the interaction between human control and the machine motions.

1. Teleoperation. A human controls each movement, each machine actuator change is specified by the operator.
2. Supervisory. A human specifies general moves or position changes and the machine decides specific movements of its actuators.
3. Task-level autonomy. The operator specifies only the task and the robot manages itself to complete it.
4. Full autonomy. The machine will create and complete all its tasks without human interaction.

ROBOTICS RESEARCH

Much of the research in robotics focuses not on specific industrial tasks, but on investigations into new types of robots, alternative ways to think about or design robots, and new ways to manufacture them but other investigations, such as MIT's cyberflora project, are almost wholly academic.

A first particular new innovation in robot design is the opensourcing of robot-projects. To describe the level of advancement of a robot, the term "Generation Robots" can be used. This term is coined by Professor Hans Moravec, Principal Research Scientist at the Carnegie Mellon University Robotics Institute in describing the near future evolution of robot technology. *First generation* robots, Moravec predicted in 1997, should have an intellectual capacity comparable to perhaps a lizard and should become available by 2010. Because the *first generation* robot would be incapable of learning, however, Moravec predicts that the *second generation* robot would be an improvement over the *first* and become available by 2020, with the intelligence maybe comparable to that of a mouse. The *third generation* robot should have the intelligence comparable to that of a monkey. Though *fourth generation* robots, robots with human intelligence, professor Moravec predicts, would become possible, he does not predict this happening before around 2040 or 2050.

The second is Evolutionary Robots. This is a methodology that uses evolutionary computation to help design robots, especially the body form, or motion and behavior controllers. In a similar way to natural evolution, a large population of robots is allowed to compete in some way, or their ability to perform a task is measured using a fitness function. Those that perform worst are removed from the population, and replaced by a new set, which have new behaviors based on those of the winners. Over time the population improves, and eventually a satisfactory robot may appear. This happens without any direct programming of the robots by the researchers. Researchers use this method both to create better robots, and to explore the nature of evolution. Because the process often requires many generations of robots to be simulated, this technique may be run entirely or mostly in simulation, then tested on real robots once the evolved algorithms are good enough. Currently, there are about 1 million industrial robots toiling around the world, and Japan is the top country having high density of utilizing robots in its manufacturing industry.

Dynamics and Kinematics

The study of motion can be divided into kinematics and dynamics. Direct kinematics refers to the calculation of end effector position, orientation, velocity, and acceleration when the corresponding joint values are known. Inverse kinematics refers to the opposite case in which required joint values are calculated for given end effector values, as done in path planning. Some special aspects of kinematics include handling of redundancy (different possibilities of performing the same movement), collision avoidance, and singularity avoidance. Once all relevant positions, velocities, and accelerations have been calculated using kinematics, methods from the field of dynamics are used to study the effect of forces upon these movements. Direct dynamics refers to the calculation of accelerations in the robot once the applied forces are known. Direct dynamics is used in computer simulations of the robot. Inverse dynamics refers to the calculation of the actuator forces necessary to create a prescribed end effector acceleration. This information can be used to improve the control algorithms of a robot.

In each area mentioned above, researchers strive to develop new concepts and strategies, improve existing ones, and improve the interaction between these areas. To do this, criteria for "optimal" performance and ways to optimize design, structure, and control of robots must be developed and implemented.

EDUCATION AND TRAINING

Robotics engineers design robots, maintain them, develop new applications for them, and conduct research to expand the potential of robotics. Robots have become a popular educational tool in some middle and high schools, as well as in numerous youth summer camps, raising interest in programming, artificial intelligence and robotics among students. First-year computer science courses at several universities now include programming of a robot in addition to traditional software engineering-based coursework. On the Technion I&M faculty an educational laboratory was established in 1994 by Dr. Jacob Rubinovitz.

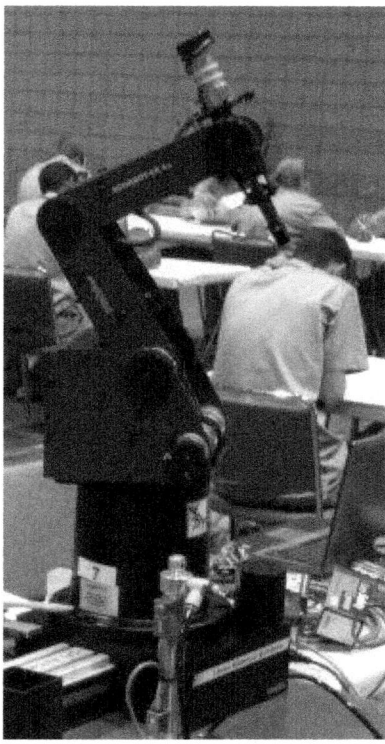

The SCORBOT-ER 4u – educational robot.

Career Training

Universities offer bachelors, masters, and doctoral degrees in the field of robotics. Vocational schools offer robotics training aimed at careers in robotics.

Certification

The Robotics Certification Standards Alliance (RCSA) is an international robotics certification authority that confers various industry- and educational-related robotics certifications.

Summer Robotics Camp

Several national summer camp programs include robotics as part of their core curriculum, including Digital Media Academy, RoboTech, and Cybercamps. In addition, youth summer robotics programs are frequently offered by celebrated museums such as the American Museum of Natural History and The Tech Museum of Innovation in Silicon Valley, CA, just to name a few. An educational robotics lab also exists at the IE & mgmnt Faculty of the Technion. It was created by Dr. Jacob Rubinovitz.

Robotics Afterschool Programs

Many schools across the country are beginning to add robotics programs to their after school curriculum. Two main programs for afterschool robotics are Botball and FIRST Robotics Competition.

EMPLOYMENT

A robot technician builds small all-terrain robots.

Robotics is an essential component in many modern manufacturing environments. As factories increase their use of robots, the number of robotics-related jobs grow and have been observed to be steadily rising.

Chapter 4
ARTIFICIAL INTELLIGENCE

Artificial intelligence (AI) is the human-like intelligence exhibited by machines or software. It is also an academic field of study. Major AI researchers and textbooks define the field as "the study and design of intelligent agents", where an intelligent agent is a system that perceives its environment and takes actions that maximize its chances of success. John McCarthy, who coined the term in 1955, defines it as "the science and engineering of making intelligent machines".

AI research is highly technical and specialised, and is deeply divided into subfields that often fail to communicate with each other. Some of the division is due to social and cultural factors : subfields have grown up around particular institutions and the work of individual researchers. AI research is also divided by several technical issues. Some subfields focus on the solution of specific problems. Others focus on one of several possible approaches or on the use of a particular tool or towards the accomplishment of particular applications.

The central problems (or goals) of AI research include reasoning, knowledge, planning, learning, natural language processing (communication), perception and the ability to move and manipulate objects. General intelligence (or "strong AI") is still among the field's long term goals. Currently popular approaches include statistical methods, computational intelligence and traditional symbolic AI. There are a large number of tools used in AI, including versions of search and mathematical optimization, logic, methods based on probability and economics, and many others. The AI field is interdisciplinary, in which a number of sciences and professions converge, including computer science, psychology, linguistics, philosophy and neuroscience, as well as other specialized field such as artificial psychology.

The field was founded on the claim that a central property of humans, intelligence—the sapience of *Homo sapiens*—"can be so precisely described that a machine can be made to simulate it." This raises philosophical issues about the nature of the mind and the ethics of creating artificial beings endowed with human-like intelligence, issues which have been addressed by myth, fiction and philosophy since antiquity. Artificial intelligence has been the subject of tremendous optimism but has also suffered stunning setbacks. Today it has become an essential part of the technology industry, providing the heavy lifting for many of the most challenging problems in computer science.

HISTORY

Thinking machines and artificial beings appear in Greek myths, such as Talos of Crete, the bronze robot of Hephaestus, and Pygmalion's Galatea. Human likenesses believed to have intelligence were built in every major civilization : animated cult images were worshiped in Egypt and Greece and humanoid automatons were built by Yan Shi, Hero of Alexandria and Al-Jazari. It was also widely believed that artificial beings had been created by Jābir ibn Hayyān, Judah Loew and Paracelsus. By the 19th and 20th centuries, artificial beings had become a common feature in fiction, as in Mary Shelley's *Frankenstein* or Karel Čapek's *R.U.R. (Rossum's Universal Robots)*. Pamela McCorduck argues that all of these are examples of an ancient urge, as she describes it, "to forge the gods". Stories of these creatures and their fates discuss many of the same hopes, fears and ethical concerns that are presented by artificial intelligence.

Mechanical or "formal" reasoning has been developed by philosophers and mathematicians since antiquity. The study of logic led directly to the invention of the programmable digital electronic computer, based on the work of mathematician Alan Turing and others. Turing's theory of computation suggested that a machine, by shuffling symbols as simple as "0" and "1", could simulate any conceivable act of mathematical deduction. This, along with concurrent discoveries in neurology, information theory and cybernetics, inspired a small group of researchers to begin to seriously consider the possibility of building an electronic brain.

The field of AI research was founded at a conference on the campus of Dartmouth College in the summer of 1956. The attendees, including John McCarthy, Marvin Minsky, Allen Newell and Herbert Simon, became the leaders of AI research for many decades. They and their students wrote programs that were, to most people, simply astonishing : computers were solving word problems in algebra, proving logical theorems and speaking English. By the middle of the 1960s, research in the U.S. was heavily funded by the Department of Defense and laboratories had been established around the world. AI's founders were profoundly optimistic about the future of the new field : Herbert Simon predicted that "machines will be capable, within twenty years, of doing any work a man can do" and Marvin Minsky agreed, writing that "within a generation ... the problem of creating 'artificial intelligence' will substantially be solved".

They had failed to recognize the difficulty of some of the problems they faced. In 1974, in response to the criticism of Sir James Lighthill and ongoing pressure from the US Congress to fund more productive projects, both the U.S. and British governments cut off all undirected exploratory research in AI. The next few years would later be called an "AI winter", a period when funding for AI projects was hard to find.

In the early 1980s, AI research was revived by the commercial success of expert systems, a form of AI program that simulated the knowledge and analytical skills of one or more human experts. By 1985 the market for AI had reached over a billion dollars. At the same time, Japan's fifth generation computer project inspired the U.S and British governments to restore funding for academic research in the field. However, beginning with the collapse of the Lisp Machine market in 1987, AI once again fell into disrepute, and a second, longer lasting AI winter began.

In the 1990s and early 21st century, AI achieved its greatest successes, albeit somewhat behind the scenes. Artificial intelligence is used for logistics, data mining, medical diagnosis and many other areas throughout the technology industry. The success was due to several factors : the increasing computational power of computers, a greater emphasis on solving specific subproblems, the creation of new ties between AI and other fields working on similar problems, and a new commitment by researchers to solid mathematical methods and rigorous scientific standards.

On 11 May 1997, Deep Blue became the first computer chess-playing system to beat a reigning world chess champion, Garry Kasparov. In 2005, a Stanford robot won the DARPA Grand Challenge by driving autonomously for 131 miles along an unrehearsed desert trail. Two years later, a team from CMU won the DARPA Urban Challenge when their vehicle autonomously navigated 55 miles in an urban environment while adhering to traffic hazards and all traffic laws. In February 2011, in a *Jeopardy!* quiz show exhibition match, IBM's question answering system, Watson, defeated the two greatest Jeopardy champions, Brad Rutter and Ken Jennings, by a significant margin. The Kinect, which provides a 3D body–motion interface for the Xbox 360 and the Xbox One, uses algorithms that emerged from lengthy AI research as does the iPhone's Siri.

GOALS

The general problem of simulating (or creating) intelligence has been broken down into a number of specific sub-problems. These consist of particular traits or capabilities that researchers would like an intelligent system to display. The traits described below have received the most attention.

Deduction, Reasoning, Problem Solving

Early AI researchers developed algorithms that imitated the step-by-step reasoning that humans use when they solve puzzles or make logical deductions. By the late 1980s and 1990s, AI research had also developed highly successful methods for dealing with uncertain or incomplete information, employing concepts from probability and economics.

For difficult problems, most of these algorithms can require enormous computational resources – most experience a "combinatorial explosion" : the amount of memory or computer time required becomes astronomical when the problem goes beyond a certain size. The search for more efficient problem-solving algorithms is a high priority for AI research.

Human beings solve most of their problems using fast, intuitive judgements rather than the conscious, step-by-step deduction that early AI research was able to model. AI has made some progress at imitating this kind of "sub-symbolic" problem solving : embodied agent approaches emphasize the importance of sensorimotor skills to higher reasoning; neural net research attempts to simulate the structures inside the brain that give rise to this skill; statistical approaches to AI mimic the probabilistic nature of the human ability to guess.

Knowledge Representation

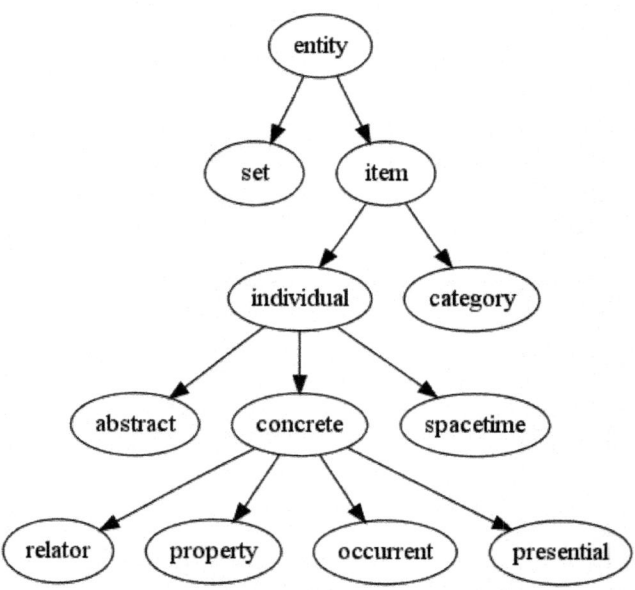

An ontology represents knowledge as a set of concepts within a domain and the relationships between those concepts.

Knowledge representation and knowledge engineering are central to AI research. Many of the problems machines are expected to solve will require extensive knowledge about the world. Among the things that AI needs to represent are : objects, properties, categories and relations between objects; situations, events, states and time; causes and effects; knowledge about knowledge (what we know about what other people know); and many other, less well researched domains. A representation of "what exists" is an ontology : the set of objects, relations, concepts and so on that the machine knows about. The most general are called upper ontologies, which attempt to provide a foundation for all other knowledge.

Among the most difficult problems in knowledge representation are :

Default reasoning and the qualification problem

Many of the things people know take the form of "working assumptions." For example, if a bird comes up in conversation, people typically picture an animal that is fist sized, sings, and flies. None of these things are true about all birds. John McCarthy identified this problem in 1969 as the qualification problem : for any commonsense rule that AI researchers care to represent, there tend to be a huge number of exceptions. Almost nothing is simply true or false in the way that abstract logic requires. AI research has explored a number of solutions to this problem.

The Breadth of Commonsense Knowledge

The number of atomic facts that the average person knows is astronomical. Research projects that attempt to build a complete knowledge base of commonsense knowledge (*e.g.*, Cyc) require enormous amounts of laborious ontological engineering — they must be built, by hand, one complicated concept at a time. A major goal is to have the computer understand enough concepts to be able to learn by reading from sources like the internet, and thus be able to add to its own ontology.

The Subsymbolic Form of Some Commonsense Knowledge

Much of what people know is not represented as "facts" or "statements" that they could express verbally. For example, a chess master will avoid a particular chess position because it "feels too exposed" or an art critic can take one look at a statue and instantly realize that it is a fake. These are intuitions or tendencies that are represented in the brain non-consciously and sub-symbolically. Knowledge like this informs, supports and provides a context for symbolic, conscious knowledge. As with the related problem of sub-symbolic reasoning, it is hoped that situated AI, computational intelligence, or statistical AI will provide ways to represent this kind of knowledge.

Planning

Intelligent agents must be able to set goals and achieve them. They need a way to visualize the future (they must have a representation of the state of the world and be able to make predictions about how their actions will change it) and be able to make choices that maximize the utility (or "value") of the available choices.

In classical planning problems, the agent can assume that it is the only thing acting on the world and it can be certain what the consequences of its actions may be. However, if the agent is not the only actor, it must periodically ascertain whether the world matches its predictions and it must change its plan as this becomes necessary, requiring the agent to reason under uncertainty.

Multi-agent planning uses the cooperation and competition of many agents to achieve a given goal. Emergent behavior such as this is used by evolutionary algorithms and swarm intelligence.

Hierarchical Control System

[Diagram: A pyramidal hierarchy with a "top level node" at the apex, connected to two "node" boxes below it, with arrows labeled "sensations, results" going up and "tasks, goals" going down. The middle nodes connect to "sensor", "actuator", and "sensor/actuator" boxes at the bottom, which interface with the "Controlled system, controlled process, or environment" via "sensations" (up) and "actions" (down).]

A hierarchical control system is a form of control system in which a set of devices and governing software is arranged in a hierarchy.

Learning

Machine learning is the study of computer algorithms that improve automatically through experience and has been central to AI research since the field's inception.

Unsupervised learning is the ability to find patterns in a stream of input. Supervised learning includes both classification and numerical regression. Classification is used to determine what category something belongs in, after seeing a number of examples of things from several categories. Regression is the attempt to produce a function that describes the relationship between inputs and outputs and predicts how the outputs should change as the inputs change. In reinforcement learning the agent is rewarded for good responses and punished for bad ones. These can be analyzed in terms of decision theory, using concepts like utility. The mathematical analysis of machine learning algorithms and their performance is a branch of theoretical computer science known as computational learning theory.

Within developmental robotics, developmental learning approaches were elaborated for lifelong cumulative acquisition of repertoires of novel skills by a robot, through autonomous self-exploration and social interaction with human teachers, and using guidance mechanisms such as active learning, maturation, motor synergies, and imitation.

Natural Language Processing (Communication)

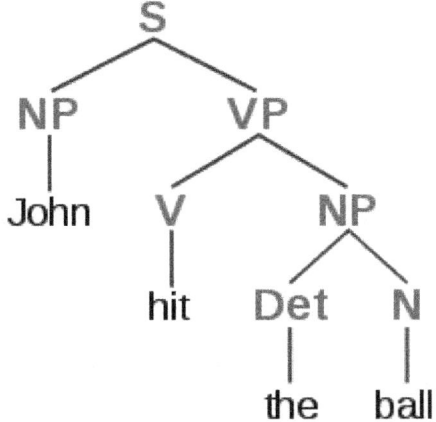

A parse tree represents the syntactic structure of a sentence according to some formal grammar.

Natural language processing gives machines the ability to read and understand the languages that humans speak. A sufficiently powerful natural language processing system would enable natural language user interfaces and the acquisition of knowledge directly from human-written sources, such as news-wire texts. Some straightforward applications of natural language processing include information retrieval (or text mining) and machine translation.

A common method of processing and extracting meaning from natural language is through semantic indexing. Increases in processing speeds and the drop in the cost of data storage makes indexing large volumes of abstractions of the users input much more efficient.

Perception

Machine perception is the ability to use input from sensors (such as cameras, microphones, tactile sensors, sonar and others more exotic) to deduce aspects of the world. Computer vision is the ability to analyze visual input. A few selected subproblems are speech recognition, facial recognition and object recognition.

Motion and Manipulation

The field of robotics is closely related to AI. Intelligence is required for robots to be able to handle such tasks as object manipulation and navigation, with subproblems of localization (knowing where you are, or finding out where other things are), mapping (learning what is around you, building a map of the environment), and motion planning (figuring out how to get there) or path planning (going from one point in space to another point, which may involve compliant motion - where the robot moves while maintaining physical contact with an object).

Long-term Goals

Among the long-term goals in the research pertaining to artificial intelligence are : (1) Social intelligence, (2) Creativity, and (3) General intelligence.

Social Intelligence

Affective computing is the study and development of systems and devices that can recognize, interpret, process, and simulate human affects. It is an interdisciplinary field spanning computer sciences, psychology, and cognitive science. While the origins of the field may be traced as far back as to early philosophical inquiries into emotion, the more modern branch of computer science originated with Rosalind Picard's 1995 paper on affective computing. A motivation for the research is the ability to simulate empathy. The machine should interpret the emotional state of humans and adapt its behaviour to them, giving an appropriate response for those emotions.

Emotion and social skills play two roles for an intelligent agent. First, it must be able to predict the actions of others, by understanding their motives and emotional states. (This involves elements of game theory, decision theory, as well as the ability to model human emotions and the perceptual skills to detect emotions.) Also, in an effort to facilitate human-computer interaction, an intelligent machine might want to be able to *display* emotions — even if it does not actually experience them itself — in order to appear sensitive to the emotional dynamics of human interaction.

Creativity

A sub-field of AI addresses creativity both theoretically (from a philosophical and psychological perspective) and practically (via specific implementations of systems that generate outputs that can be considered creative, or systems that identify and assess creativity). Related areas of computational research are Artificial intuition and Artificial thinking.

General Intelligence

Many researchers think that their work will eventually be incorporated into a machine with *general* intelligence (known as strong AI), combining all the skills above and exceeding human abilities at most or all of them. A few believe that anthropomorphic features like artificial consciousness or an artificial brain may be required for such a project.

Many of the problems above may require general intelligence to be considered solved. For example, even a straightforward, specific task like machine translation requires that the machine read and write in both languages (NLP), follow the author's argument (reason), know what is being talked about (knowledge), and faithfully reproduce the author's intention (social intelligence). A problem like machine translation is considered "AI-complete". In order to solve this particular problem, you must solve all the problems.

APPROACHES

There is no established unifying theory or paradigm that guides AI research. Researchers disagree about many issues. A few of the most long standing questions that have remained unanswered are these : should artificial intelligence simulate natural intelligence by studying psychology or neurology? Or is human biology as irrelevant to AI research as bird biology is to aeronautical engineering? Can intelligent behavior be described using simple, elegant principles (such as logic or optimization)? Or does it necessarily require solving a large number of completely unrelated problems? Can intelligence be reproduced using high-level symbols, similar to words and ideas? Or does it require "sub-symbolic" processing? John Haugeland, who coined the term GOFAI (Good Old-Fashioned Artificial Intelligence), also proposed that AI should more properly be referred to as synthetic intelligence, a term which has since been adopted by some non-GOFAI researchers.

Cybernetics and Brain Simulation

In the 1940s and 1950s, a number of researchers explored the connection between neurology, information theory, and cybernetics. Some of them built machines that used electronic networks to exhibit rudimentary intelligence, such as W. Grey Walter's turtles and the Johns Hopkins Beast. Many of these researchers gathered for meetings of the Teleological Society at Princeton University and the Ratio Club in England. By 1960, this approach was largely abandoned, although elements of it would be revived in the 1980s.

Symbolic

When access to digital computers became possible in the middle 1950s, AI research began to explore the possibility that human intelligence could be reduced to symbol manipulation. The research was centered in three institutions : Carnegie Mellon University, Stanford and MIT, and each one developed its own style of research. John Haugeland named these approaches to AI "good old fashioned AI" or "GOFAI". During the 1960s, symbolic approaches had achieved great success at simulating high-level thinking in small demonstration programs. Approaches based on cybernetics or neural networks were abandoned or pushed into the background. Researchers in the 1960s and the 1970s were convinced that symbolic approaches would eventually succeed in creating a machine with artificial general intelligence and considered this the goal of their field.

Cognitive Simulation

Economist Herbert Simon and Allen Newell studied human problem-solving skills and attempted to formalize them, and their work laid the foundations of the field of artificial intelligence, as well as cognitive science, operations research and management science. Their research team used the results of psychological experiments to develop programs that simulated the techniques that people used to solve problems. This tradition, centered at Carnegie Mellon University would eventually culminate in the development of the Soar architecture in the middle 1980s.

Logic-based

Unlike Newell and Simon, John McCarthy felt that machines did not need to simulate human thought, but should instead try to find the essence of abstract reasoning and problem solving, regardless of whether people used the same algorithms. His laboratory at Stanford (SAIL) focused on using formal logic to solve a wide variety of problems, including knowledge representation, planning and learning. Logic was also the focus of the work at the University of Edinburgh and elsewhere in Europe which led to the development of the programming language Prolog and the science of logic programming.

"Anti-logic" or "Scruffy"

Researchers at MIT (such as Marvin Minsky and Seymour Papert) found that solving difficult problems in vision and natural language processing required ad-hoc solutions – they argued that there was no simple and general principle (like logic) that would capture all the aspects of intelligent behavior. Roger Schank described their "anti-logic" approaches as "scruffy" (as opposed to the "neat" paradigms at CMU and Stanford). Commonsense knowledge bases (such as Doug Lenat's Cyc) are an example of "scruffy" AI, since they must be built by hand, one complicated concept at a time.

Knowledge-based

When computers with large memories became available around 1970, researchers from all three traditions began to build knowledge into AI applications. This "knowledge revolution" led to the development and deployment of expert systems (introduced by Edward Feigenbaum), the first truly successful form of AI software. The knowledge revolution was also driven by the realization that enormous amounts of knowledge would be required by many simple AI applications.

Sub-symbolic

By the 1980s progress in symbolic AI seemed to stall and many believed that symbolic systems would never be able to imitate all the processes of human cognition, especially perception, robotics, learning and pattern recognition. A number of researchers began to look into "sub-symbolic" approaches to specific AI problems.

Bottom-up, embodied, situated, behavior-based or nouvelle AI

Researchers from the related field of robotics, such as Rodney Brooks, rejected symbolic AI and focused on the basic engineering problems that would allow robots to move and survive. Their work revived the non-symbolic viewpoint of the early cybernetics researchers of the 1950s and reintroduced the use of control theory in AI. This coincided with the development of the embodied mind thesis in the related field of cognitive science : the idea that aspects of the body (such as movement, perception and visualization) are required for higher intelligence.

Computational Intelligence

Interest in neural networks and "connectionism" was revived by David Rumelhart and others in the middle 1980s. These and other sub-symbolic approaches, such as fuzzy systems and evolutionary computation, are now studied collectively by the emerging discipline of computational intelligence.

Statistical

In the 1990s, AI researchers developed sophisticated mathematical tools to solve specific subproblems. These tools are truly scientific, in the sense that their results are both measurable and verifiable, and they have been responsible for many of AI's recent successes. The shared mathematical language has also permitted a high level of collaboration with more established fields (like mathematics, economics or operations research). Stuart Russell and Peter Norvig describe this movement as nothing less than a "revolution" and "the victory of the neats." Critics argue that these techniques are too focused on particular problems and have failed to address the long term goal of general intelligence. There is an ongoing debate about the relevance and validity of statistical approaches in AI, exemplified in part by exchanges between Peter Norvig and Noam Chomsky.

Integrating the Approaches

Intelligent Agent Paradigm

An intelligent agent is a system that perceives its environment and takes actions which maximize its chances of success. The simplest intelligent agents are programs that solve specific problems. More complicated agents include human beings and organizations of human beings (such as firms). The paradigm gives researchers license to study isolated problems and find solutions that are both verifiable and useful, without agreeing on one single approach. An agent that solves a specific problem can use any approach that works – some agents are symbolic and logical, some are sub-symbolic neural networks and others may use new approaches. The paradigm also gives researchers a common language to communicate with other fields — such as decision theory and economics — that also use concepts of abstract agents. The intelligent agent paradigm became widely accepted during the 1990s.

Agent Architectures and Cognitive Architectures

Researchers have designed systems to build intelligent systems out of interacting intelligent agents in a multi-agent system. A system with both symbolic and sub-symbolic components is a hybrid intelligent system, and the study of such systems is artificial intelligence systems integration. A hierarchical control system provides a bridge between sub-symbolic AI at its lowest, reactive levels and traditional symbolic AI at its highest levels, where relaxed time constraints permit planning and world modelling. Rodney Brooks' subsumption architecture was an early proposal for such a hierarchical system.

TOOLS

In the course of 50 years of research, AI has developed a large number of tools to solve the most difficult problems in computer science. A few of the most general of these methods are discussed below.

Search and Optimization

Many problems in AI can be solved in theory by intelligently searching through many possible solutions : Reasoning can be reduced to performing a search. For example, logical proof can be viewed as searching for a path that leads from premises to conclusions, where each step is the application of an inference rule. Planning algorithms search through trees of goals and subgoals, attempting to find a path to a target goal, a process called means-ends analysis. Robotics algorithms for moving limbs and grasping objects use local searches in configuration space. Many learning algorithms use search algorithms based on optimization.

Simple exhaustive searches are rarely sufficient for most real world problems : the search space (the number of places to search) quickly grows to astronomical numbers. The result is a search that is too slow or never completes. The solution, for many problems, is to use "heuristics" or "rules of thumb" that eliminate choices that are unlikely to lead to the goal (called "pruning the search tree"). Heuristics supply the program with a "best guess" for the path on which the solution lies. Heuristics limit the search for solutions into a smaller sample size.

A very different kind of search came to prominence in the 1990s, based on the mathematical theory of optimization. For many problems, it is possible to begin the search with some form of a guess and then refine the guess incrementally until no more refinements can be made. These algorithms can be visualized as blind hill climbing : we begin the search at a random point on the landscape, and then, by jumps or steps, we keep moving our guess uphill, until we reach the top. Other optimization algorithms are simulated annealing, beam search and random optimization.

Evolutionary computation uses a form of optimization search. For example, they may begin with a population of organisms (the guesses) and then allow them to mutate and recombine, selecting only the fittest to survive each generation (refining the guesses). Forms of evolutionary computation include swarm intelligence algorithms (such as ant colony or particle swarm optimization) and evolutionary algorithms (such as genetic algorithms, gene expression programming, and genetic programming).

Logic

Logic is used for knowledge representation and problem solving, but it can be applied to other problems as well. For example, the satplan algorithm uses logic for planning and inductive logic programming is a method for learning.

Several different forms of logic are used in AI research. Propositional or sentential logic is the logic of statements which can be true or false. First-order logic also allows the use of quantifiers and predicates, and can express facts about objects, their properties, and their relations with each other. Fuzzy logic, is a version of first-order logic which allows the truth of a statement to be represented as a value between 0 and 1, rather than simply True (1) or False (0). Fuzzy systems can be used for uncertain reasoning and have been widely used in modern industrial and consumer product control systems. Subjective logic models uncertainty in a different and more explicit manner than fuzzy-logic : a given binomial opinion satisfies belief + disbelief + uncertainty = 1 within a Beta distribution. By this method, ignorance can be distinguished from probabilistic statements that an agent makes with high confidence.

Default logics, non-monotonic logics and circumscription are forms of logic designed to help with default reasoning and the qualification problem. Several extensions of logic have been designed to handle specific domains of knowledge, such as : description logics; situation calculus, event calculus and fluent calculus (for representing events and time); causal calculus; belief calculus; and modal logics.

Probabilistic Methods for Uncertain Reasoning

Many problems in AI (in reasoning, planning, learning, perception and robotics) require the agent to operate with incomplete or uncertain information. AI researchers have devised a number of powerful tools to solve these problems using methods from probability theory and economics.

Bayesian networks are a very general tool that can be used for a large number of problems : reasoning (using the Bayesian inference algorithm), learning (using the expectation-maximization algorithm), planning (using decision networks) and perception (using dynamic Bayesian networks). Probabilistic algorithms can also be used for filtering, prediction, smoothing and finding explanations for streams of data, helping perception systems to analyze processes that occur over time (*e.g.*, hidden Markov models or Kalman filters).

A key concept from the science of economics is "utility" : a measure of how valuable something is to an intelligent agent. Precise mathematical tools have been developed that analyze how an agent can make choices and plan, using decision theory, decision analysis, information value theory. These tools include models such as Markov decision processes, dynamic decision networks, game theory and mechanism design.

Classifiers and Statistical Learning Methods

The simplest AI applications can be divided into two types : classifiers ("if shiny then diamond") and controllers ("if shiny then pick up"). Controllers do however also classify conditions before inferring actions, and therefore classification forms a central part of many AI systems. Classifiers are functions that use pattern matching to determine a closest match. They can be tuned according to examples, making them very attractive for use in AI. These examples are known as observations or patterns. In supervised learning, each pattern belongs to a certain predefined class. A class can be seen as a decision that has to be made. All the observations combined with their class labels are known as a data set. When a new observation is received, that observation is classified based on previous experience.

A classifier can be trained in various ways; there are many statistical and machine learning approaches. The most widely used classifiers are the neural network, kernel methods such as the support vector machine, k-nearest neighbor algorithm, Gaussian mixture model, naive Bayes classifier, and decision tree. The performance of these classifiers have been compared over a wide range of tasks. Classifier performance depends greatly on the characteristics of the data to be classified. There is no single classifier that works best on all given problems; this is also referred to as the "no free lunch" theorem. Determining a suitable classifier for a given problem is still more an art than science.

Neural Networks

The study of artificial neural networks began in the decade before the field AI research was founded, in the work of Walter Pitts and Warren McCullough. Other important early researchers were Frank Rosenblatt, who invented the perceptron and Paul Werbos who developed the backpropagation algorithm.

The main categories of networks are acyclic or feedforward neural networks (where the signal passes in only one direction) and recurrent neural networks (which allow feedback). Among the most popular feedforward networks are perceptrons, multi-layer perceptrons and radial basis networks. Among recurrent networks, the most famous is the Hopfield net, a form of attractor network, which was first described by John Hopfield in 1982. Neural networks can be applied to the problem of intelligent control (for robotics) or learning, using such techniques as Hebbian learning and competitive learning.

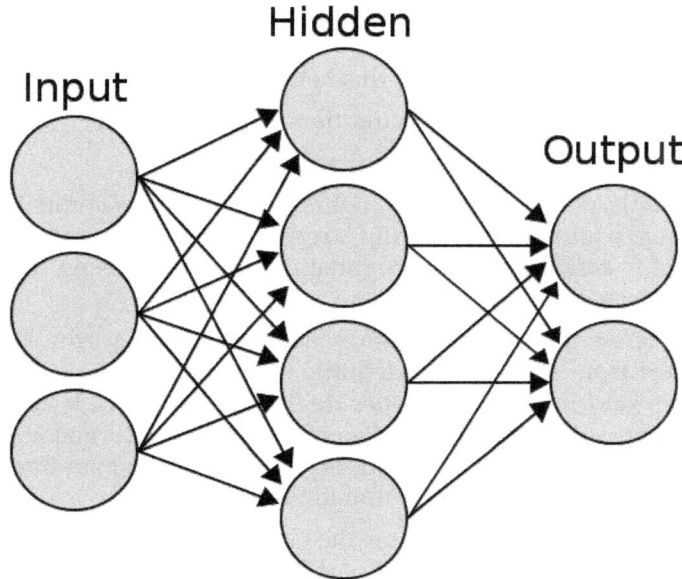

A neural network is an interconnected group of nodes, akin to the vast network of neurons in the human brain.

Hierarchical temporal memory is an approach that models some of the structural and algorithmic properties of the neocortex.

Control Theory

Control theory, the grandchild of cybernetics, has many important applications, especially in robotics.

Languages

AI researchers have developed several specialized languages for AI research, including Lisp and Prolog.

EVALUATING PROGRESS

In 1950, Alan Turing proposed a general procedure to test the intelligence of an agent now known as the Turing test. This procedure allows almost all the major problems of artificial intelligence to be tested. However, it is a very difficult challenge and at present all agents fail.

Artificial intelligence can also be evaluated on specific problems such as small problems in chemistry, hand-writing recognition and game-playing. Such tests have been termed subject matter expert Turing tests. Smaller problems provide more achievable goals and there are an ever-increasing number of positive results.

One classification for outcomes of an AI test is :

1. Optimal : it is not possible to perform better.
2. Strong super-human : performs better than all humans.
3. Super-human : performs better than most humans.
4. Sub-human : performs worse than most humans.

For example, performance at draughts (*i.e.* checkers) is optimal, performance at chess is super-human and nearing strong super-human and performance at many everyday tasks (such as recognizing a face or crossing a room without bumping into something) is sub-human.

A quite different approach measures machine intelligence through tests which are developed from *mathematical* definitions of intelligence. Examples of these kinds of tests start in the late nineties devising intelligence tests using notions from Kolmogorov complexity and data compression. Two major advantages of mathematical definitions are their applicability to nonhuman intelligences and their absence of a requirement for human testers.

A derivative of the Turing test is the Completely Automated Public Turing test to tell Computers and Humans Apart (CAPTCHA). as the name implies, this helps to determine that a user is an actual person and not a computer posing as a human. In contrast to the standard Turing test, CAPTCHA administered by a machine and targeted to a human as opposed to being administered by a human and targeted to a machine. A computer asks a user to complete a simple test then generates a grade for that test. Computers are unable to solve the problem, so correct solutions are deemed to be the result of a person taking the test. A common type of CAPTCHA is the test that requires the typing of distorted letters, numbers or symbols that appear in an image undecipherable by a computer.

APPLICATIONS

Artificial intelligence techniques are pervasive and are too numerous to list. Frequently, when a technique reaches mainstream use, it is no longer considered artificial intelligence; this phenomenon is described as the AI effect. An area that artificial intelligence has contributed greatly to is Intrusion detection.

Competitions and Prizes

There are a number of competitions and prizes to promote research in artificial intelligence. The main areas promoted are : general machine intelligence, conversational behavior, data-mining, robotic cars, robot soccer and games.

Platforms

A platform (or "computing platform") is defined as "some sort of hardware architecture or software framework (including application frameworks), that allows software to run." As Rodney Brooks pointed out many years ago, it is not just the artificial intelligence software that defines the AI features of the platform, but rather the actual platform itself that affects the AI that results, *i.e.*, there needs to be work in AI problems on real-world platforms rather than in isolation.

A wide variety of platforms has allowed different aspects of AI to develop, ranging from expert systems, albeit PC-based but still an entire real-world system, to various robot platforms such as the widely available Roomba with open interface.

PHILOSOPHY

Artificial intelligence, by claiming to be able to recreate the capabilities of the human mind, is both a challenge and an inspiration for philosophy. Are there limits to how intelligent machines can be? Is there an essential difference between human intelligence and artificial intelligence? Can a machine have a mind and consciousness? A few of the most influential answers to these questions are given below.

Turing's "Polite Convention"

We need not decide if a machine can "think"; we need only decide if a machine can act as intelligently as a human being. This approach to the philosophical problems associated with artificial intelligence forms the basis of the Turing test.

The Dartmouth Proposal

"Every aspect of learning or any other feature of intelligence can be so precisely described that a machine can be made to simulate it." This conjecture was printed in the proposal for the Dartmouth Conference of 1956, and represents the position of most working AI researchers.

Newell and Simon's Physical Symbol System Hypothesis

"A physical symbol system has the necessary and sufficient means of general intelligent action." Newell and Simon argue that intelligences consist of formal operations on symbols. Hubert Dreyfus argued that, on the contrary, human expertise depends on unconscious instinct rather than conscious symbol manipulation and on having a "feel" for the situation rather than explicit symbolic knowledge.

Gödel's Incompleteness Theorem

A formal system (such as a computer program) cannot prove all true statements. Roger Penrose is among those who claim that Gödel's theorem limits what machines can do.

Searle's Strong AI Hypothesis

"The appropriately programmed computer with the right inputs and outputs would thereby have a mind in exactly the same sense human beings have minds."

John Searle counters this assertion with his Chinese room argument, which asks us to look *inside* the computer and try to find where the "mind" might be.

The Artificial Brain Argument

The brain can be simulated. Hans Moravec, Ray Kurzweil and others have argued that it is technologically feasible to copy the brain directly into hardware and software, and that such a simulation will be essentially identical to the original.

PREDICTIONS AND ETHICS

Many thinkers have speculated about the future of artificial intelligence technology and society. The existence of an artificial intelligence that rivals or exceeds human intelligence raises difficult ethical issues, and the potential power of the technology inspires both hopes and fears.

If research into Strong AI produced sufficiently intelligent software, it might be able to reprogram and improve itself. The improved software would be even better at improving itself, leading to recursive self-improvement. The new intelligence could thus increase exponentially and dramatically surpass humans.

Hyper-intelligent software may not necessarily decide to support the continued existence of mankind, and would be extremely difficult to stop. This topic has also recently begun to be discussed in academic publications as a real source of risks to civilization, humans, and planet Earth.

One proposal to deal with this is to ensure that the first generally intelligent AI is 'Friendly AI', and will then be able to control subsequently developed AIs. Some question whether this kind of check could really remain in place.

Martin Ford, author of *The Lights in the Tunnel : Automation, Accelerating Technology and the Economy of the Future,* and others argue that specialized artificial intelligence applications, robotics and other forms of automation will ultimately result in significant unemployment as machines begin to match and exceed the capability of workers to perform most routine and repetitive jobs. Ford predicts that many knowledge-based occupations – and in particular entry level jobs – will be increasingly susceptible to automation via expert systems, machine learning and other AI-enhanced applications. AI-based applications may also be used to amplify the capabilities of low-wage offshore workers, making it more feasible to outsource knowledge work.

Joseph Weizenbaum wrote that AI applications can not, by definition, successfully simulate genuine human empathy and that the use of AI technology in fields such as customer service or psychotherapy was deeply misguided. Weizenbaum was also bothered that AI researchers (and some philosophers) were willing to view the human mind as nothing more than a computer program (a position now known as computationalism). To Weizenbaum these points suggest that AI research devalues human life.

Many futurists believe that artificial intelligence will ultimately transcend the limits of progress. Ray Kurzweil has used Moore's law (which describes the relentless exponential improvement in digital technology) to calculate that desktop computers will have the same processing power as human brains by the year 2029. He also predicts that by 2045 artificial intelligence will reach a point where it is able to improve *itself* at a rate that far exceeds anything conceivable in the past, a scenario that science fiction writer Vernor Vinge named the "singularity".

Robot designer Hans Moravec, cyberneticist Kevin Warwick and inventor Ray Kurzweil have predicted that humans and machines will merge in the future into cyborgs that are more capable and powerful than either. This idea, called transhumanism, which has roots in Aldous Huxley and Robert Ettinger, has been illustrated in fiction as well, for example in the manga *Ghost in the Shell* and the science-fiction series *Dune*. In the 1980s artist Hajime Sorayama's Sexy Robots series were painted and published in Japan depicting the actual organic human form with life-like muscular metallic skins and later "the Gynoids" book followed that was used by or influenced movie makers including George Lucas and other creatives. Sorayama never considered these organic robots to be real part of nature but always unnatural product of the human mind, a fantasy existing in the mind even when realized in actual form. Almost 20 years later, the first AI robotic pet, AIBO, came available as a companion to people. AIBO grew out of Sony's Computer Science Laboratory (CSL). Famed engineer Toshitada Doi is credited as AIBO's original progenitor : in 1994 he had started work on robots with artificial intelligence expert Masahiro Fujita, at CSL. Doi's, friend, the artist Hajime Sorayama, was enlisted to create the initial designs for the AIBO's body. Those designs are now part of the permanent collections of Museum of Modern Art and the Smithsonian Institution, with later versions of AIBO being used in studies in Carnegie Mellon University. In 2006, AIBO was added into Carnegie Mellon University's "Robot Hall of Fame".

Political scientist Charles T. Rubin believes that AI can be neither designed nor guaranteed to be benevolent. He argues that "any sufficiently advanced benevolence may be indistinguishable from malevolence." Humans should not assume machines or robots would treat us favorably, because there is no *a priori* reason to believe that they would be sympathetic to our system of morality, which has evolved along with our particular biology (which AIs would not share).

Edward Fredkin argues that "artificial intelligence is the next stage in evolution", an idea first proposed by Samuel Butler's "Darwin among the Machines" , and expanded upon by George Dyson in his book of the same name in 1998.

ROBOTS AND ARTIFICIAL INTELLIGENCE

Artificial intelligence (AI) is arguably the most exciting field in robotics. It's certainly the most controversial : Everybody agrees that a robot can work in an assembly line, but there's no consensus on whether a robot can ever be intelligent.

Like the term "robot" itself, artificial intelligence is hard to define. Ultimate AI would be a recreation of the human thought process -- a man-made machine with our intellectual abilities. This would include the ability to learn just about anything, the ability to reason, the ability to use language and the ability to formulate original ideas. Roboticists are nowhere near achieving this level of artificial intelligence, but they have made a lot of progress with more limited AI. Today's AI machines can replicate some specific elements of intellectual ability.

Computers can already **solve problems** in limited realms. The basic idea of AI problem-solving is very simple, though its execution is complicated. First, the AI robot or computer gathers facts about a situation through sensors or human input. The computer compares this information to stored data and decides what the information signifies. The computer runs through various possible actions and predicts which action will be most successful based on the collected information. Of course, the computer can only solve problems it's programmed to solve -- it doesn't have any generalized analytical ability. Chess computers are one example of this sort of machine.

Some modern robots also have the ability to **learn** in a limited capacity. Learning robots recognize if a certain action (moving its legs in a certain way, for instance) achieved a desired result (navigating an obstacle). The robot stores this information and attempts the successful action the next time it encounters the same situation. Again, modern computers can only do this in very limited situations. They can't absorb any sort of information like a human can. Some robots can learn by mimicking human actions. In Japan, roboticists have taught a robot to dance by demonstrating the moves themselves.

Some robots can **interact socially**. Kismet, a robot at M.I.T's Artificial Intelligence Lab, recognizes human body language and voice inflection and responds appropriately. Kismet's creators are interested in how humans and babies interact, based only on tone of speech and visual cue. This low-level interaction could be the foundation of a human-like learning system.

Kismet and other humanoid robots at the M.I.T. AI Lab operate using an unconventional control structure. Instead of directing every action using a central computer, the robots control lower-level actions with lower-level computers. The program's director, Rodney Brooks, believes this is a more accurate model of human intelligence. We do most things automatically; we don't decide to do them at the highest level of consciousness.

The real challenge of AI is to understand how natural intelligence works. Developing AI isn't like building an artificial heart -- scientists don't have a simple, concrete model to work from. We do know that the brain contains billions and billions of neurons, and that we think and learn by establishing electrical connections between different neurons. But we don't know exactly how all of these connections add up to higher reasoning, or even low-level operations. The complex circuitry seems incomprehensible.

Because of this, AI research is largely theoretical. Scientists hypothesize on how and why we learn and think, and they experiment with their ideas using robots. Brooks and his team focus on humanoid robots because they feel that being able to experience the world like a human is essential to developing human-like intelligence. It also makes it easier for people to interact with the robots, which potentially makes it easier for the robot to learn.

Just as physical robotic design is a handy tool for understanding animal and human anatomy, AI research is useful for understanding how natural intelligence works. For some roboticists, this insight is the ultimate goal of designing robots. Others envision a world where we live side by side with intelligent machines and use a variety of lesser robots for manual labor, health care and communication. A number of robotics experts predict that robotic evolution will ultimately turn us into cyborgs -- humans integrated with machines. Conceivably, people in the future could load their minds into a sturdy robot and live for thousands of years!

In any case, robots will certainly play a larger role in our daily lives in the future. In the coming decades, robots will gradually move out of the industrial and scientific worlds and into daily life, in the same way that computers spread to the home in the 1980s.

Chapter 5

MACHINE VISION

Machine vision (MV) is the technology and methods used to provide imaging-based automatic inspection and analysis for such applications as automatic inspection, process control, and robot guidance in industry. The scope of MV is broad. MV is related to, though distinct from, computer vision.

APPLICATIONS

The primary uses for machine vision are automatic inspection and industrial robot guidance. Common machine vision applications include quality assurance, sorting, material handling, robot guidance, and optical gauging.

METHODS

Machine vision methods are defined as both the process of defining and creating an MV solution, and as the technical process that occurs during the operation of the solution. Here the latter is addressed. As of 2006, there was little standardization in the interfacing and configurations used in MV. This includes user interfaces, interfaces for the integration of multi-component systems and automated data interchange. Nonetheless, the first step in the MV sequence of operation is acquisition of an image, typically using cameras, lenses, and lighting that has been designed to provide the differentiation required by subsequent processing. MV software packages then employ various digital image processing techniques to extract the required information, and often make decisions (such as pass/fail) based on the extracted information.

Imaging

While conventional (2D visible light) imaging is most commonly used in MV, alternatives include imaging various infrared bands, line scan imaging, 3D imaging of surfaces and X-ray imaging. Key divisions within MV 2D visible light imaging are monochromatic *vs.* color, resolution, and whether or not the imaging process is simultaneous over the entire image, making it suitable for moving processes. The most commonly used method for 3D imaging is scanning based triangulation which utilizes motion of the product or image during the imaging process. Other 3D methods used for machine vision are time of flight, grid based and stereoscopic.

The imaging device (*e.g.* camera) can either be separate from the main image processing unit or combined with it in which case the combination is generally called a smart camera or smart sensor. When separated, the connection may be made to specialized intermediate hardware, a frame grabber using either a standardized or custom interface. MV implementations also have used digital cameras capable of direct connections (without a framegrabber) to a computer via FireWire, USB or Gigabit Ethernet interfaces.

Though the vast majority of machine vision applications are solved using two-dimensional imaging, machine vision applications utilizing 3D imaging are growing niche within the industry. One method is grid array based systems using pseudorandom structured light system as employed by the Microsoft Kinect system circa 2012. Another method of generating a 3D image is to use laser triangulation, where a laser is projected onto the surfaces of an object and the deviation of the line is used to calculate the shape. In machine vision this is accomplished with a scanning motion, either by moving the workpiece, or by moving the camera & laser imaging system. Stereoscopic vision is used in special cases involving unique features present in both views of a pair of cameras.

Image Processing

After an image is acquired, it is processed. Machine vision image processing methods include

- Stitching/Registration : Combining of adjacent 2D or 3D images.
- Filtering (*e.g.* morphological filtering)
- Thresholding : Thresholding starts with setting or determining a gray value that will be useful for the following steps. The value is then used to separate portions of the image, and sometimes to transform each portion of the image simply black and white based on whether it is below or above that grayscale value.
- Pixel counting : counts the number of light or dark pixels
- Segmentation : Partitioning a digital image into multiple segments to simplify and/or change the representation of an image into something that is more meaningful and easier to analyze.
- Edge detection : finding object edges
- Color Analysis : Identify parts, products and items using color, assess quality from color, and isolate features using color.

- Blob discovery & manipulation : inspecting an image for discrete blobs of connected pixels (*e.g.* a black hole in a grey object) as image landmarks. These blobs frequently represent optical targets for machining, robotic capture, or manufacturing failure.
- Neural net processing : weighted and self-training multi-variable decision making
- Pattern recognition including template matching. Finding, matching, and/or counting specific patterns. This may include location of an object that may be rotated, partially hidden by another object, or varying in size.
- Barcode, Data Matrix and "2D barcode" reading
- Optical character recognition : automated reading of text such as serial numbers
- Gauging/Metrology : measurement of object dimensions (*e.g.* in pixels, inches or millimeters)
- Comparison against target values to determine a "pass or fail" or "go/no go" result. For example, with code or bar code verification, the read value is compared to the stored target value. For gauging, a measurement is compared against the proper value and tolerances. For verification of alpha-numberic codes, the OCR'd value is compared to the proper or target value. For inspection for blemishes, the measured size of the blemishes may be compared to the maximums allowed by quality standards.

Outputs

A common output from machine vision systems is pass/fail decisions. These decisions may in turn trigger mechanisms that reject failed items or sound an alarm. Other common outputs include object position and orientation information from robot guidance systems. Additionally, output types include numerical measurement data, data read from codes and characters, displays of the process or results, stored images, alarms from automated space monitoring MV systems, and process control signals.

VISION GUIDED ROBOTIC SYSTEMS

Introduction

A Vision Guided Robot System comprises three core systems including robotic system, vision system, and component bulk handling system (hopper or conveyor system).

The vision system determines the position of randomly fed products onto a recycling conveyor system. The vision system and control software gives the robot exact coordinates of the components, which are spread out randomly beneath the camera field of vision, enabling the robot arm(s) to move to a selected component and pick from the conveyor belt. The conveyor, normally, stops under the camera where the position of the parts is determined. If the cycle time is short it is also possible to pick a component without stopping the conveyor. This is achieved by fitting an encoder to the conveyor and tracking the component through the vision software.

This functionality is usually referred to as vision guided robotics (VGR). It is a fast growing technology and a way to reduce manpower and retain production, especially in countries with high manufacturing overheads and labour costs.

Vision Systems for Robot Guidance

A vision system comprises a camera and microprocessor or computer, with associated software. This is a very wide definition that can be used to cover many different types of systems which aim to solve a large variety of different tasks. Vision systems can be implemented in virtually any industry for any purpose. It can be used for quality control to check dimensions, angles, colour or surface structure-or for the recognition of an object as used in VGR systems.

A camera can be anything from a standard compact camera system with integrated vision processor to more complex laser sensors and high resolution high speed cameras. Combinations of several cameras to build up 3D images of an object are also available.

Limitations of a Vision System

There are always difficulties of integrated vision system to match the camera with the set expectations of the system, in most cases this is caused by lack of knowledge on behalf of the integrator or machine builder. Many vision systems can be applied successfully to virtually any production activity, as long as the user knows exactly how to set up system parameters. This set-up, however, requires a large amount of knowledge by the integrator and the number of possibilities can make the solution complex. Lighting in industrial environments can be another major downfall of many vision systems.

VGR Systems Benefits

Traditional automation means serial production with large batch sizes and limited flexibility. Complete automation lines are usually built up around a single product or possibly a small family of similar products that can run in the same production line. If a component is changed or if a complete new product is introduced, this usually causes large changes in the automation process-in most cases new component fixtures are required with time consuming set up procedures. If components are delivered to the process by traditional hoppers and vibrating feeders, new bowl feeder tooling or additional bowl feeder tops are required. It may be that different product must be manufactured on the same process line, the cost for pallets, fixtures and bowl feeders can often be a large part of the investment. Other areas to be considered are space constraints, storage of change parts, spare components, and changeover time between products.

VGR systems can run side-by-side with very little mechanical set up, in the most extreme cases a gripper change is the only requirement, and the need to position components to set pick-up position is eliminated. With its vision system and control software, it is possible for the VGR system to handle different types of components. Parts with various geometry, can be fed in any random orientation to the system and be picked and placed without any mechanical changes to the machine, resulting in quick changeover times. Other features and benefits of VGR system are :

- Switching between products and batch runs is software controlled and very fast, with no mechanical adjustments.
- High residual value, even if production is changed.
- Short lead times, and short payback periods
- High machinery efficiency, reliability, and flexibility
- Possibility to integrate a majority of secondary operations such as deburring, clean blowing, washing, measuring and so on.
- Reduces manual work

Chapter 6

WELDING ROBOT APPLICATIONS

ROBOT ARC WELDING

Robot welding means welding that is performed and controlled by robotic equipment. In general equipment for automatic arc welding is designed differently from that used for manual arc welding. Automatic arc welding normally involves high duty cycles, and the welding equipment must be able to operate under those conditions. In addition, the equipment components must have the necessary features and controls to interface with the main control system.

A special kind of electrical power is required to make an arc weld. The special power is provided by a welding machine, also known as a *power source*. All arc welding processes use an *arc welding gun* or *torch* to transmit welding current from a welding cable to the electrode. They also provide for shielding the weld area from the atmosphere.

The nozzle of the torch is close to the arc and will gradually pick up spatter. A *torch cleaner* (normally automatic) is often used in robot arc welding systems to remove the spatter. All of the continous electrode wire arc processes require an *electrode feeder* to feed the consumable electrode wire into the arc.

Welding fixtures and *workpiece manipulators* hold and position parts to ensure precise welding by the robot. The productivity of the robot welding cell is speeded up by having an automatically rotating or switching fixture, so that the operator can be fixing one set of parts while the robot is welding another.

To be able to guarantee that the electrode tip and the tool frame are accurately known with respect to each other, the calibration process of the *TCP* (Tool Center Point) is important. An automatic TCP calibration device facilitates this time consuming task.

Arc Welding Robot

During the short time that industrial welding robots have been in use, the jointed arm or revolute type has become by far the most popular. For welding it has almost entirely replaced the other types except for the Cartesian, which is used for very large and very small robots. The reason for the popularity of the jointed arm type is that it allows the welding torch to be manipulated in almost the same fashion as a human being would manipulate it. The torch angle and travel angle can be changed to make good quality welds in all positions. Jointed arm robots also allow the arc to weld in areas that are difficult to reach. Even so, a robot cannot provide the same manipulative motion as a human being, although it can come extremely close. In addition, jointed arm robots are the most compact and provide the largest work envelope relative to their size.Usually arc welding robots have five or six free programmable arms or axes.

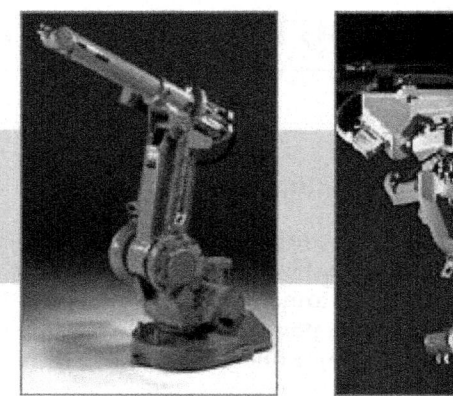
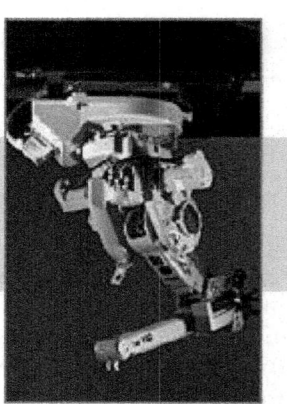

Off-the-shelf programmable robot arms are today available from different suppliers such as ABB, FANUC, PANASONIC, KUKA, MOTOMAN.

Arc Welding Power Sources

A welding power source must deliver controllable current at a voltage according to the requirements of the welding process. Normally, the power required is from 10 to 35 V and from 5 to 500 A. The various welding processes and procedures have specific arc characteristics that demand specific outputs of the welding machine.

Automatic arc welding machines may require power sources more complex than those used for semi-automated welding. An automatic welding machine usually electronically communicates with the power source to control the welding power program for optimum performance. A power source for arc welding is designed to provide electric power of the proper values and characteristics to maintain a stable arc suitable for welding.

There are three types of arc welding power sources, distinguished according to their static characteristics output curve. The *constant-power* (CP) is the conventional type of power source that has been used for many years for shielded metal arc welding using stick electrodes. It can be used for submerged arc welding and gas tungsten arc welding. The *constant-voltage* (CV) power source is the type normally used for gas metal arc and flux cored arc welding using small-diameter electrode wire. The *constant-current* (CC) power source is normally used for gas tungsten arc and plasma arc welding.

The selection of a welding power source is based on
1. The process or processes to be used
2. The amount of current required
3. The power available at the job site
4. Economic factors and convenience.

Welding Torch

A welding torch is used in an automatic welding system to direct the welding electrode into the arc, to conduct welding power to the electrode, and to provide shielding of the arc area. There are many types of welding torches, and the choice depends on the welding process, the welding process variation, welding current, electrode size and shielding medium.

Welding torches can be categorized according to the way in which they are cooled. They may be water-cooled with circulating cooling water or air-cooled with ambient air. A torch can be used for a consumable electrode welding process such as gas metal arc or flux cored arc welding, and shielding gas may or may not be employed.

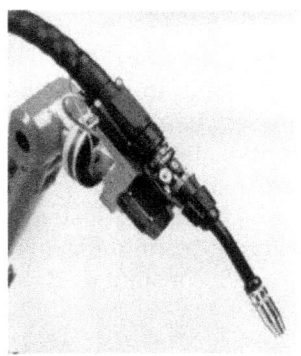

A torch can be described according to whether it is a straight torch or has a bend in its barrel. A torch with a bend is often used for robotic arc welding applications to provide access for the weld.

The major function of the torch is to deliver the welding current to the electrode. For concumable electrode process this means transferring the current to the electrode as the electrode moves through the torch.

A second major task of the torch is to deliver the shielding gas, if one is used, to the arc area. Gas metal arc welding uses a shielding gas that may be an active gas usually carbon dioxide or a mixture of an inert gas, normally argon, with CO_2 or oxygen.

The welding torch is mounted to the robot flange with a matching mounting arm. Preferably an anti collision clutch is used to prevent damages on expensive weld equipment in case of sticking electrode and crashes during installation and start-up.

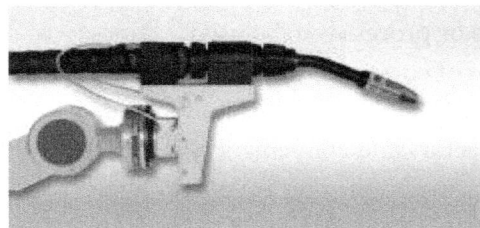

Wire Feeder

Wire feeders are used to add filler metal during robotic welding. This allows flexibility in establishing various welding wire feed rates to suit specific requirements for an assembly. Normally, the wire feeder for robotic welding is mounted on the robot arm, separate from the power supply. For robotic welding, a control interface between the robot controller, the power supply and wire feeder is needed. The wire feeding system must be matched to the welding process and the type of power source being used.

There are two basic types of wire feeders. The first type is used for the consumable electrode wire process and is known as an *electrode wire feeder*. The electrode is part of the welding circuit, and the melted metal from the electrode crosses the arc to become the weld deposit. There are two different types of electrode wire feeders. The constant-power power source requires a voltage-sensing wire feed system in which the feed rate may be changing continously. The constant-voltage system requires a constant feed rate during the welding operation.

The second type of wire feeder is known as a *cold wire feeder* and is especially used for gas tungsten arc welding. The electrode is not part of the circuit, and the filler wire fed into the arc area melts from the heat of the arc and becomes the weld metal.

Workpiece Fixation and Positioning

In order to join parts successfully in a robotic welding application, individual parts must be aligned precisely and held securely in place while the welding is proceeding. An important consideration, then, is the design of a fixture which holds the individual parts in the proper alignment. The tool must allow for quick and easy loading, it must hold the parts in place securely until they are welded together and must allow the welding gun unrestricted access to each weld point.

One starting point for positioning the workpiece for robotic welding may be the fixture already used for manual welding even though specialized positioners are used to improve the versatility and to extend the range of robotic arc welding systems. The usable portion of a robot work envelope can be limited becuse the welding torch mounting method does not allow the torch to reach the joint properly. Special positioners eliminate some of these limitations by making the workpiece more accessible to the robot welding torch.

The positioners used with robots also have to be more accurate than required for manual or semiautomatic welding. In addition the robot positioner controls must be compatible and controllable by the robot controller in order to have simultaneous coordinated motion of several axes while welding.

However, loading and unloading stationary jigs of the robot cell can be time consuming and impractical. It is often more efficient to have two or more fixtures on a revolving workpiece positioner, despite a higher initial cost. With a revolving table for instance, the operator can load and unload while the robot is welding. Obviously, this speeds up the process and keeps the robot welding as much of the time as possible.

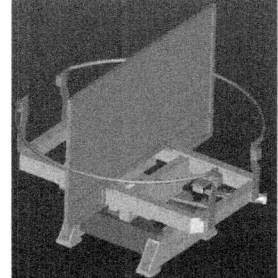

Torch Cleaner

Periodic cleaning of arc welding guns is required for proper and reliable operation of robotic arc welding equipment. The high duty cycle of an automatic operation may require automated gun cleaning. Systems are available that spray an antispatter agent into the nozzle of the gun. Additionally, tools that ream the nozzle to remove accumulated spatter and cut the wire are available. The cleaning system is automatically activated at required intervals by the welding control system.

TCP-Calibration Unit

End-of-arm sensor and tool centre point calibration is a critical aspect of successful system implementation. End-of-arm sensing, in the context of robotic welding, is used to detect the actual position of the seam on the workpiece with respect to the robot tool frame.

Analysis of the profile data yields the relative position of the the seam with respect to the sensor reference frame. If the sensor reference frame pose is known with respect to the end-frame of the robot, and the tool frame pose is known with respect to the end-frame, then the sensor data may be used to accurately position the tool centre point (TCP) with respect to the workpiece.

While end-of-arm sensor based control would appear to solve both robot accuracy and workpiece position error problems, this is only so if the sensor frame, end frame, and tool frame are accurately known with respect to each other.

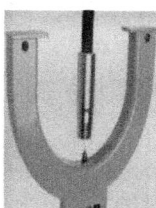

Should the sensor be accidentally knocked out of position, the robot system becomes a highly consistent scrap production facility. Indeed, this very concern has been one of the reasons why some companies that would benefit from a sensor based correction system have been reluctant to implement such a system. What is required is not only a technique that enables the frames to be automatically calibrated, but that also enables the system to quickly determine if recalibration is necessary. This second capability is perhaps the more important in practice, since it can be reasonably assumed that any calibration error will be caused by an unanticipated event that could occur during any welding cycle.

ROBOT SPOT WELDING

Automatic welding imposes specific demands on resistance welding equipment. Often, equipment must be specially designed and welding procedures developed to meet robot welding requirements.

The spot welding robot is the most imortant component of a robotized spotwelding installation. Welding robots are available in various sizes, rated by payload capacity and reach. Robots are also classified by the number of axes. A spot welding gun applies approriate pressure and current to the sheets to be welded. There are different types of welding guns, used for different applications, available. An automatic *weld-timer* initiates and times the duration of current.

During the resistance welding process the welding electrodes are exposed to severe heat and pressure. In time, these factors begin to deform (mushroom) the electrodes. To restore the shape of the electrodes, an automatic *tip-dresser* is used.

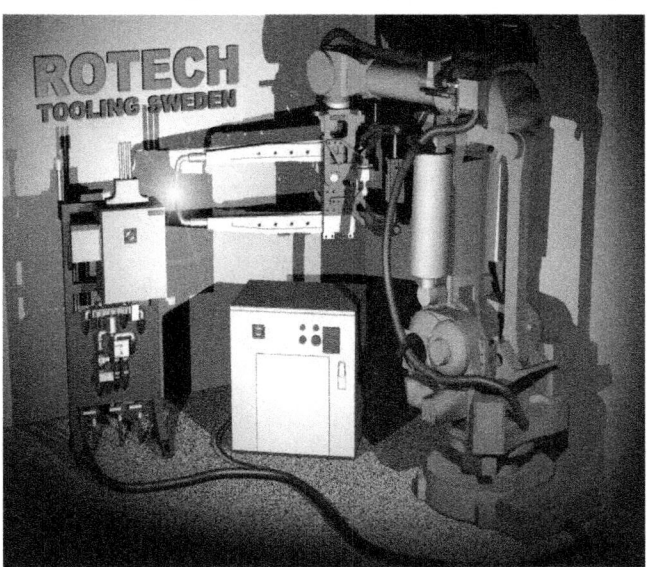

One problem when welding with robots is that the cables and hoses used for current and air *etc.* tend to limit the capacity of movement of the robot wrist. A solution to this problem is the *swivel*, which permits passage of compressed air, cooling water, electric current and signals within a single rotating unit. The swivel unit also enables off-line programming as all cables and hoses can be routed along defined paths of the robot arm.

Spot Welding Robot

A robot can repeatedly move the welding gun to each weld location and position it perpendicular to the weld seam. It can also replay programmed welding schedules. A manual welding operator is less likely to perform as well because of the weight of the gun and monotony of the task.

Spot welding robots should have six ore more axes of motion and be capable of approaching points in the work envelope from any angle. This permits the robot to be flexible in positioning a welding gun to weld an assembly. Some movements that are awkward for an operator, such as positioning the welding gun upside down, are easily performed by a robot.

Spot Welding Guns

Spot welding guns are normally designed to fit the assembly. Many basic types of guns are available, the two most commonly used being the direct acting type, generally known as a "C"-type gun, where the operating cylinder is connected directly to the moving electrode, and the "X"-type (also known as "Scissors" or "Pinch") where the operating cylinder is remote from the moving electrode, the force being applied to it by means of a lever arm. C guns are generally the cheapest and the most commonly used. There are many variations available in each basic type with regard to the shape and style of the frame and arms, and also the duty for which the gun is designed with reference to welding pressure and current.

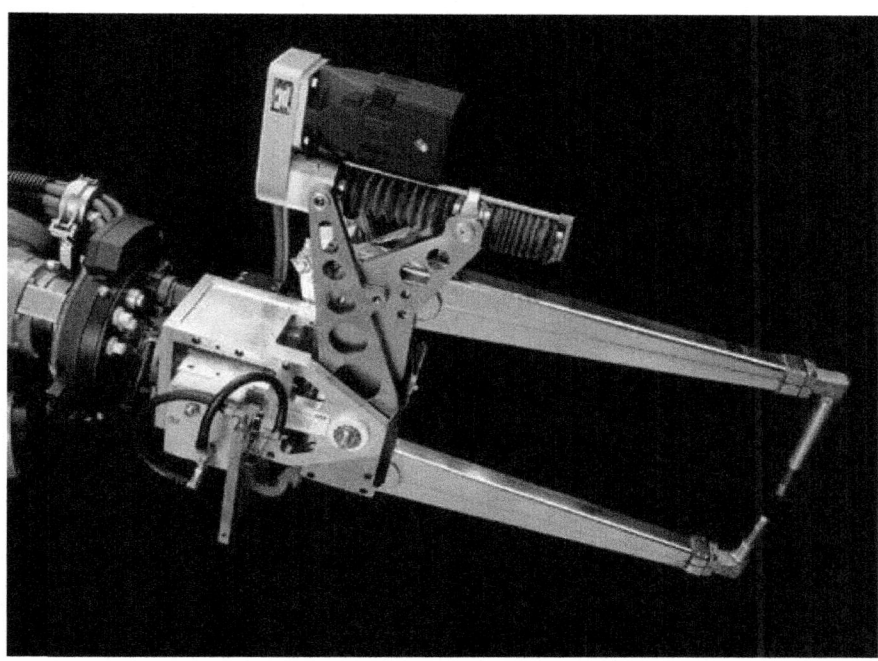

Pneumatic guns are usually preferred because they are faster, and they apply a uniform electrode force. Hydraulic spot welding guns are normally used where space is limited or where high electrode forces are required.

Weld Timer

An automated spot welding cell needs control equipment to initiate and time the duration of current. A spot weld timer (weld control unit) automatically controls welding time when spot welding. It also may control the current magnitude as well as sequence and time of other parts of the welding cycle.

Electrode Tip Dresser

The function of the electrodes is to conduct the current and to withstand the high pressures in order to maintain a uniform contact area and to ensure the continued proper relationship between selected current and pressure. Uniform contacting areas should therefore be maintained.

Good weld quality is essential and depends, to a considerable degree, upon uniformity of the electrode contact surface. This surface tends to be deformed (mushroomed) with each weld. Primary causes for mushrooming are too soft electrode material, too high welding pressure, too small electrode contact surface, and most importantly, too high welding current. These conditions cause excessive heat build-up and softening of electrode tips. Welding of today´s coated materials also tends to contaminate the face of the electrodes.

As the electrode deforms, the weld control is called upon to "step" up the welding current in order to compensate for "mushroomed" weld tips. Eventually, the production line will have to be shut down in order to replace the electrodes or to manually go in and hand dress the electrodes. This process will improve the weld cycle but in either case, the line is stopped and time is lost. Furthermore the deformed electrodes have caused unnecessary high consumption of energy and electrodes.

In automatic tip dressing, a tip dresser is mounted on the line where it can be accessed by the welding robot. The robot is programmed to dress the electrodes at regular time intervals. The dressing can be done after each working cycle, after every second cycle, and so on. It depends upon how many spot-welds are done in each cycle. For welding in galvanized sheet, dressing after about 25 spot-welds is recommended. The dressing takes approximately 1 to 2 seconds, and is perfomed when the work pieces are loaded, unloaded and transported. Maintaining proper electrode geometry minimizes production downtime and utility costs and increases weld efficiency.

Spot Welding Swivel

A major advancement in resistance spot welding is the swivel. This unit permits passage of compressed air, cooling water, electric current and signals through different channels within a single rotating unit.

This invention greatly improves total efficiency of robotic spot-weld installations. Electrical connection between swivel and transformer is minimal thus permitting maximum utilization of access to spot-weld areas.

Basic Advantages Are :

- Less work space needed -No mass of cables and hoses hanging from the robot arm, resulting in floorspace economy.
- Improved accessability - Since no limitation on the robot wrist caused by any cables or hoses.
- Improved safety - Greatly improved safety factors through reduction of air, electric and water lines; now limited to quick-connect piping, and hoses within robot arm.

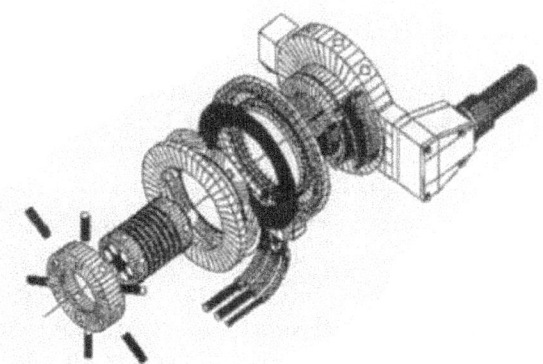

- Saving in capital equipment - Compact weld-gun assembly accessable to areas formly blocked by transformer, cables, and controlboxes. More welds per station means big savings through fewer work stations and less capital equipment.
- Reduced try-out costs - No un-defined cables exist on the robot, which reduces programming time to minimum. True off-line programming is now a reality.

The swivel, which fits directly onto the weld-gun fixture plate without any hoses or cables, ensures the highest quality condition of the spot-weld. No electrical degeneration on cables and no hoses that wear.

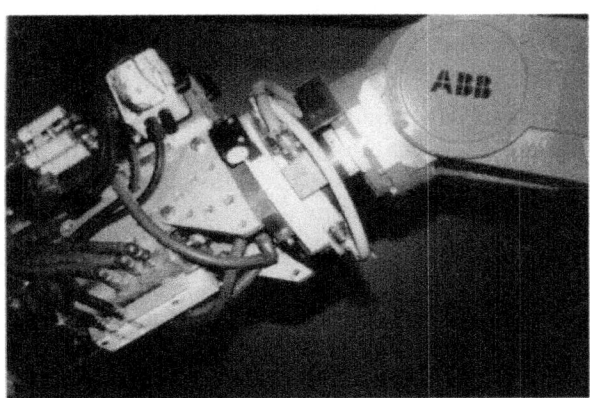

WELDING SAFETY

Welding is an established manufacturing process with known potential hazards. Potential safety hazards associated with arc welding include arc radiation, air contamination, electrical shock, fire and explosion, compressed gases, and other hazards. Robots were originally designed to perform the job functions of a human. They were designed to relieve humans of the drudgery of unpleasant, fatiguing, or repetitive tasks and also to remove humans from a potentially hazardous environment. In this regard, robots can replace humans in the performance of dangerous jobs and are considered beneficial for preventing industrial accidents. On the other hand, robots have caused fatal accidents.

The introduction of robots requires appropriate safety features in order to protect both those working directly with the robot and others in the workshop who may not be aware of its potential dangers. This can be provided in a number of ways.

One of the best solutions for robot safety is to purchase a complete welding cell from a robotic integrator . A complete cell includes barriers, all necessary safety devices, and a method of loading and unloading the workstation.

Each robot installation must be carefully planned from safety viewpoint to eliminate hazards. When the robot is in operation it is necessary that people remain outside the work envelope. Barriers or fences should be in place around the robot. All doors and maintenance openings must be protected by safety switches, and the weld areas must be safe guarded so that the power is immediately removed from the robot when a door is opened.. Emergency stop buttons should be placed on all operator panels, robot cabinets and robot programming panels. Barriers must be designed to completely surround the robot and eliminate the possibility of people climbing over or under to get inside the barrier. Signal lights must be arranged on the robot or in the robot area to indicate that the robot is powered.

WELDING PROCESSES

Welding is the most economical and efficient way to join metals permanently. Welding is used to join all of the commercial metals and to join metals of different types and strengths.

A weld is produced either by heating the materials to the welding temperature with or without the application of pressure alone with or without the use of filler metal. There are different kinds of welding processes who all use different sources of heat, for instance arc welding which uses an electric arc as a heat source. Another commonly used welding process is spot welding (resistance welding).

Welding is considered to be the most complex of all manufacturing technologies. In order to transform welding from a manual operation to an automated production process, it is necessary to understand the scientific principles involved.

THINGS TO KNOW ABOUT ROBOTIC WELDING SYSTEMS

Implementing a robotic welding system isn't something that happens on a whim — at least not successfully. Converting to this technology can help companies gain greater productivity, improve quality and reduce costs in the welding operation, but the process requires thorough planning to gain those results. Working closely with a robotic integrator is a good step to ensure every aspect of the implementation is carefully orchestrated and that the robotic welding system works properly for the given application — in reality, not just theory. Before adding a robotic welding system, it's helpful to know some key factors that can maximize the return on investment (ROI) in the technology and also help prevent potential problems.

Part Repeatability is Critical to Successful Automation

The quality of part produced by a robotic welding system depends on the quality of the part that enters the weld cell. That's why it's not uncommon to hear the phrase "garbage in, garbage out" when it comes to robotic welding systems — if the part entering the welding cell is flawed, the subsequent weld will be, too.

Before adding a robotic welding system, it's helpful to know some key factors that can maximize the return on investment (ROI) in the technology and also help prevent potential problems.

To protect against poor weld quality, it is critical to have simple, consistent parts that allow the robot to execute the weld in the same location, repeatedly. Having a blueprint or electronic CAD drawing is helpful for confirming that repeatability. Robotic integrators can review the blueprint or they may want to create a software simulation that assesses the suitability of the part for the robotic welding system. After the assessment, they can advise of any adjustments that need to be made prior to implementation.

Proper fixturing is also critical to achieving part repeatability, regardless of whether the application is high volume/low variety or low volume/high variety. Parts that meet the exact specifications can easily be welded incorrectly if they are not held in an exact position during the process. Many robot manufacturers offer vision systems to aid in part recognition and to ensure that the weld path can be altered in real time if part fit-up issues exist. These systems usually work very well, but may cost more.

Training is Essential

Robotic welding systems require a properly trained operator to oversee them. A skilled welding operator or an individual with previous robotic welding management experience is a good candidate for the job. Again, a trusted robotic integrator is an excellent resource to provide the necessary training, which should cover proper programming, troubleshooting and preventive maintenance. As a best practice, companies should also consider ongoing training support to keep the operator's knowledge of the system up to date. In many cases, robotic OEMs offer online tutorials, troubleshooting information and/or additional on-site training as aftercare support.

Additional Safety Equipment may be Necessary

Many facilities already have fume extraction systems in place for manual welding operations, but converting to a robotic welding system may require additional equipment to help maintain a healthy work environment. With the increased production brought forth by a robotic welding system, there is also an increase in fume generation. Given the stringent regulations and recommendations from OSHA (Occupational Safety and Health Administration) and other safety regulatory bodies, proper equipment is necessary to maintain compliance.

For larger facilities with higher production robotic welding applications, a centralized fume extraction system is a good option. These systems involve the installation of ductwork throughout the facility and the placement of fume extraction hoods over the welding cell. Smaller shops with fewer robotic welding cells may want to consider a less expensive portable fume extraction system. Operators can wheel these systems right next to the welding cell and adjust the extendable arm toward the robot to suction the fumes. It is also critical that the proper cage and screens are in place around the robotic welding system to protect employees from the welding arc and moving parts within the cell.

Weld Data Monitoring and/or Peripherals Can Help Improve Results

Adding weld data monitoring capabilities and/or peripherals into a robotic welding system can help improve weld quality and productivity. Achieving these results, however, requires an additional up-front investment.

The addition of peripherals, particularly a nozzle cleaning station can improve weld quality and productivity, as well as minimize the loss of shielding gas coverage that could lead to poor weld quality and rework.

Weld data monitoring (whether integrated in a power source or via a third party) allows companies to track the parameters of individual welds, determine the cause of weld defects and identify general inefficiencies in order to rectify those problems and optimize the process for peak quality and productivity. This equipment requires the purchase of software and computers, as well as the establishment and maintenance of an Ethernet network throughout the facility. Companies will also need tech-savvy individuals to review the data and make the necessary adjustments to the robotic welding system according to the data provided.

Similarly, the addition of peripherals — particularly a nozzle cleaning station (also called a reamer or spatter cleaner) can improve weld quality and productivity. By cleaning spatter from the inside of the welding consumables on the front end of the GMAW welding gun, this peripheral helps extend consumable life, reduces downtime for changeover during production and also reduces the cost for replacing consumables. Nozzle cleaning stations also help minimize the loss of shielding gas coverage that could lead to poor weld quality and rework.

Proper Maintenance can help Protect the Investment in Automation

Preventive maintenance of the entire robotic welding system, including the robotic GMAW (gas metal arc welding) gun, consumables and cables is an important step in protecting the investment in this technology. Neglecting maintenance can easily lead to unscheduled downtime, poor quality parts and/or costly repairs. It may even lead to failures that require equipment replacements.

Scheduling time to check connections throughout the system, clean fixturing (to prevent debris that may affect part fit-up) and check TCP (tool center point) helps ensure that the robotic welding system continues to operate within its proper parameters. Certain maintenance can occur in between shifts — cleaning off the robot or changing consumables, for example — while other activities like greasing the robot's joints may occur less frequently and during a longer scheduled stop. Companies need to assess their individual needs and plan the preventive maintenance schedule accordingly. For larger companies, hiring a maintenance crew to take care of preventive maintenance may be desirable.

Communication is Key to Proper Weld Quality And Cost Savings

Retrofitting robotic welding systems is a common practice among many companies, particularly those investing in automation for the first time or for smaller shops requiring only one or two weld cells. It's significantly less expensive to purchase a used robot than a new one. When retrofitting a robot, however, it is absolutely essential that it is capable of communicating with the selected power source if companies are to have the entire robotic welding system function properly. New power sources feature software that may not be immediately compatible with a robot that is older, or in some cases, the robot may need a specific robotic GMAW gun that isn't readily available at a welding distributor or possibly even discontinued.

For this reason, it is critical to contact an experienced robotic integrator who can both recommend and help set up all components in the retrofitted robotic welding system. The investment in this assistance can help ensure the proper functioning of the equipment and the long-term cost savings sought by implementing the system. Not to mention, it can also save a lot of frustration and downtime.

Robots Can do More than Just Weld

Robots rely on the input of the operator to execute a given task. That task, however, doesn't have to be limited to just welding or to welding the same part every time. Operators can program the robot to weld multiple parts over the course of a single shift, enhancing the versatility of the robotic welding system and positioning the company to produce additional output. Operators can also program robots to move parts so that a particular unit is not sitting idle when it isn't tasked with welding; there are components that offer gripping capabilities and can be installed in addition to a welding gun. Companies may even have a tool dock that allows the robot to be fitted with a different tool and proceed with its work. Some companies with multiple robots may also benefit from installing a vision system in order to check on the work of the others, ensuring that part fit-up is optimal and that the robot is correctly placing welds.

Given that the goal of any robotic welding system is uptime, having the versatility to use a robot for multiple tasks can contribute meaningfully to the other advantages of this technology – increased productivity, improved quality, decreased costs — and may help give companies a real competitive edge.

WHAT YOU MUST KNOW ABOUT ROBOTIC WELDING

In the fabrication and manufacturing world, quality and productivity are everything. To remain competitive, companies need to look continually for ways to increase throughput and minimize defects, while also keeping costs low for parts and labor. In many cases, turning to robotic welding is a means to achieve those goals — for both the smaller job shop and larger manufacturing facilities.

In many cases, robotic welding operations can help companies achieve greater productivity, better quality and lower costs. Planning the operation carefully and paying close attention to details after implementation are both critical, however, to achieving success with this technology.

The decision to implement a robotic weld cell, however, takes a good deal of consideration and planning if the system is to function in the most efficient, productive and profitable manner. And it requires a significant investment.

Fortunately, the long-term benefits of a robotic welding operation can be very positive. For companies who have already invested in robotic welding, but are looking to improve or better understand their operations, or for those considering the investment, it is critical to consider some key factors about the technology. Here, we will explore "what you must know about robotic welding" to make the most of the process.

There's More to the Payback on a Robotic Welding System than Just Speed

Justifying the cost of a robotic weld cell comes down to the ability to gain (and prove) a payback on the investment. Typically, that payback comes in the form of greater productivity and higher-quality welds (which minimize instances of costly and time-consuming rework), but there are other contributing factors to the return on the investment (ROI) in this technology. Robotic welding also offers the advantage of lower energy and labor costs, and in many cases lower material costs due to fewer instances of overwelding. Overwelding is a common and costly occurrence in semi-automatic welding. A weld bead that is 1/8-inch larger than necessary can double filler metal costs, but a robot can reduce those costs by only putting down as much material as necessary. Plus, robotic welding systems use bulk filler metals (600-pound drums, for example) that companies can often purchase at a greater discount.

For companies just considering the investment in robotic welding, it is important to consider how to calculate the payback. Assess the current part cycle times and compare those to the potential cycle times of a robot. A trusted robotic welding integrator or OEM can often help with this calculation. During this process, also consider the possibility of reallocating existing labor to other parts of the welding operation, where these individuals can add value to the process. Remember, up to 75 percent of the cost in a semi-automatic welding operation is labor. If there is the opportunity to use that labor elsewhere to increase part production, the payback on the investment in robotic welding will increase.

Most companies — particularly smaller ones or those with frequent production changes — seek a payback on the robotic welding investment of no greater than 12 to 15 months. That time frame is entirely possible to achieve with proper up-front planning of the part blueprints, fixturing and general setup of the system. In some cases, companies may be able to justify a longer payback period if they know that their production needs will remain relatively static for longer periods of time.

Parts and Product Flow Need to be Consistent

The output from a robotic welding cell is only as good as the parts fed into it. In order to gain the advantages of these systems, it is critical to have accurate, repeatable part designs. Gaps, poor fit-up or poor joint access all prevent a robot from completing its job correctly.

The best part designs for a robotic welding application are simple ones that allow the robot to execute the same weld repeatedly. High-volume applications with low-variety parts are especially poised to gain the advantages of robotic welding. Companies should try to avoid part designs that require intricate tooling or clamping to hold it in place, as both can hinder the efficiency of the robot and also add to the up-front cost of the operation. That said, in some cases, companies may still be able to gain a good payback on the investment in tooling for slightly more complex parts, but they will need to weigh out the pros and cons of that cost ahead of time.

The robotic MIG welding gun, as well as the consumables can impact productivity and profitability in a robotic welding application. Companies need to be certain to select a gun with the right amperage and duty cycle, and use consumables appropriate for the application.

Companies also need to be certain to assess their overall welding operation for consistent process flow. Bottlenecks upstream can easily slow down the movement of parts into the robotic work cell and the ability of the system to function to its full capacity. A robot that sits idle costs time and money. Some companies may need to reconfigure operations or set up a flexible cell that can manage quick tool and fixture changes in order to minimize bottlenecks in the process flow. It is also important to have adequate labor to supply the robot with parts.

Again, companies should consider tapping the knowledge of a robotic welding integrator for advice and assistance to optimize process flow.

The MIG Guns and Consumables on the Robot can Impact Productivity and Profitability

The robotic MIG gun and consumables on a robot together are responsible for directing the current to the arc to complete the weld, making them integral components in the whole system. To gain the best quality and to avoid expensive downtime for maintenance, repairs or replacement, companies need to select a robotic MIG gun that is suitable for the amperage, duty cycle and cooling capacity needed in the application. Using a robotic MIG gun that offers inadequate cooling or amperage can cause performance issues and lead to premature failure — both factors that increase costs and downtime. Likewise, using a robotic MIG gun that offers higher amperages than necessary raises the total cost of ownership, as typically the cost of a robotic MIG gun increases directly in proportion to its amperage.

Companies also need to select their consumables — contact tips, nozzles, retaining heads (diffusers) and liners — carefully and manage them properly in order to gain optimal productivity and lower costs.

Heavy- or extended-duty contact tips and nozzles, for example, are a good choice to withstand the heat of higher amperage applications and can help minimize downtime for changeover. Conversely, lower amperage applications or ones with shorter arc times may be better suited to standard-duty consumables, which also tend to cost less.

As with robotic MIG guns, carefully matching the type of consumables to the application can keep companies from having to address premature failures and/or accrue costly downtime (not to mention, lapses in production). It can also keep them from overpaying for consumables that may be too much for the application. Companies should also consider the mode of welding when selecting consumables, as technology such as Pulsed welding tends to be especially harsh on consumables and often requires heavy-duty options to withstand the heat of the arc for longer periods of time.

Peripherals can help Improve the Return on Investment in a Robotic Welding Operation

Peripherals refer to any additional equipment integrated into the robotic welding system to maximize its performance. They include items such as a nozzle cleaning station (sometimes called a reamer or spatter cleaner), anti-spatter sprayers, wire cutters and neck alignment tools. Unfortunately, some companies downplay the value of peripherals, viewing them as an unnecessary cost, and don't realize that they can play an important role in reducing downtime and rework, improving quality and increasing productivity. Consider a nozzle cleaning station, for example. As its name implies, this peripheral cleans the nozzle of dirt, debris and spatter, typically during routine pauses in the robotic welding operation. This cleaning action helps prevent shielding gas coverage loss that could lead to weld defects, expensive rework and lost productivity. The equipment also helps the front-end consumables last longer — and longer consumable life means less downtime for changeover and less expense for replacements. The addition of an anti-spatter sprayer further improves consumable life and performance by adding an anti-spatter compound that serves as a protective barrier against spatter buildup and other contaminants.

In the long run, the up-front investment in peripherals such as these can lead to measurable savings and provide a better return on investment by aiding the robot in doing what it does best : complete consistent, high-quality welds for longer periods of time than a semi-automatic welding operation.

Peripherals, such as a nozzle cleaning station or reamer can often help companies gain a better return on the investment in a robotic welding operation. They can lower downtime and help maintain higher quality welds.

Having Skilled Operators with Proper Training to Oversee the Robotic Weld Cell is Critical.

Robotic welding operations require ongoing supervision and maintenance, and that job needs to be completed by a skilled operator who has undergone the proper training. When considering an investment in robotic welding, companies should take care to evaluate the available pool of talent. As a rule, skilled welding operators and/or employees with prior robotic welding experience are the best candidates to supervise the weld cell. After the proper training, which a robotic integrator or OEM can typically provide, these employees can provide the necessary operating and troubleshooting skills to ensure the maximum uptime in the robotic welding cell.

As part of the routine training, it is absolutely necessary for the operators who will be overseeing the robot to be able to schedule and perform routine preventive maintenance on the system. Implementing preventive maintenance helps minimize unnecessary downtime and keep the robotic welding system running more smoothly. If problems can be solved before they arise and the robotic welding equipment made to last longer, it can protect the company's investment, and ensure the productivity and profitability sought by this equipment in the first place.

Companies should consider vetting robotic welding integrators to determine the availability and costs associated with the training of personnel. Typically training lasts one to three weeks, depending on the certification level desired, and continuing tutorials are often available.

In the end, careful planning, good equipment selection and proper training are all "must-knows" for managing a profitable and productive robotic welding operation. So whether a company is new to robotic welding or trying to improve an existing operation, knowing some key factors can go a long way in helping to gain a competitive edge and to make the most out of the investment.

TROUBLESHOOTING ROBOTIC WELDING

In an ideal world, companies would never experience downtime or costs related to troubleshooting problems in the weld cell. Unfortunately, that just isn't the reality. And like any welding process, robotic welding is no exception to problems caused by equipment failures or human error. However, given that most companies invest in welding automation to increase throughput and profitability, there is a lot at stake when something goes wrong in the process.

Most often, when a problem occurs with a robotic welding system, it's valuable to ask first : What has recently changed in the process? Has the operator recently reprogrammed the robot? Or was the system restarted after a long shutdown? What about the consumables or robotic MIG gun — has anything changed with them?

Quite often, looking at the most recently changed variable in the process can help narrow down the point of trouble. The issue may be something as simple as a loose contact tip or more complex like an incorrect tool center point (TCP). Whichever the case, it's important to have good troubleshooting skills to help narrow down the focus and get the robotic welding system back on line sooner.

Poor Consumable Performance and/or Premature Failure

The longevity of consumables — nozzles, contact tips, diffusers and liners — in a robotic welding application depends in part on the material being welded and the welding parameters. For example, high-amperage, high-deposition-rate applications tend to be harsher on consumables than those with lower amperages. However, if it appears that consumables for the given application aren't lasting their usual length of time or they are performing poorly, there could be multiple causes.

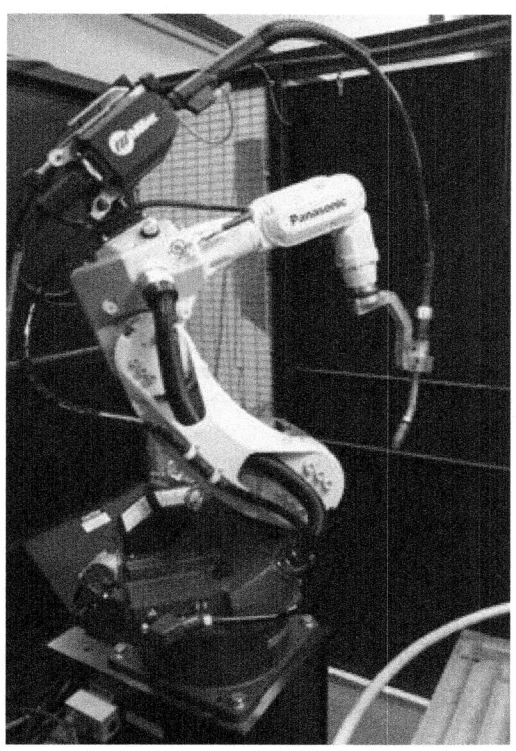

When a problem occurs with a robotic welding system, it is critical to identify the problem as quickly and accurately as possible. Look first at the variables that have most recently changed, as these may be the culprits.

In many cases, a loose connection between consumables can be the culprit. Loose connections increase electrical resistance, causing the consumables to generate additional heat that can shorten their lifespan and/or cause them to perform poorly. Be certain to tighten consumables properly upon installation, per the manufacturer's instructions, and check them periodically during routine pauses in welding. For companies that weld thick materials or long welds, it is especially important to make sure that consumables are tightened properly, as the rework for quality issues caused by poorly performing ones can generate much more costly rework than an application that produces multiple smaller parts.

Issues with the contact tip are also not uncommon, particularly burnbacks. These are often the result of a liner being trimmed too short. Welding operators should follow the manufacturer's instructions for trimming and installation, and when possible use a liner gauge to confirm the correct liner length.

If the robotic welding system utilizes a nozzle cleaning station (also called a reamer) and consumable issues occur, such as spatter build-up, check to see that this equipment is working properly. Also be certain that the reamer is cleaning the consumables at a frequency that is appropriate for the application. It may be necessary to increase the frequency of cleaning and/or anti-spatter spray application throughout the programmed welding cycle.

If weld defects — like porosity or lack of fusion — are occurring frequently, it might also be indicative of an issue with the consumables.

Premature Cable Failure

Premature power cable failure can occur in both through-arm robotic welding systems, where the cable feeds through the arm of the robot, or in standard robotic welding systems (also referred to as over-the-arm). The power cable may become kinked or worn, causing the failure — or in extreme cases, it may even snap.

If any of these situations occur, it is important consider the path the robot is programmed to follow, as well as the length of the power cable being used. First, be certain that the robot's movements have not been programmed to be too fast or abrupt. Aggressive movements can cause the power cable to snap. Or in some cases, it may cause it to flop around, allowing the power cable to rub against the robot or tooling, or catch on components — both instances that can lead to premature failure.

Also, check that the power cable being used is not too short for the application or too long. If it is too short, the power cable will stretch beyond its capacity during routine robotic movements, leading to greater wear. Conversely, if the power cable is too long it may be prone to kinking or becoming pinched by the robot's arm.

Poor Wire Feeding

Poor wire feeding in a robotic welding application can lead to equally poor weld quality. Issues with the liner, including debris build-up, can often cause the problem. Be certain to change out the liner during routine maintenance to prevent debris build-up from the welding wires and the environment. Blowing compressed air through the liner also helps. Ideally, consider using a robotic MIG gun with an "air blast" feature, which blows the air through the liner during a scheduled time in the robotic program (for example, during a reaming or cleaning cycle).

An improperly functioning wire feeder — specifically the drive rolls — can also cause poor wire feeding. Over time, these components can become worn and may not guide the welding wire properly. Or the drive rolls may not be tightened correctly. Inspect the drive rolls for signs of wear and replace them as necessary.

Welding operators can also determine whether the drive rolls are the problem through a process of elimination. Namely, by conducting a "two finger" test — disengage the drive rolls, grasp the welding wire and pull it through the gun. It should be able to pull easily through. If it does then it's possible that the drive rolls are the cause of the poor wire feeding. If the wire does not pull through easily, it indicates a problem outside of the wire feeder and drive rolls, such as debris in the liner or another such restriction within the robotic MIG gun. It may even be the result of having too small of a contact tip in place.

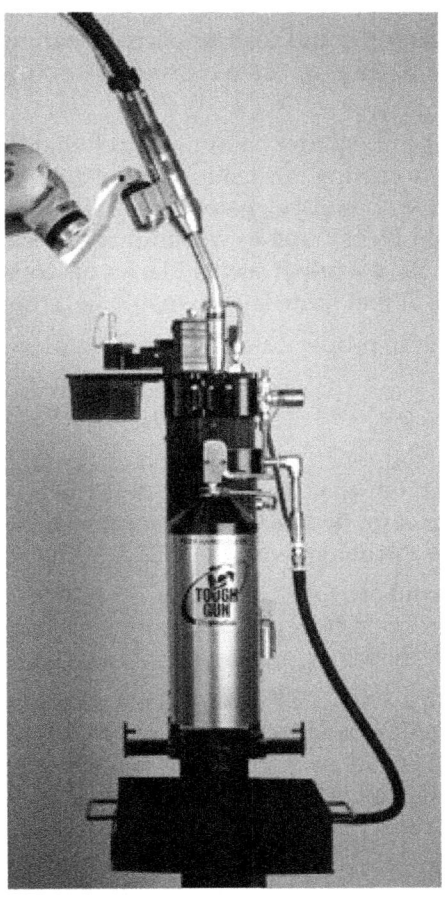

Having a properly functioning reamer can help extend consumable life. Should any problems occur with the equipment, check that the reamer is positioned accurately and is applying the correct mount of anti-spatter solution.

Welding operators should also look for kinks in the power cable, as these can also lead to wire feeding problems.

Poorly Performing Peripherals

Peripherals — in particular, reamers — can help companies optimize their robotic welding performance and extend the life of their consumables. If a welding operator notices that there is an excessive build-up of spatter on the consumables, however, it may indicate a problem with the reamer.

There are typically three reasons for a reamer to function poorly. The first relates to the taught position of the robotic MIG gun nozzle in relation to the reamer. That is, where the robot clamps to the reamer. The position should be exactly perpendicular to the cutting blade on the reamer. Any misalignment of the nozzle during cleaning could lead to partial cleaning of the nozzle and excessive spatter build-up. As a first step in troubleshooting, check that the taught position is correct.

Secondly, if using anti-spatter solution, check that the spray location is correct. Is the solution fully coating the inside of the nozzle? If not, adjust the location accordingly. The nozzle should be coated until it is slightly damp on the inside and the outside should be covered to within three-quarters of an inch from the bottom of the nozzle. And while it seems like an obvious troubleshooting step : Always be sure to check that there is anti-spatter solution in the sprayer!

Lastly, be certain the proper cutting blade is in place and that it is sharp.

Trouble with TCP?

In addition to speed, one of the greatest advantages of a robotic welding system is the repeatability that it provides, and the subsequent quality of the welds. If a welding operator begins to notice inconsistent welds or welds that are off-location, it may be a problem with the TCP.

TCP is the focal point of a tool. In the case of a robotic welding system, it refers to the location of the robotic MIG gun and how it corresponds with the position of the welding wire in the joint (gun-to-work distance).

Most often, issues with TCP occur after a collision, during which the neck of the robotic MIG gun becomes bent. To rectify the problem, welding operators should use a neck-checking fixture or neck alignment tool to make sure the neck is bent to the proper angle. It is also important to check that the neck is installed correctly. If the neck isn't fully seated, it may extend too far and lead to TCP problems. To protect against future issues, it may also be helpful to program a TCP check to verify the proper position. Welding operators, however, shouldn't assume that welds that are off-location are always caused by an incorrect TCP. In some cases, they can be the result of improper fixturing, fixturing that allows the part to move or a loose robot base. Or there may be a variation in the part itself.

To differentiate between a TCP problem and other problems that could cause off-location welds, first take the neck off the robot, implement a TCP check via the robotic program and verify that everything is on-location. If everything checks out properly, the problem is likely a part or position variation.

Final Considerations

When something goes wrong in a robotic welding system, it is critical to identify the problem as quickly and accurately as possible. Not only can swift troubleshooting ensure that the operation returns to producing quality, repeatable parts, but it can also help prevent unnecessary costs for replacing components that may not need replacing. Always start with the simplest solutions first and consider keeping a checklist for setup and maintenance procedures. Having a quick reference point can help facilitate the troubleshooting process by identifying potential variables that have changed during the course of routine operations.

WELDING IN TODAY'S AUTOMOTIVE INDUSTRY

Worldwide, companies serving the automotive industry have faced a unique set of challenges in the last several years, including changes in material types, a lack of skilled labor and initiatives by OEMs to decrease the weight of vehicles. Still, as the economy continues to rebound, each must find ways to maintain their productivity and profitability — often with fewer employees than before the recent recession.

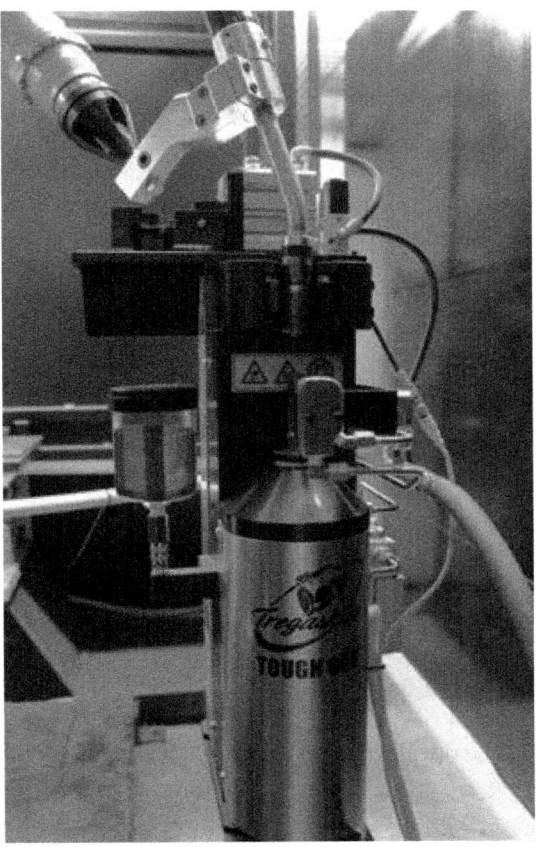

A large part of maintaining productivity in an automotive application is ensuring high levels of uptime in the robotic welding operations and maximizing net throughput. It is equally important to find ways to minimize errors and obtain predictive weld data that helps anticipate problems in the operation.

A large part of maintaining that productivity is to ensure high levels of uptime in the robotic welding operations in order to maximize net throughput. It is equally important to find ways to minimize errors and obtain predictive weld data to help anticipate problems in the operation. Conventional issues like spatter, burn-through and poor part fit-up often hinder these attempts, as can the need to manage large amounts of inventory and contend with downtime to service welding equipment. That's why it's so important, too, for companies to find equipment that minimizes the total cost of ownership.

Unfortunately, there is no single answer to these challenges. There are, however, some considerations that may help reduce automotive suppliers' pains and assist in other interrelated parts of the process.

Best practice meetings : When possible, suppliers in the automotive industry should work with original equipment manufacturers (OEMs) and vendors or welding distributors who can engage regularly in best practice meetings. These meetings can occur by conference call, webinar or in person, and can help determine what practices in the welding operation are working most effectively and what areas need improvement. "Open issues" can be prioritized in order to determine time-phased solutions.

These meetings can be especially helpful to companies with multiple locations (even globally), since they help identify opportunities for changes that could positively affect other facilities. They are also an excellent platform for brainstorming error-proofing ideas and serve to open communication among the parties involved in the success of a company's welding operation. Ultimately, the goal is to spread an assessment of the operation to a broader peer group, extending the company's core competencies to gain solutions from others' input.

Streamline vendors : Automotive suppliers, particularly those with multiple locations, may want to consider purchasing their robotic gas metal arc welding (GMAW) guns, peripherals, consumables and other welding supplies from a single-source vendor via a welding distributor. Having multiple vendors may appear to provide cost savings on the surface; however, a per-item approach can actually increase the total spend. Instead, by single sourcing a product line, a company is better poised to maximize their purchasing power with one vendor and gain loyalty discounts. The vendor may also be more inclined to aid in new efficiencies and groundbreaking technologies. Plus, a trusted single-source vendor can often help automotive suppliers assess their total weld spend, streamline inventory and reduce costly paperwork. The goal is to work with a vendor who can "own the arc," providing assistance throughout the whole welding operation by assessing predictive data and offering suggestions for ongoing improvements.

"Co-opetition" : If you already work with several welding vendors, co-opetition is your next best option to maintaining an effective welding operation and in some cases can occur as part of best practice meetings. This term refers, in short, to cooperation that occurs between the various equipment manufacturers who are building the end user's welding solution. Sometimes these companies have competitive product overlap. For example, the manufacturer of the robotic

GMAW gun or welding wire may be in direct competition with the company whose power sources are in an automotive supplier's weld cell. Even so, finding equipment manufacturers who are willing to work together to address problems in the welding operation is key to resolving issues when they arise.

A problem with the contact tip, for example, is usually a barometer of other things happening in the process. In short, it is very often a symptom of a problem, as opposed to the root cause. Having partners who are willing to put aside competitive differences for the good of resolving problems like these is important to gaining good welding performance. If this co-opetition is not feasible, companies may want to consider moving to a single-source vendor.

Equipment standardization: Recent increases in demand for production have caused some automotive suppliers, especially those in North America, to make capital investments that they previously postponed during the recession. When possible, standardizing on a single brand and style of welding power source, robotic controller, and GMAW gun and consumables during this investment can streamline inventory and maintenance procedures, thereby lowering management costs. It can also help companies avoid long lead times associated with specialty products and improve access to spare parts.

For companies in an organic growth mode with new programs and/or greenfield operations, this standardization can help in long-term equipment redeployment to other facilities, as well as streamline the learning curve among employees, and improve adoption rates and costs.

For companies that are in acquisition mode, however, this standardization may not be feasible. Instead, these suppliers should, at a minimum, consider standardizing on a single brand and style of robotic GMAW guns and consumables to minimize inventory. Doing so can also reduce the risk of improper consumable installation, which often leads to unscheduled downtime to rectify.

Appropriate welding technology : Many automotive suppliers rely on tandem- welding operations as a means to generate greater productivity. Companies can use this process for line production in the cells housing the majority of the welds. The benefit is that these operations require less floor space and can simultaneously improve throughput.

Advancements in single arc pulsed technology have also proven very efficient in providing faster travel speeds and minimizing spatter. This single arc technology, which effectively lowers the average amperage level during welding (by regularly switching the current between high peak amperages and low background amperages), is also quite easy to operate. Given the reduction in workforce in the automotive industry, combined with an overall shortage of skilled labor, this less complex (but highly efficient) technology has already proven beneficial for many automotive suppliers.

Companies should work with an appropriate welding distributor or robotic integrator to assess the individual application in order to determine the most appropriate welding technology.

Error-proofing: In addition to standardizing equipment when possible, using welding products that minimize the opportunity for human errors is an important part of keeping the welding process flowing. For example, nozzle detection can eliminate the potential of excessive rework or scrap. Avoiding errors in equipment installation is also critical, as missing or incorrectly installed components on the front end of a robotic GMAW gun can cause it to become electrically alive, leading to premature failure and poor welding performance.

Preventive maintenance: Even though preventive maintenance or PM may have become a commonplace buzzword in recent years, the fundamentals are still critical to providing good welding performance and reducing unscheduled downtime in the automotive industry. Companies should take care to inspect all connections in the ground cables, feeding assembly, wire feeder, GMAW gun and consumables on a regularly scheduled basis. Replacing worn components during scheduled downtime (at the beginning of a shift, for example) can help prevent problems during production. On some welding robots, "predictive maintenance" technology is available to send alerts when consumables need to be changed.

Built-in buffers: As is typical in automotive "just-in-time" applications, suppliers want to reduce work in progress (WIP) — maintaining only strategically determined micro-inventories — and keep parts flowing (Takt time). To continue that workflow but still allow for any instances of stoppage in a robotic welding cell, suppliers may consider building a buffer into production. For example, if a company has a production line of 40 welding robots, breaking that line into fifths (five sections of eight robots), allows them to address any instances of failure while causing a stoppage of only eight robots instead of shutting down production on all 40. That buffer can mean a significant difference in terms of lost production and money.

And while no single one of these considerations can ensure the levels of productivity and profitability to which automotive suppliers strive as production demands increase, they can be a step in the right direction. Automotive suppliers should consider working with a trusted welding equipment manufacturer and vendor to discuss a plan for assessing their robotic welding operation and identifying opportunities for improvement.

CONSUMABLES FOR ROBOTIC WELDING

What You Should Know to Improve Performance and Reduce Costs

When you invest in automation, the goal is to gain productivity and quality improvements that set your welding operation apart from the competition and help increase your bottom line. To achieve success with an automated welding system, however, you need to ensure that the parts you are welding are consistent and repeatable, confirm that your welding operation has good workflow and have properly trained welding operators to oversee the system. You also need the right equipment for the job.

In addition to working with a reliable robotic integrator to select and implement the robot, you should also take care to select the right robotic MIG gun and consumables — contact tips, nozzles, liners and retaining heads — for the application. The consumables, in particular, are an easily overlooked part of an automated welding system, but they can have a measurable impact on downtime and day-to-day costs. Consider these suggestions for getting the best performance from these components.

Mind Your Extensions and Connections

The contact-tip-to-nozzle relationship for an automated welding system varies according to the application, but it still has an impact on the welding performance and quality you achieve. Applications that have complex joints or tooling often require an extended contact-tip-to-nozzle relationship. This relationship provides greater access into more complex joints and can help you better accommodate for complex tooling. You should be mindful that this relationship also makes your contact tip more prone to spatter accumulation and may reduce the tip life due to it being more exposed to the heat of the arc. The application of an anti-spatter compound can offer some protection against such situations, but you will also need to monitor your contact tips regularly for signs of wear. Remember, preventive maintenance is better than downtime for resolving problems. Change over your contact tips before issues occur.

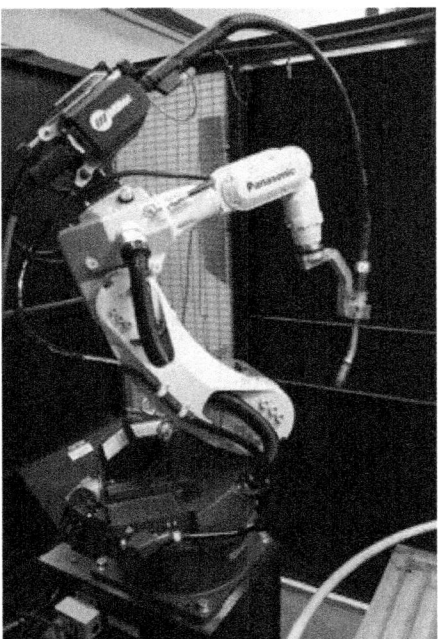

Consumables are an easily overlooked part of an automated welding system, but they can have a measurable impact on downtime and day-to-day costs. Be certain to carefully select and maintain them to get the best performance and minimize downtime.

Using heavy or extended-life heavy duty contact tips composed of chrome zirconium is also a good option for gaining longer performance. Chrome zirconium contact tips are harder and more durable than copper ones, and while they offer slightly less conductivity, the difference is negligible. Typically, you can identify these types of contact tips by the machined groove at the base of the thread.

Checking your contact tips, retaining heads (or diffusers) and nozzles for good connections can also have a measurable impact on your welding performance. Solid connections help ensure reliable electrical conductivity and minimize heat, which in turn provides more consistent weld quality and helps your consumables last longer. Look for consumables that are designed to thread together and mate securely, too, as these can further increase their longevity.

The Impact of Welding Wires on Contact Tip Selection

The welding wires you use can impact the performance of your contact tips and it can also affect what size you should use. Larger drums of wires — 500 to 1,000 pounds — are commonly used for automated welding systems to minimize changeover; however, the wire in these drums tends to have less of a cast and/or helix than wire that feeds off of a smaller spool. As a result, the wire often feeds through the contact tip relatively straight, making little or no contact with it.

The effect is twofold : one, it minimizes the electrical conductivity necessary to create a good arc and a sound weld; and two, it can cause the welding wire to contact the part being welded and arc back into the contact tip, thereby creating a burnback. This condition automatically creates downtime to change over the contact tip. As a solution, consider undersizing your contact tips particularly if you are using a solid wire. For example, a .040-inch (1 mm) diameter contact tip could work for a .045-inch wire. Check with a trusted robotic integrator or welding distributor if you are using metal-cored wires, as undersizing them is not always feasible due to their tubular construction.

You should also consider the impact that the wire you are using has on the longevity of your contact tip. For example, non-copper-coated solid wires tend to wear contact tips more quickly than copper-coated ones. The copper on a copper-coated wire acts like a lubricant to improve feedability and can often extend consumable life. It may be worthwhile to factor in the higher up-front cost of these wires compared to the increased cost of purchasing more contact tips for use with a non-copper-coated wire, as well as the downtime for changeover.

What is Your Mode of Welding?

Automated welding systems require consumables that are capable of withstanding longer periods of welding — and most often higher amperages — than a semi-automatic application. The specific mode of transfer for (GMAW) or (MIG) welding you use can also impact the type of consumables you require. For example, pulsed welding programs in which the power source "pulses" between low background currents and high peaks, are especially harsh on consumables due to the higher levels of heat that the process generates. They tend to cause the contact tip to erode more quickly and therefore require more frequent changeover.

Solid connections between your contact tip and retaining head help ensure reliable electrical conductivity and minimize heat. The result is more consistent weld quality and longer-lasting consumables.

You should carefully monitor your contact tip usage if using such a welding program so that you can determine how often the contact tips need to be replaced. Changing over these consumables before they experience problems can help prevent issues like loss of electrical conductivity, burnbacks or excessive spatter accumulation, the latter of which tends to occur when the contact tip becomes too hot and the consumable material softens. Use the time during routine pauses in production for contact tip changeover to avoid interrupting arc-on time. You should also consider using heavy-duty contact tips for higher heat applications. Again, contact tips made of chrome zirconium are a good choice.

Selecting the Right Nozzle ... and Maintaining It

Typically, the tooling on your automated welding system dictates the type of nozzle that you will need to use. Bottleneck, straight or tapered nozzles are common choices since they are narrower than standard nozzles and can provide better access around tooling or into complex joints. Still, always consider the duty cycle and amperage of your application when deciding which nozzle to use. The more tapered a nozzle, typically the thinner it is and the less able it is to withstand higher amperage or higher-duty-cycle applications. If your automated welding system welds at higher amperages (300 amps or greater) and has high levels of arc-on time, it may be a good idea to select a heavy-duty style since these have thicker walls and insulators and are more able to resist heat. Nozzles composed of copper are also a good option, as are those featuring high-temperature fiberglass insulators. Work with your robotic integrator or welding distributor to make the right nozzle selection. Remember that you need to be sure to select one that provides access to the joint, but that is not so narrow (especially in relation to the contact tip) that you compromise shielding gas coverage or unnecessarily shorten the consumables' life.

A Nozzle Cleaning Station, or Reamer, cleans the robotic gun nozzle of spatter and clears away debris in the retaining head that accum lates during the welding process. It can also help extend the life of your nozzles, retaining heads and contact tips.

For all styles and types of nozzles, it is always recommended that you employ a nozzle cleaning station or reamer to help maintain them. A nozzle cleaning station cleans the robotic gun and nozzle of spatter and clears away debris in the retaining head that accumulates during the welding process. These stations can also be outfitted with a sprayer that applies a water- or oil-based anti-spatter compound to protect the nozzle, retaining head, and workpiece from spatter after it has been cleaned. The nozzle cleaning station should be placed close to your robot so it is easily accessible. Also, you should program your robot to use it in between cycles — during part loading or tool transfer — so as not to interrupt your welding operation. It should only take a few seconds for the nozzle cleaning station to complete its job.

Other Considerations

As a general rule, it is best to select consumables that are well-machined and have smooth, round surfaces, as these are less prone to collecting spatter and tend to last longer. It is also important that you use the heaviest-duty consumables for your application that will still allow you access to tooling. Doing so can help extend their life.

Keep in mind that you also need to pay attention to your retaining head selection and the liners that you use in your robotic MIG gun. The retaining head should match your nozzle and contact tip appropriately and offer a secure connection so that you obtain the best conductivity. Also, always trim and install liners according to the manufacturer's recommendation, using a liner gauge to determine the appropriate length. A liner that is too short or too long can cause wire-feeding problems that require downtime to rectify.

As with any part of an automated welding system, the goal is to keep your consumables in working order so that you spend more time reaping the benefits of the process and less time troubleshooting problems.

HANDLING, INSTALLING, AND MAINTAINING GMAW CONSUMABLES

Tips to Follow and Pitfalls to Avoid

When it comes to welding, many variables can influence productivity and quality. The power source, filler metals, and consumables all factor into the equation and require special attention during the selection process. You must manage these variables properly to ensure their longevity and to help minimize downtime for maintenance and repair.

For MIG consumables in particular, several pitfalls exist that can shorten their lifespan. Taking the time to learn tips for keeping them clean and lasting longer can positively affect productivity, quality, and the bottom line.

Handling, installing and maintaining consumables properly can minimize downtime and costs.

The Heat Factor

The welding process generates heat that significantly affects the cleanliness and longevity of MIG consumables. Processes like pulsed MIG and other high-amperage applications tend to subject consumables to high heat levels, as do those that generate a lot of reflective heat. As the consumables heat up during welding, the material becomes soft, making the surface area much more prone to spatter accumulation.

To avoid this problem, you must determine the best consumables for each application and manage them properly throughout the course of a welding shift. For example, high-amperage applications (above 300 amps) most often benefit from using heavy-duty consumables because they have greater mass and are more capable of dissipating heat. However, if the welding procedure requires you to change the contact tip frequently, a standard-duty contact tip may suffice.

Your goal should be to determine which consumables — heavy or standard duty — are most capable of withstanding the duty cycle and heat of the application. A reliable welding integrator often can help you make this determination.

Using Anti-Spatter Solution

When used sparingly, anti-spatter compound can help keep MIG consumables clean in both semiautomatic and robotic welding applications.

In a semiautomatic application, dip only the front 1.5 in. of the nozzle into the anti-spatter compound. Submerging the entire nozzle can saturate its fiberglass insulator and potentially plug up the gas holes on the diffuser. This buildup may cause premature nozzle failure or unbalanced gas coverage that can lead to weld porosity.

In robotic applications, use the minimum amount of anti-spatter compound required for the application. Too much anti-spatter can build up on the consumables or cause the nozzle to become clogged with debris, leading to poor gas coverage, inconsistent electrical conductivity, or shortened consumable life.

Another important way to combat spatter is to inspect the nozzle for buildup on a regular basis and clean it with a soft wire brush or spatter-cleaning tool as needed.

Using the recommended amount of anti-spatter compound, maintaining good connections, and selecting the right consumables for the application can help prevent the spatter buildup shown here.

Storing and Handling Consumables

Always keep MIG consumables in their original packaging until they are ready for use. Opening them and placing them in a bin can lead to scratches or dents that allow spatter to adhere and will ultimately shorten the products' life. Similarly, removing contact tips or diffusers from their packaging and storing them in open or dirty containers can cause dirt and oil to accumulate in the threads, which can impede their properly seating together.

Keep storage containers for new consumables separate from those for discarded ones to avoid selecting an old contact tip or nozzle that may have dents or scratches and be prone to spatter accumulation. Always wear clean gloves when handling or replacing contact tips, nozzles, and diffusers to prevent dirt, oil, or other contaminants from adhering to them.

Establishing and Maintaining Good Connections

Installing MIG consumables correctly and inspecting them periodically for good connections minimizes the chance of poor conductivity and the spatter accumulation or premature failure that can result. Always follow the MIG consumable manufacturer's suggestions for installing contact tips and gas diffusers. Use a pair of channel-lock pliers or other recommended installation tools to install tips and diffusers. Never use wire cutters or side cutters, as too much pressure from these tools can damage the inside diameter of the contact tip. These tools also tend to scratch the surface of the consumables, leaving marks that attract spatter.

A good rule of thumb is to hand-tighten the contact tip until it is fully seated into the diffuser, then grip the contact tip with an appropriate tool as close to the base as possible, tightening it one-quarter to one-half turn past finger tight. This procedure helps ensure a good connection, minimizing electrical resistance, overheating, and damage to the consumables, as well as excessive spatter accumulation. Follow the same procedure for installing and tightening the diffuser so that it fully connects with the neck.

Some contact tips can be installed and held in place by hand-tightening the nozzle. Check the manufacturer's recommendation for proper installation instructions.

Inspect consumable connections regularly to ensure that they are secure.

Trimming Liners Correctly

A liner that is trimmed and installed improperly can cause a host of wire feeding problems that require downtime to rectify. It also affects MIG consumables' performance, cleanliness, and longevity. Cutting a liner too short causes the liner to misalign with or in the gas diffuser. A misaligned liner will feed the wire off-center, and the contact can fail prematurely as a result.

Debris often builds up between the liner and the retaining head when the liner is too short, causing wire feeding issues and poor weld quality. In some cases the gap that is present between the gas diffuser and liner when a liner has been cut too short will cause the welding wire to catch, shaving off a tiny portion of the wire. The small shavings can plug up the contact tip and cause it to fail quickly.

A liner that's too long can kink, which again leads to wire feeding issues that shorten the life of the contact tip. Always be sure to remove any burrs or sharp edges after cutting a liner to ensure smooth and consistent welding wire feeding.

Always consult with the liner manufacturer's recommendation for proper trimming and installation instructions. Also be sure to wear gloves when handling the liner, and avoid dragging it on the ground to keep debris away from the MIG gun. Debris can contaminate the weld and hinder consumable performance.

Minding the Contact Tip Position and Nozzle Size

The position of the contact tip (extended or recessed) affects consumable lifespan and cleanliness. The nozzle used in conjunction with a specific contact tip and the wire size also makes a difference. The farther the contact tip extends from the nozzle and the closer it is to the arc, the more prone it is to damage from reflective heat by way of spatter accumulation and burnbacks. A recessed contact tip can help prevent these problems while also providing better shielding gas coverage.

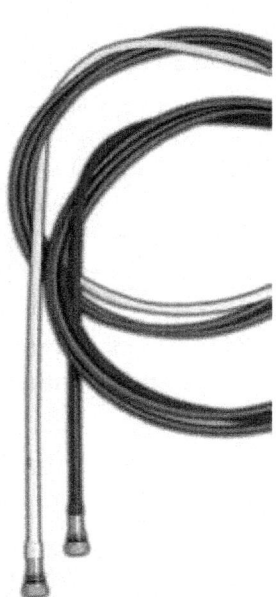

Always consult with the liner manufacturer's recommendation for proper trimming and installation instructions. Also be sure to wear gloves when handling the liner to avoid contaminating it.

For applications that require access into restricted areas, it is important to select a nozzle that provides that access but isn't tapered so much that it impedes the space around the contact tip. If there isn't enough space for shielding gas to flow out of the nozzle, the shielding gas could hit the workpiece and begin jetting back or swirling. This action pulls oxygen into the weld pool and increases the risk for spatter. As the bore size on the nozzle decreases, there is less mass to that portion of the consumable, increasing the risk for heat absorption and spatter adherence.

Things to Remember

As a general rule, select the largest consumable that will work for the application while still providing necessary joint access. Larger consumables are more able to resist heat and spatter buildup, and they often last longer as a result.

Selecting consumables with the right material for the application is important too. For example, brass nozzles tend to resist spatter well and are good for lower-amperage applications (100 to 300 amps), whereas copper nozzles are better for high-amperage applications (more than 300 amps) or for those with longer arc-on time.

Lastly, always pay attention to the manner in which you manage consumables. Using the same consumables throughout the welding operation can help you to maintain consistent performance and troubleshoot problems more quickly when they occur. The result can be longer-lasting, cleaner consumables that provide more reliable performance and quality.

UNDERSTANDING MIG WELDING NOZZLES

Nozzles play an important part in the MIG welding process. These components are responsible for directing the shielding gas to the weld pool and protecting it from contamination. Without proper gas flow, the final weldment can be prone to problems like excessive spatter and porosity that cause downtime for rework. Also, having the wrong nozzle for an application can cause overheating and lead to premature consumable failure.

Nozzles that feature a fiberglass insulator and brass insert, as shown in this cut-away, can help extend the life of the consumable. The brass insert, in particular, helps maintain the inner diameter of the nozzle and reduce wear.

Unfortunately, like other MIG welding consumables, the importance of selecting the right nozzle is often overlooked. In any welding application, the right shape and style of nozzle, however, can have a significant impact on the quality, productivity and overall cost of the welding operation. Knowing how to store and handle nozzles properly can also help improve their overall performance. Consider these tips to get the best results.

Selecting the Right Shape of Nozzle

There are several shapes of nozzles available, including straight, bottleneck and short or long taper nozzles. Straight nozzles typically have larger inside diameters (*e.g.*, 3/4 inch), but don't offer as good of joint access. If greater joint access is critical, a bottleneck nozzle may be the better option. These nozzles are particularly good for automated welding applications. A common inside diameter for a bottleneck nozzle is 1/2 inch.

Short and long taper nozzles are also common choices for gaining good joint access. Note, that long taper nozzles typically have a smaller inside diameters and may collect spatter more readily. When possible, using a short taper nozzle can help prevent such a problem.

When selecting a nozzle, it is important to find one that provides the best joint access for the application. It is also imperative that the nozzle allows for the proper gas flow to the weld puddle in order to keep contaminants away. The best choice is to use as large of a nozzle as possible that still allows access to the weld joint. Doing so helps ensure the greatest shielding gas flow. Larger nozzles are also less prone to collecting spatter compared to those with smaller inside diameters.

Selecting the Best Material

Nozzles are typically available in heavy-duty or standard styles, and in slip-on or thread-on varieties. Heavy-duty nozzles have thicker walls, as well as thicker insulators, and are designed for use in applications ranging from 400 to 600 amps. Due to their heavier construction, these nozzles resist heat better than standard varieties. Standard nozzles tend to have a thinner wall and are better for 100- to 300-amp applications. Slip-on nozzles, as their name implies, simply slip on to the front end of the MIG gun. These nozzles are quite prevalent in the industry, compared to thread-on nozzles that need to be twisted to install, and they offer the advantage of being able to change over more quickly. A note of caution : when installing slip-on nozzles, be certain that they are fully seated on the retaining head to prevent shielding gas leaks that could lead to poor weld quality.

Nozzles are typically available in brass or copper, although chrome-plated nozzles are also available. Brass nozzles tend to resist spatter well and are good for lower-amperage applications (100 to 300 amps), whereas copper nozzles are better for high-amperage applications (above 300 amps) or for those with longer arc-on time.

For high-amperage water-cooled applications, there are also nozzles available that circulate coolant around the nozzles, but these tend to be much more expensive.

Proper Storage, Handling and Maintenance

It is important to handle, store and maintain nozzles properly to gain consistent welding performance and prevent premature failure. Selecting high quality nozzles can help these consumables last longer, too.

Checking the nozzle periodically for spatter build-up, as seen here, and cleaning it properly can help extend the life of the consumable. Adding anti-spatter can also help prevent build-up.

Look for nozzles that are engineered with a smooth surface finish and edges, as these resist spatter build-up compared to nozzles that have an uneven surface or burrs on the edges. Nozzles that have some mass to them are also more desirable than lighter or thinner ones since they tend to resist heat better. Also, consider purchasing nozzles that feature a brass insert. This insert helps the nozzle maintain its inner diameter, and prevents the nozzle from rocking and wearing prematurely. The addition of a high-temperature fiberglass insulator can also help extend nozzle life. Finally, look for heavy-duty crimping on the nozzle — the crimping holds the layers together and is an indication that the nozzle has been built for longevity.

When storing nozzles, keep them in their original packaging, usually a small plastic bag. Removing them from that packaging and placing them in a bin can lead to scratches or dents that allow spatter to adhere and will ultimately shorten the life of the nozzle. Use gloves when handling nozzles or replacing nozzles to prevent dirt, oil or other contaminants from adhering to them and inadvertently entering the weld puddle.

Periodically inspect the nozzle for spatter build-up and clean it using the tool recommended by the manufacturer as needed and/or consider using an anti-spatter compound to protect against spatter.

As with any front-end consumable, nozzles play an important role in maintaining good weld quality and can have a measurable impact on productivity and costs, too. Take the time to select the right ones for each application and maintain them properly. Careful selection and maintenance can minimize downtime and keep your welding operation running more smoothly in the long run.

PREVENTIVE ROBOTIC MIG GUN MAINTENANCE : THE WHOS, WHENS, WHYS AND HOWS

Companies invest in welding automation to increase productivity, improve quality and reduce costs. Any unnecessary downtime can quickly interfere with obtaining those goals. But what about small amounts of scheduled downtime for maintenance? In most cases, a well-planned, efficient preventive maintenance (PM) program can yield positive results. Not only does it help ensure reliable throughput, but a properly executed PM program can also lower labor costs, reduce waste and minimize rework. It may even expedite the return on investment (ROI) in the automated welding system.

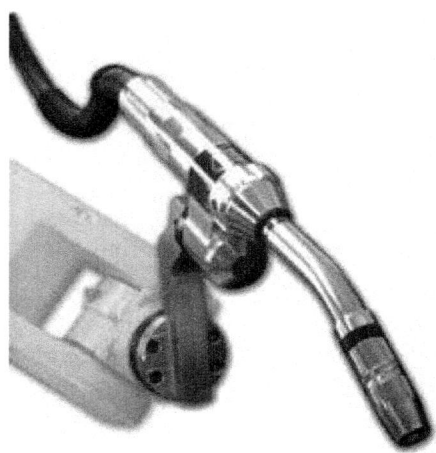

Regular maintenance of the robotic MIG gun can help provide a positive return on an automated welding investment.

Caring properly for the whole of an automated welding system is imperative, of course, but so too is maintaining the robotic MIG gun. In fact, the robotic MIG gun and its consumables are frequently overlooked components in the system. They are also relatively easy to maintain, and doing so can positively contribute to the efficiency of the entire welding operation.

Pm Program Basics : The Whos and Whens

All companies, regardless of their size or arc count, can benefit from regular maintenance of their robotic MIG guns and consumables. The scope of the PM program, however, will vary according to each company's application. For example, a company with higher-risk applications — those with large, thick parts; long cycle times and/or expensive rework — generally require more frequent care of the equipment than companies that weld smaller, less expensive parts. They simply stand to lose more (in both downtime and money) should something go wrong in the welding process.

Most of the maintenance on a robotic MIG gun can be completed shift-by-shift with minimal off-line time. Welding engineers, welding supervisors, tool and die employees or members of the maintenance staff are all viable candidates to oversee the process. All personnel involved, however, need to be properly trained to identify potential problems in the weld cell and learn how to prevent them. They should also be aware that "in-process" maintenance does not constitute the whole of a PM program. Some activities may need to take place off-shift due to their complexity and the time needed to complete them.

Taking Action : The Whys and Hows

There are several key components to a good PM program for robotic MIG guns. Before starting any task, it is important to have the correct tools for the job. For example, be sure to have the proper adjustable or crescent wrench for changing diffusers or retaining heads, as well as the recommended pliers, welpers or tip installation tools for installing contact tips. Keep a sharp pair of side cutters on hand, too, to trim the robotic MIG gun liner. These tools help prevent burrs on the liner that can wear or drag on the welding wire.

After establishing that the proper tools are in place to support the PM program, consider the following practices.

Secure Connections on a Regular Basis

During pauses in production — when the robot finishes welding a part or during routine contact tip changeover, for example — check for clean, secure connections between the MIG gun neck, the diffuser or retaining heads and the contact tip. Also, check that the nozzle is secure and any seals around it are in good condition.

Having tight connections from the neck through the contact tip helps ensure a solid electrical flow throughout the components and minimizes heat build-up that could cause premature failure, poor arc stability, quality issues and/or rework. It also reduces the opportunity for burnbacks, which can lead to unplanned downtime for changeover. Look for changes in consumable colors, too, as those are a good indication that they are loose and require tightening.

Prevent Spatter Build-up

Spatter build-up can cause excessive heat in the consumables and MIG guns, block shielding gas flow, and increase costs for inventory and downtime to change over nozzles, diffusers and contact tips. Visually inspect consumables on a regular basis for signs of spatter, replacing them as needed. Also, consider adding a nozzle cleaning station (also called a reamer or spatter cleaner) to the weld cell. Like its name implies, a nozzle cleaning station removes spatter (and other debris) that builds up in the nozzle and diffuser. Using this equipment in conjunction with a sprayer that applies an anti-spatter compound can further protect against spatter accumulation.

Mind the Liner

Track how long it takes for the liner in the robotic MIG gun to become worn or fouled, and schedule a replacement as needed. Replacing the liner prior to a failure prevents unplanned downtime to remedy wire feeding or quality problems later. Also, always cut the liner according to the manufacturer's recommendation to prevent kinking and poor wire feeding that can lead to premature contact tip failure and/or arc instability.

Periodically, release the drive rolls and check the force required to pull the welding wire from the feeder through the robotic MIG gun. Excessive drag indicates that there is a build-up of debris in the liner and it needs to be replaced. It is best to perform this task in between shifts, as opposed to during contact tip changeover, as it tends to take more time.

Assess the Welding Cable and Power Pin

Check regularly that the welding cable leads are properly secured and assess the condition of the welding cable on the robotic MIG gun. Look for signs of wear and be certain that the cable is not rubbing against any part of the robot's metal casting, as that friction can cause the cable to loosen or become damaged. A worn spot on the robot (*e.g.*, the absence of paint) or on the tooling is a good indication that the cable is rubbing against it. Rectifying the situation will likely involve repositioning the tooling or a cable management device and may need to occur while the robot is off-line. Still, a quick in-process inspection that identifies the issue can flag it for a later, proactive solution.

Parting Thoughts on PM Programs

Preventive maintenance programs don't have to be complicated — only effective. Most of the robotic MIG gun maintenance discussed here can be completed on a shift-by-shift basis with minimal interference to cycle times and with minimal labor costs. The scope and frequency of a PM program will vary from company to company, of course, but carefully executed maintenance activities can help companies better realize the potential of their automated welding operation. And it can reduce costs by preventing problems, instead of being forced to resolve them.

HOW PERIPHERALS CAN MAXIMIZE YOUR ROBOTIC WELDING PERFORMANCE

Robotic welding systems introduce speed, accuracy and repeatability into the fabrication process. The repeatability of these systems can increase productivity and reduce welding production costs, thereby maximizing the return on the investment in automation. When compared to semi-automatic welding operations, a robot has the ability to perform the same or more tasks, with extreme precision and a lower labor cost per part.

Still, achieving these results isn't subject to chance. It's the result of careful equipment purchases, software programming and operator training. In many cases, it can also be the result of complementary equipment called peripherals.

Peripherals are any equipment integrated into the robotic welding process with the objective of maximizing its effectiveness and protecting the overall equipment investment. In short, these devices (in most cases) can add significantly to the ROI a company achieves with its welding robot. The key to successful peripheral selection and usage, like that of any other equipment, is simply a matter of education.

Get a Grip

All robotic welding systems require some form of collision detection in order to reduce the damage to both the robot and the welding system in the event of an impact. Impacts occur for several reasons. These include a robotic gun colliding with an incorrectly positioned work piece or tooling that has been left out of position, or striking an item that has inadvertently been left in the weld cell.

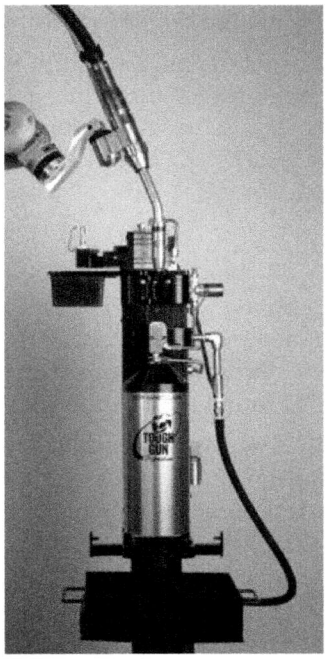

A nozzle cleaning station can lengthen the life of the robotic MIG gun and its consumables, and save companies money for extra parts.

Depending on the type of collision detection utilized by the particular robot manufacturer, it will require either a shock sensor or safety clutch as protection. In cases where collision detection is integral to the robot, a solid mounting arm can be used.

The sensitivity of a clutch or shock sensor can be calibrated to accommodate the robotic welding gun's mass and moment of inertia. The function of a clutch is both mechanical and electrical. The clutch first recognizes the physical impact of the torch on a solid surface, which sends an electrical signal back to the robot controller, causing the system to stop. This action prevents damage to the robot and the robotic gun. It also alerts the welding operator overseeing the operation that there is an incorrect variable in the weld cell.

Some robotic systems are capable of monitoring current rates and/or torque via robot collision detection software that stops the robot in the event of an impact. In this situation, a solid arm mount would be used in lieu of a clutch. As its name implies, a solid arm mount is just that : solid. It does not provide electrical feedback during an impact, but rather relies on the software to stop the robot during an impact.

Both clutches and solid arm mounts require mounting arms that attach them to the robotic MIG gun. A mounting arm is generally composed of a durable aluminum alloy that can resist breakage during an impact. Its purpose is to hold the robotic MIG gun in a specified position (even after a welding operator replaces the gun), so the robot can repeat the same weld throughout the welding process.

Clutches and solid arm mounts are also quite robust and require little to no maintenance to keep them operating to their fullest. However, should a company feel that maintenance or repairs are necessary for one of these peripherals, maintenance personnel should contact their welding distributor, integrator or robotic equipment manufacturer for advice.

Making the Cut

For companies whose robotic welding applications require consistent welding wire stick-out (the distance the wire extends from the end of the contact tip) when the arc initiates, a wire cutter is recommended. Note, consistent welding wire stick-out is not required for all applications.

Again, as its name implies, a wire cutter cuts the welding wire to a specified length or stick-out and/or it also removes any balling at the end of the wire. In doing so, this peripheral helps provide smooth arc starts. It also helps attain reliable, repeatable welds, as many companies who own automated systems program the robot to seam track, or find the joint, with touch sensing. This touch sensing depends on the robotic MIG gun having a consistent length of wire with which to locate the correct spot and begin welding. Most wire cutters are designed to cut a range of different types of welding wire, including stainless steel, flux-cored and metal-cored, usually up to 1/16-inch diameter. They can often be mounted on a nozzle station or remotely located to be used as needed.

Inspected and Ready to Weld

Another key peripheral is a neck (or gooseneck) inspection fixture. A neck inspection fixture tests the tolerance of a robotic MIG gun's neck to the tool center point so it can be readjusted after an impact or after bending due to routine welding. Most inspection fixtures will accommodate standard necks for that particular brand of robotic gun. They are designed with a precision-tooled steel base to withstand the harsh robotic welding environment and also to guarantee accuracy after long-term use.

The advantage of adding a neck inspection fixture to a robotic weld cell is two-fold. One, it ensures the neck meets the specifications to which the robotic welding system has been programmed. Once the tolerance has been determined, a trained welding operator simply adjusts the neck accordingly. This adjustment helps prevent costly rework due to missing weld joints. Accurate neck adjustments also prevent the downtime necessary to reprogram the robot to meet the welding specifications with the existing bent neck.

Secondly, a neck inspection fixture can save companies time, money and confusion when exchanging necks from one robotic MIG gun to another. This is especially advantageous for companies that maintain a large number of welding robots. Welding operators can simply remove a bent neck and change it with a spare that has already been inspected and adjusted, and put the robot back in service immediately. The damaged neck can then be set aside for inspection while the robot is still online. This again lowers downtime and also helps companies save money for extra parts.

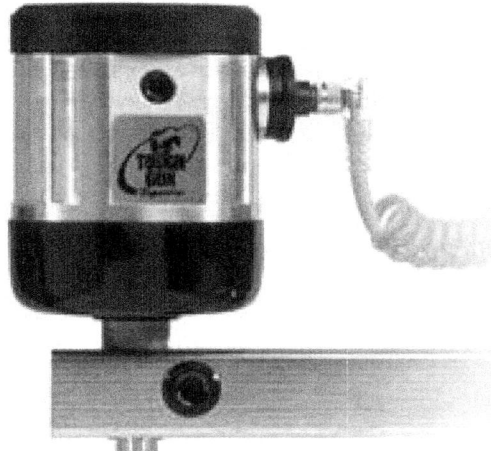

Clutches offer protection in the event of an impact by signaling to the robot to stop welding.

Cleaned, Sprayed and Spatter-Free

One of the most important peripherals a company should consider for its robotic welding system is a nozzle cleaning station, also called a reamer. This peripheral can be used by itself or in conjunction with a sprayer that applies anti-spatter compound.

A nozzle cleaning station cleans the robotic gun nozzle of spatter and/or clears away debris in the diffuser that accumulates during the welding process. If a sprayer has been mounted on the nozzle cleaning station, it will apply a water- or oil-based anti-spatter compound to protect the nozzle, diffuser and work piece from spatter after it has been cleaned.

Again, there are several benefits a nozzle cleaning station can have on the robotic welding process. First, by minimizing the accumulation of spatter and debris in the nozzle, it helps lengthen the life of the robotic gun consumables (nozzle, contact tip and diffuser), and of the robotic gun itself. This longer equipment life translates into less downtime and labor for component changeover and also less cost for equipment—both factors contribute positively to a company's ROI of its robotic welding system. A clean nozzle also helps provide better weld quality and reduce problems that could lead to rework.

To achieve all of these benefits, it's important to consider two factors : one, the location of the nozzle cleaning station, and two, the timing of its use. Ideally, the nozzle cleaning station should be placed in close proximity to the welding robot so that it is easily accessible when cleaning is necessary. As well, the nozzle cleaning process should be programmed so that the function occurs in-between cycles—during part loading or tooling transfer. In this case, the cleaning time would not be added to the overall cycle time per part, as a typical nozzle cleaning station needs only a matter of three or four seconds to complete the job.

If a company attaches a sprayer to its nozzle cleaning station, it should be certain to use only the minimum amount of anti-spatter compound required for the application. Excessive anti-spatter usage can lead to unnecessary costs and the compound may build up on the nozzle, the welding robot and the parts being welded. In the long term, a high spray volume could cause additional problems that are just as bad as spatter build-up itself.

Finally, implementing a preventative maintenance plan for a nozzle cleaning station is imperative to gaining long-lasting results from the equipment. And it's easy. Simply clean off the peripheral, wiping it free of dirt, debris and/or spatter, on a weekly basis to prevent malfunctions, and in turn, quality issues in the robotic weld cell.

No Peripheral Decision

The decision for a company, large or small, to invest in robotic welding equipment is significant. It requires time, knowledge and a trusted relationship with a robotic welding equipment manufacturer and/or integrator to find the right system for the application. The same holds true for peripherals. And although these devices do add to the initial cost of automating, they can lead to measurable cost savings and profits in the long term. Remember, the goal in robotic welding is repeatability and increased productivity, any additional equipment that can help achieve that result is worth the consideration.

Chapter 7

ROBOT END EFFECTOR

In robotics, an **end effector** is the device at the end of a robotic arm, designed to interact with the environment. The exact nature of this device depends on the application of the robot.

In the strict definition, which originates from serial robotic manipulators, the end effector means the last link (or end) of the robot. At this endpoint the tools are attached. In a wider sense, an end effector can be seen as the part of a robot that interacts with the work environment. This does not refer to the wheels of a mobile robot or the feet of a humanoid robot which are also not end effectors—they are part of the robot's mobility.

End effectors may consist of a gripper or a tool. The gripper can be of two, three or even five fingers.

The end effectors that can be used as tools serve various purposes, such as spot welding in an assembly, spray painting where uniformity of painting is necessary, and for other purposes where the working conditions are dangerous for human beings. Surgical robots have end effectors that are specifically manufactured for the purpose.

MECHANISM OF GRIPPING

A common form of robotic grasping is force closure.

Generally, the gripping mechanism is done by the grippers or mechanical fingers. Generally only two-finger grippers are used for industrial robots as they tend to be built for specific tasks and can therefore be less complex.

The fingers are also replaceable whether or not the gripper itself is replaced. There are two mechanisms of gripping the object in between the fingers (for the sake of simplicity, the following explanations consider only two finger grippers).

Shape of the Gripping Surface

The shape of the gripping surface of the fingers can be chosen according to the shape of the objects that are to be manipulated. For example, if a robot is designed to lift a round object, the gripper surface shape can be a concave impression of it to make the grip efficient, or for a square shape the surface can be a plane.

Force Required to Grip the Object

Though there are numerous forces acting over the body that has been lifted by the robotic arm, the main force acting there is the frictional force. The gripping surface can be made of a soft material with high coefficient of friction so that the surface of the object is not damaged. The robotic gripper must withstand not only the weight of the object but also acceleration and the motion that is caused due to frequent movement of the object. To find out the force required to grip the object, the following formula is used

$$F = \mu W n$$

where :

F is the force required to grip the object,

μ is the coeffecient of friction,

n is the number of fingers in the gripper and

W is the weight of the object.

But the above equation is incomplete. The direction of the movement also plays an important role over the gripping of the object. For example, when the body is moved upwards, against the gravitational force, the force required will be more than towards the gravitational force. Hence, another term is introduced and the formula becomes :

$$F = \mu W n g$$

Here, the value of g should not be taken as the acceleration due to gravity. In fact, here g stands for multiplication factor. The value of g ranges from 1 to 3. When the body is moved in the horizontal direction then the value is taken to be 2, when moved against the gravitational force then 3 and along the gravitational force, *i.e.*, downwards, 1.

EXAMPLES

The end effector of an assembly line robot would typically be a welding head, or a paint spray gun. A surgical robot's end effector could be a scalpel or others tools used in surgery. Other possible end effectors are machine tools, like a drill or milling cutters. The end effector on the space shuttle's robotic arm uses a pattern of wires which close like the aperture of a camera around a handle or other grasping point.

When referring to robotic prehension there are four general categories of robot grippers, these are :

1. Impactive – jaws or claws which physically grasp by direct impact upon the object.
2. Ingressive – pins, needles or hackles which physically penetrate the surface of the object (used in textile, carbon and glass fibre handling).
3. Astrictive – suction forces applied to the objects surface (whether by vacuum, magneto- or electro-adhesion).
4. Contigutive – requiring direct contact for adhesion to take place (such as glue, surface tension or freezing).

ROBOTIC ARM

The SSRMS while deploying a payload from the cargo bay of the Space Shuttle

A **robotic arm** is a type of mechanical arm, usually programmable, with similar functions to a human arm; the arm may be the sum total of the mechanism or may be part of a more complex robot. The links of such a manipulator are connected by joints allowing either rotational motion (such as in an articulated robot) or translational (linear) displacement. The links of the manipulator can be considered to form a kinematic chain. The terminus of the kinematic chain of the manipulator is called the end effector and it is analogous to the human hand.

Robotic Hand

The end effector, or robotic hand, can be designed to perform any desired task such as welding, gripping, spinning *etc.*, depending on the application. For example robot arms in automotive assembly lines perform a variety of tasks such as welding and parts rotation and placement during assembly. In some circumstances, close emulation of the human hand is desired, as in robots designed to conduct bomb disarmament and disposal.

Types

6 Axis Articulated Robots from KUKA

- Cartesian robot / Gantry robot : Used for pick and place work, application of sealant, assembly operations, handling machine tools and arc welding. It's a robot whose arm has three prismatic joints, whose axes are coincident with a Cartesian coordinator.
- Cylindrical robot : Used for assembly operations, handling at machine tools, spot welding, and handling at diecasting machines. It's a robot whose axes form a cylindrical coordinate system.
- Spherical robot / Polar robot (such as the Unimate) : Used for handling at machine tools, spot welding, diecasting, fettling machines, gas welding and arc welding. It's a robot whose axes form a polar coordinate system.
- SCARA robot : Used for pick and place work, application of sealant, assembly operations and handling machine tools. This robot features two parallel rotary joints to provide compliance in a plane.
- Articulated robot : Used for assembly operations, diecasting, fettling machines, gas welding, arc welding and spray painting. It's a robot whose arm has at least three rotary joints.
- Parallel robot : One use is a mobile platform handling cockpit flight simulators. It's a robot whose arms have concurrent prismatic or rotary joints.
- Anthropomorphic robot : Similar to the robotic hand Luke Skywalker receives at the end of The Empire Strikes Back. It is shaped in a way that resembles a human hand, *i.e.* with independent fingers and thumbs.

Notable Robotic Arms

In space the Space Shuttle Remote Manipulator System also known as Canadarm or SSRMS and its successor Canadarm2 are examples of multi degree of freedom robotic arms that have been used to perform a variety of tasks such as inspections of the Space Shuttle using a specially deployed boom with cameras and sensors attached at the end effector and satellite deployment and retrieval manoeuvres from the cargo bay of the Space Shuttle. The *Curiosity* rover on the planet Mars also uses a robotic arm.

Chapter 8

ROBOTICS TECHNOLOGY

Most robots of today are nearly deaf and blind. Sensors can provide some limited feedback to the robot so it can do its job. Compared to the senses and abilities of even the simplest living things, robots have a very long way to go.

The sensor sends information, in the form of electronic signals back to the controller. Sensors also give the robot controller information about its surroundings and lets it know the exact position of the arm, or the state of the world around it. Sight, sound, touch, taste, and smell are the kinds of information we get from our world. Robots can be designed and programmed to get specific information that is beyond what our 5 senses can tell us. For instance, a robot sensor might "see" in the dark, detect tiny amounts of invisible radiation or measure movement that is too small or fast for the human eye to see.

Here are some things sensors are used for :

Physical Property	Technology
Contact	Bump, Switch
Distance	Ultrasound, Radar, Infra Red
Light Level	Photo Cells, Cameras
Sound Level	microphones
Strain	Strain Gauges
Rotation	Encoders
Magnetism	Compasses
Smell	Chemical
Temperature	Thermal, Infra Red
Inclination	Inclinometers, Gyroscope
Pressure	Pressure Gauges
Altitude	Altimeters

Sensors can be made simple and complex, depending on how much information needs to be stored. A switch is a simple on/off sensor used for turning the robot on and off. A human retina is a complex sensor that uses more than a hundred million photosensitive elements (rods and cones). Sensors provide information to the robots brain, which can be treated in various ways. For example, we can simply *react* to the sensor output : if the switch is open, if the switch is closed, go.

LEVELS OF PROCESSING

To figure out if the switch is open or closed, you will need to measure the voltage going through the circuit, that's electronics. Now lets say that you have a microphone and you want to recognize a voice and separate it from noise; that's signal processing. Now you have a camera, and you want to take the pre-processed image and now you need to figure out what those objects are, perhaps by comparing them to a large library of drawings; that's computation. Sensory data processing is a very complex thing to try and do but the robot needs this in order to have a "brain". The brain has to have analog or digital processing capabilities, wires to connect everything, support electronics to go with the computer, and batteries to provide power for the whole thing, in order to process the sensory data. Perception requires the robot to have sensors (power and electronics), computation (more power and electronics, and connectors.

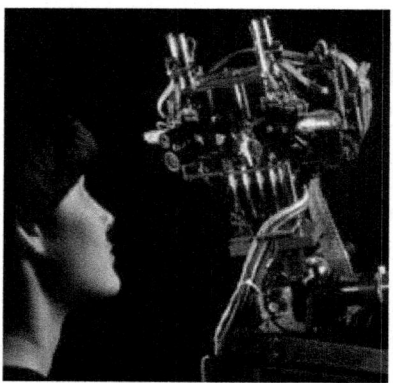

SWITCH SENSORS

Switches are the simplest sensors of all. They work without processing, at the electronics (circuit) level. Their general underlying principle is that of an open vs. closed circuit. If a switch is open, no current can flow; if it is closed, current can flow and be detected. This simple principle can (and is) used in a wide variety of ways.

Switch sensors can be used in a variety of ways :
- contact sensors : detect when the sensor has contacted another object (*e.g.*, triggers when a robot hits a wall or grabs an object; these can even be whiskers)
- limit sensors : detect when a mechanism has moved to the end of its range

- shaft encoder sensors : detects how many times a shaft turns by having a switch click (open/close) every time the shaft turns (*e.g.*, triggers for each turn, allowing for counting rotations)

There are many common switches : button switches, mouse switches, key board keys, phone keys, and others. Depending on how a switch is wired, it can be normally open or normally closed. This would of course depend on your robot's electronics, mechanics, and its task. The simplest yet extremely useful sensor for a robot is a "bump switch" that tells it when it's bumped into something, so it can back up and turn away. Even for such a simple idea, there are many different ways of implementation.

LIGHT SENSORS

Switches measure physical contact and light sensors measure the amount of light impacting a photocell, which is basically a resistive sensor. The resistance of a photocell is low when it is brightly illuminated, *i.e.*, when it is very light; it is high when it is dark. In that sense, a light sensor is really a "dark" sensor. In setting up a photocell sensor, you will end up using the equations we learned above, because you will need to deal with the relationship of the photocell resistance photo, and the resistance and voltage in your electronics sensor circuit. Of course since you will be building the electronics and writing the program to measure and use the output of the light sensor, you can always manipulate it to make it simpler and more intuitive. What surrounds a light sensor affects its properties. The sensor can be shielded and positioned in various ways. Multiple sensors can be arranged in useful configurations and isolate them from each other with shields.

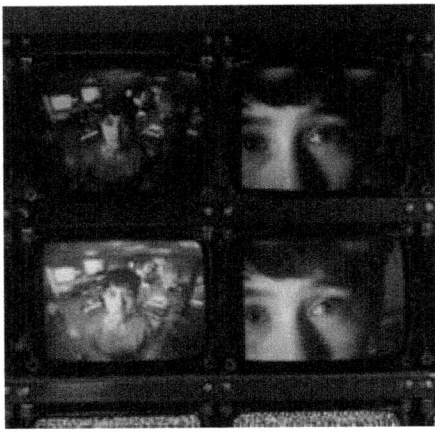

Just like switches, light sensors can be used in many different ways :
- Light sensors can measure :
    o light intensity (how light/dark it is)
    o differential intensity (difference between photocells)

- o break-beam (change/drop in intensity)
- Light sensors can be shielded and focused in different ways
- Their position and directionality on a robot can make a great deal of difference and impact

POLARIZED LIGHT

"Normal" light emanating from a source is non-polarized, which means it travels at all orientations with respect to the horizon. However, if there is a polarizing filter in front of a light source, only the light waves of a given orientation of the filter will pass through. This is useful because now we can manipulate this remaining light with other filters; if we put it through another filter with the same characteristic plane, almost all of it will get through. But, if we use a perpendicular filter (one with a 90-degree relative characteristic angle), we will block all of the light. Polarized light can be used to make specialized sensors out of simple photocells; if you put a filter in front of a light source and the same or a different filter in front of a photocell, you can cleverly manipulate what and how much light you detect.

RESISTIVE POSITION SENSORS

We said earlier that a photocell is a resistive device. We can also sense resistance in response to other physical properties, such as *bending*. The resistance of the device increases with the amount it is bent. These bend sensors were originally developed for video game control (for example, Nintendo Powerglove), and are generally quite useful. Notice that repeated bending will wear out the sensor. Not surprisingly, a bend sensor is much less robust than light sensors, although they use the same underlying resistive principle.

POTENTIOMETERS

These devices are very common for manual tuning; you have probably seen them in some controls (such as volume and tone on stereos). Typically called *pots*, they allow the user to manually adjust the resistance. The general idea is that the device consists of a movable tap along two fixed ends. As the tap is moved, the resistance changes. As you can imagine, the resistance between the two ends is fixed, but the resistance between the movable part and either end varies as the part is moved. In robotics, *pots* are commonly used to sense and tune position for sliding and rotating mechanisms.

BIOLOGICAL ANALOGS

- All of the sensors we described exist in biological systems
- Touch/contact sensors with much more precision and complexity in all species

- Bend/resistance receptors in muscles

Reflective Optosensors

We mentioned that if we use a light bulb in combination with a photocell, we can make a break-beam sensor. This idea is the underlying principle in reflective optosensors : the sensor consists of an emitter and a detector. Depending of the arrangement of those two relative to each other, we can get two types of sensors :

Reflectance sensors (the emitter and the detector are next to each other, separated by a barrier; objects are detected when the light is reflected off them and back into the detector)

Break-beam sensors (the emitter and the detector face each other; objects are detected if they interrupt the beam of light between the emitter and the detector)

The emitter is usually made out of a light-emitting diode (an LED), and the detector is usually a photodiode/phototransistor.

Note that these are not the same technology as resistive photocells. Resistive photocells are nice and simple, but their resistive properties make them slow; photodiodes and photo-transistors are much faster and therefore the preferred type of technology.

What can you do with this simple idea of light reflectivity? Quite a lot of useful things :

- object presence detection
- object distance detection
- surface feature detection (finding/following markers/tape)
- wall/boundary tracking
- rotational shaft encoding (using encoder wheels with ridges or black & white color)
- bar code decoding

Note, however, that light reflectivity depends on the color (and other properties) of a surface. A light surface will reflect light better than a dark one, and a black surface may not reflect it at all, thus appearing invisible to a light sensor. Therefore, it may be harder (less reliable) to detect darker objects this way than lighter ones. In the case of object distance, lighter objects that are farther away will seem closer than darker objects that are not as far away. This gives you an idea of how the physical world is partially-observable. Even though we have useful sensors, we do not have complete and completely accurate information.

Another source of noise in light sensors is ambient light. The best thing to do is subtract the ambient light level out of the sensor reading, in order to detect the actual change in the reflected light, not the ambient light. How is that done? By taking two (or more, for higher accuracy) readings of the detector, one with the emitter on, and one with it off, and subtracting the two values from each other. The result is the ambient light level, which can then be subtracted from future readings. This process is called sensor *calibration*. Of course, remember that ambient light levels can change, so the sensors may need to be calibrated repeatedly.

BREAK-BEAM SENSORS

We already talked about the idea of break-beam sensors. In general, any pair of compatible emitter-detector devices can be used to produce such a sensors :

- an incandescent flashlight bulb and a photocell
- red LEDs and visible-light-sensitive photo-transistors
- or infra-red IR emitters and detectors

Shaft Encoding

Shaft encoders measure the angular rotation of an axle providing position and/or velocity info. For example, a speedometer measures how fast the wheels of a vehicle are turning, while an odometer measures the number of rotations of the wheels.

In order to detect a complete or partial rotation, we have to somehow mark the turning element. This is usually done by attaching a round disk to the shaft, and cutting notches into it. A light emitter and detector are placed on each side of the disk, so that as the notch passes between them, the light passes, and is detected; where there is no notch in the disk, no light passes.

If there is only one notch in the disk, then a rotation is detected as it happens. This is not a very good idea, since it allows only a low level of resolution for measuring speed : the smallest unit that can be measured is a full rotation. Besides, some rotations might be missed due to noise.

Usually, many notches are cut into the disk, and the light hits impacting the detector are counted. An alternative to cutting notches in the disk is to paint the disk with black (absorbing, non-reflecting) and white (highly reflecting) wedges, and measure the reflectance. In this case, the emitter and the detector are on the same side of the disk.

In either case, the output of the sensor is going to be a wave function of the light intensity. This can then be processes to produce the speed, by counting the peaks of the waves.

Note that shaft encoding measures both position and rotational velocity, by subtracting the difference in the position readings after each time interval. Velocity, on the other hand, tells us how fast a robot is moving, or if it is moving at all. There are multiple ways to use this measure :

- measure the speed of a driven (active) wheel
- use a passive wheel that is dragged by the robot (measure forward progress)

We can combine the position and velocity information to do more sophisticated things :

- move in a straight line
- rotate by an exact amount

Note, however, that doing such things is quite difficult, because wheels tend to slip (effector noise and error) and slide and there is usually some slop and backlash in the gearing mechanism. Shaft encoders can provide feedback to correct the errors, but having some error is unavoidable.

Quadrature Shaft Encoding

So far, we've talked about detecting position and velocity, but did not talk about direction of rotation. Suppose the wheel suddenly changes the direction of rotation; it would be useful for the robot to detect that.

An example of a common system that needs to measure position, velocity, and direction is a computer mouse. Without a measure of direction, a mouse is pretty useless. How is direction of rotation measured?

Quadrature shaft encoding is an elaboration of the basic break-beam idea; instead of using only one sensor, two are needed. The encoders are aligned so that their two data streams coming from the detector and one quarter cycle (90-degrees) out of phase, thus the name "quadrature". By comparing the output of the two encoders at each time step with the output of the previous time step, we can tell if there is a direction change. When the two are sampled at each time step, only one of them will change its state (*i.e.*, go from on to off) at a time, because they are out of phase. Which one does it determines which direction the shaft is rotating. Whenever a shaft is moving in one direction, a counter is incremented, and when it turns in the opposite direction, the counter is decremented, thus keeping track of the overall position.

Other uses of quadrature shaft encoding are in robot arms with complex joints (such as rotary/ball joints; think of your knee or shoulder), Cartesian robots (and large printers) where an arm/rack moves back and forth along an axis/gear.

Modulation and Demodulation of Light

We mentioned that ambient light is a problem because it interferes with the emitted light from a light sensor. One way to get around this problem is to emit *modulated light*, *i.e.*, to rapidly turn the emitter on and off. Such a signal is much easier and more reliably detected by a *demodulator*, which is tuned to the particular frequency of the modulated light. Not surprisingly, a detector needs to sense several on-flashes in a row in order to detect a signal, *i.e.*, to detect its frequency. This is a small point, but it is important in writing demodulator code.

The idea of modulated IR light is commonly used; for example in household remote controls.

Modulated light sensors are generally more reliable than basic light sensors. They can be used for the same purposes : detecting the presence of an object measuring the distance to a nearby object.

Infrared Proximity Detector

Infra Red (IR) Sensors

Infra red sensors are a type of light sensors, which function in the infra red part of the frequency spectrum. IR sensors consist are active sensors : they consist of an emitter and a receiver. IR sensors are used in the same ways that visible light sensors are that we have discussed so far : as break-beams and as reflectance sensors. IR is preferable to visible light in robotics (and other) applications because it suffers a bit less from ambient interference, because it can be easily modulated, and simply because it is not visible.

IR Communication

Modulated infra red can be used as a serial line for transmitting messages. This is is fact how IR modems work. Two basic methods exist :
- bit frames (sampled in the middle of each bit; assumes all bits take the same amount of time to transmit)
- bit intervals (more common in commercial use; sampled at the falling edge, duration of interval between sampling determines whether it's a 0 or 1)

Ultrasonic Distance Sensing

As we mentioned before, ultrasound sensing is based on the time-of-flight principle. The emitter produces a sonar "chirp" of sound, which travels away from the source, and, if it encounters barriers, reflects from them and returns to the receiver (microphone). The amount of time it takes for the sound beam to come back is tracked (by starting a timer when the "chirp" is produced, and stopping it when the reflected sound returns), and is used to compute the distance the sound traveled. This is possible (and quite easy) because we know how fast sound travels; this is a constant, which varies slightly based on ambient temperature.

At room temperature, sound travels at 1.12 feet per millisecond. Another way to put it that sound travels at 0.89 milliseconds per foot. This is a useful constant to remember.

The process of finding one's location based on sonar is called *echolocation*. The inspiration for ultrasound sensing comes from nature; bats use ultrasound instead of vision (this makes sense; they live in very dark caves where vision would be largely useless). Bat sonars are extremely sophisticated compared to artificial sonars; they involve numerous different frequencies, used for finding even the tiniest fast-flying prey, and for avoiding hundreds of other bats, and communicating for finding mates.

Specular Reflection

A major disadvantage of ultrasound sensing is its susceptibility to *specular reflection* (specular reflection means reflection from the outer surface of the object). While the sonar sensing principle is based on the sound wave reflecting from surfaces and returning to the receiver, it is important to remember that the sound wave will not necessarily bounce off the surface and "come right back." In fact, the direction of reflection depends on the incident angle of the sound beam and the surface. The smaller the angle, the higher the probability that the sound will merely "graze" the surface and bounce off, thus not returning to the emitter, in turn generating a false long/far-away reading. This is often called specular reflection, because smooth surfaces, with specular properties, tend to aggravate this reflection problem. Coarse surfaces produce more irregular reflections, some of which are more likely to return to the emitter. (For example, in our robotics lab on campus, we use sonar sensors, and we have lined one part of the test area with cardboard, because it has much better sonar reflectance properties than the very smooth wall behind it.)

In summary, long sonar readings can be very inaccurate, as they may result from false rather than accurate reflections. This must be taken into account when programming robots, or a robot may produce very undesirable and unsafe behavior.

Nonetheless, sonar sensors have been successfully used for very sophisticated robotics applications, including terrain and indoor mapping, and remain a very popular sensor choice in mobile robotics.

The first commercial ultrasonic sensor was produced by Polaroid, and used to automatically measure the distance to the nearest object (presumably which is being photographed). These simple Polaroid sensors still remain the most popular off-the-shelf sonars (they come with a processor board that deals with the analog electronics). Their standard properties include :

- 32-foot range
- 30-degree beam width
- sensitivity to specular reflection
- shortest distance return

Polaroid sensors can be combined into phased arrays to create more sophisticated and more accurate sensors.

One can find ultrasound used in a variety of other applications; the best known one is ranging in submarines. The sonars there have much more focused and have longer-range beams. Simpler and more mundane applications involve automated "tape-measures", height measures, burglar alarms, *etc.*

Machine Vision

So far, we have talked about relatively simple sensors. They were simple in terms of processing of the information they returned. Now we turn to machine vision, *i.e.*, to cameras as sensors.

Cameras, of course, model biological eyes. Needless to say, all biological eyes are more complex than any camera we know today, but, as you will see, the cameras and machine vision systems that process their perceptual information, are not simple at all! In fact, machine vision is such a challenging topic that it has historically been a separate branch of Artificial Intelligence.

The general principle of a camera is that of light, scattered from objects in the environment (those are called the *scene*), goes through an opening ("iris", in the simplest case a *pin hole*, in the more sophisticated case a *lens*), and impinging on what is called the *image plane*. In biological systems, the image plane is the *retina*, which is attached to numerous rods and cones (photosensitive elements) which, in turn, are attached to nerves which perform so-called "early vision", and then pass information on throughout the brain to do "higher-level" vision processing. As we mentioned before, a very large percentage of the human (and other animal) brain is dedicated to visual processing, so this is a highly complex endeavor.

In cameras, instead of having photosensitive rhodopsin and rods and cones, we use silver halides on photographic film, or silicon circuits in charge-coupled devices (CCD) cameras. In all cases, some information about the incoming light (*e.g.*, intensity, color) is detected by these photosensitive elements on the image plane.

In machine vision, the computer must make sense out of the information it gets on the image plane. If the camera is very simple, and uses a tiny pin hole, then some computation is required to compute the projection of the objects from the environment onto the image plane . If a lens is involved (as in vertebrate eyes and real cameras), then more light can get in, but at the price of being focused; only objects a particular range of distances from the lens will be in focus. This range of distances is called the camera's *depth of field*.

The image plane is usually subdivided into equal parts, called *pixels*, typically arranged in a rectangular grid. In a typical camera there are 512 by 512 pixels on the image plane (for comparison, there are 120×10^6 rods and 6×10^6 cones in the eye, arranged hexagonally). Let's call the projection on the image plane the *image*.

The brightness of each pixel in the image is proportional to the amount of light directed toward the camera by the surface patch of the object that projects to that pixel. (This of course depends on the reflectance properties of the surface patch, the position and distribution of the light sources in the environment, and the amount of light reflected from other objects in the scene onto the surface patch.) As it turns out, brightness of a patch depends on two kinds of reflections, one being specular (off the surface, as we saw before), and the other being diffuse (light that penetrates into the object, is absorbed, and then re-emitted). To correctly model light reflection, as well as reconstruct the scene, all these properties are necessary.

Let us suppose that we are dealing with a black and white camera with a 512 x 512 pixel image plane. Now we have an image, which is a collection of those pixels, each of which is an intensity between white and black. To find an object in that image (if there is one, we of course don't know *a priori*), the typical first step ("early vision") is to do *edge detection*, *i.e.*, find all the edges. How do we recognize them? We define edges as curves in the image plane across which there is significant change in the brightness.

A simple approach would be to look for sharp brightness changes by differentiating the image and look for areas where the magnitude of the derivative is large. This almost works, but unfortunately it produces all sorts of spurious peaks, *i.e.*, noise. Also, we cannot inherently distinguish changes in intensities due to shadows from those due to physical objects. But let's forget that for now and think about noise. How do we deal with noise?

We do *smoothing*, *i.e.*, we apply a mathematical procedure called *convolution*, which finds and eliminates the isolated peaks. Convolution, in effect, applies a *filter* to the image. In fact, in order to find arbitrary edges in the image, we need to convolve the image with many filters with different orientations. Fortunately, the relatively complicated mathematics involved in edge detection has been well studied, and by now there are standard and preferred approaches to edge detection.

Once we have edges, the next thing to do is try to find objects among all those edges. *Segmentation* is the process of dividing up or organizing the image into parts that correspond to continuous objects. But how do we know which lines correspond to which objects, and what makes an object? There are several cues we can use to detect objects :

1. We can have stored *models* of line-drawings of objects (from many possible angles, and at many different possible scales!), and then compare those with all possible combinations of edges in the image. Notice that this is a very computationally intensive and expensive process. This general approach, which has been studied extensively, is called *model-based vision*.

2. We can take advantage of *motion*. If we look at an image at two consecutive time-steps, and we move the camera in between, each continuous solid objects (which obeys physical laws) will move as one, *i.e.*, its brightness properties will be conserved. This hives us a hint for finding objects, by subtracting two images from each other. But notice that this also depends on knowing well how we moved the camera relative to the scene (direction, distance), and that nothing was moving in the scene at the time. This general approach, which has also been studied extensively, is called *motion vision*.

3. We can use stereo (*i.e.*, *binocular stereopsis*, two eyes/cameras/points of view). Just like with motion vision above, but without having to actually move, we get two images, which we can subtract from each other, if we know what the *disparity* between them should be, *i.e.*, if we know how the two cameras are organized/positioned relative to each other.
4. We can use *texture*. Patches that have uniform texture are consistent, and have almost identical brightness, so we can assume they come from the same object. By extracting those we can get a hint about what parts may belong to the same object in the scene.
5. We can also use *shading* and *contours* in a similar fashion. And there are many other methods, involving *object shape* and projective invariants, *etc.*

Note that all of the above strategies are employed in biological vision. It's hard to recognize unexpected objects or totally novel ones (because we don't have the models at all, or not at the ready). Movement helps catch our attention. Stereo, *i.e.*, two eyes, is critical, and all carnivores use it (they have two eyes pointing in the same direction, unlike herbivores). The brain does an excellent job of quickly extracting the information we need for the scene.

Machine vision has the same task of doing real-time vision. But this is, as we have seen, a very difficult task. Often, an alternative to trying to do all of the steps above in order to do *object recognition*, it is possible to simplify the vision problem in various ways :

1. Use color; look for specifically and uniquely colored objects, and recognize them that way (such as stop signs, for example)
2. Use a small image plane; instead of a full 512 x 512 pixel array, we can reduce our view to much less, for example just a line (that's called a *linear CCD*). Of course there is much less information in the image, but if we are clever, and know what to expect, we can process what we see quickly and usefully.
3. Use other, simpler and faster, sensors, and combine those with vision. For example, IR cameras isolate people by body-temperature. Grippers allow us to touch and move objects, after which we can be sure they exist.
4. Use information about the environment; if you know you will be driving on the road which has white lines, look specifically for those lines at the right places in the image. This is how first and still fastest road and highway robotic driving is done.

Those and many other clever techniques have to be employed when we consider how important it is to "see" in real-time. Consider highway driving as an important and growing application of robotics and AI. Everything is moving so quickly, that the system must perceive and act in time to react protectively and safely, as well as intelligently.

Now that you know how complex vision is, you can see why it was not used on the first robots, and it is still not used for all applications, and definitely not on simple robots. A robot can be extremely useful without vision, but some tasks demand it. As always, it is critical to think about the proper match between the robot's sensors and the task.

EFFECTORS

An *effector* is any device that affects the environment. Robots control their effectors, which are also known as end effectors. Effectors include legs, wheels, arms, fingers, wings and fins. Controllers cause the effectors to produce desired effects on the environment. An *actuator* is the actual mechanism that enables the effector to execute an action. Actuators typically include electric motors, hydraulic or pneumatic cylinders, *etc*. The terms effector and actuator are often used interchangeably to mean "whatever makes the robot take an action." This is not really proper use. Actuators and effectos are not the same thing. And we'll try to be more precise in the class. Most simple actuators control a single *degree of freedom*, *i.e.*, a single motion (*e.g.*, up-down, left-right, in-out, *etc*.). A motor shaft controls one rotational degree of freedom, for example. A sliding part on a plotter controls one translational degree of freedom. How many degrees of freedom (DOF) a robot has is going to be very important in determining how it can affect its world, and therefore how well, if at all, it can accomplish its task. Just as we said many times before that sensors must be matched to the robot's task, similarly, *effectors must be well matched to the robot's task* also.

In general, a free body in space as 6 DOF : three for translation (x,y,z), and three for orientation/rotation (roll, pitch, and yaw). We'll go back to DOF in a bit. You need to know, for a given effector (and actuator/s), how many DOF are available to the robot, as well as how many total DOF any given robot has. If there is an actuator for every DOF, then all of the DOF are controllable. Usually not all DOF are controllable, which makes robot control harder. A car has 3 DOF : position (x,y) and orientation (theta). But only 2 DOF are controllable : driving : through the gas pedal and the forward-reverse gear; steering : through the steering wheel. Since there are more DOF than are controllable, there are motions that cannot be done, like moving sideways (that's why parallel parking is hard). We need to make a distinction between what an actuator does (*e.g.*, pushing the gas pedal) and what the robot does as a result (moving forward). A car can get to any 2D position but it may have to follow a very complicated trajectory. Parallel parking requires a discontinuous trajectory w.r.t. velocity, *i.e.*, the car has to stop and go. When the number of controllable DOF is equal to the total number of DOF on a robot, it is holonomic. If the number of controllable DOF is smaller than total DOF, the robot is non-holonomic. If the number of controllable DOF is larger than the total DOF, the robot is redundant. A human arm has 7 DOF (3 in the shoulder, 1 in the elbow, 3 in the wrist), all of which can be controlled. A free object in 3D space (*e.g.*, the hand, the finger tip) can have at most 6 DOF! So there are redundant ways of putting the hand at a particular position in 3D space. This is the core of why manipulations is very hard!

Two basic ways of using effectors :
- to move the robot around =>locomotion
- to move other object around =>manipulation

These divide robotics into two mostly separate categories :
- mobile robotics
- manipulator robotics

In contrast to locomotion, where the body of the robot is moved to get to a particular position and orientation, a manipulator moves itself typically to get the *end effector* (*e.g.*, the hand, the finger, the fingertip) to the desired 3D position and orientation. So imagine having to touch a specific point in 3D space with the tip of your index finger; that's what a typical manipulator has to do. Of course, largely manipulators need to grasp and move objects, but those tasks are extensions of the basic reaching above. The challenge is to get there efficiently and safely. Because the end effector is attached to the whole arm, we have to worry about the whole arm; the arm must move so that it does not try to violate its own *joint limits* and it must not hit itself or the rest of the robot, or any other obstacles in the environment. Thus, doing autonomous manipulation is very challenging. Manipulation was first used in tele-operation, where human operators would move artificial arms to handle hazardous materials. It turned out that it was quite difficult for human operators to learn how to tele-operate complicated arms (such as duplicates of human arms, with 7 DOF). One alternative today is to put the human arm into an exo-skeleton , in order to make the control more direct. Using joy-sticks, for example, is much harder for high DOF. Why is this so hard? Because even as we saw with locomotion, there is typically no direct and obvious link between what the effector needs to do in physical space and what the actuator does to move it. In general, the correspondence between actuator motion and the resulting effector motion is called *kinematics*. In order to control a manipulator, we have to know its kinematics (what is attached to what, how many joints there are, how many DOF for each joint, *etc*.). We can formalize all of this mathematically, and get an equation which will tell us how to convert from, say, angles in each of the joints, to the Cartesian positions of the end effector/point. This conversion from one to the other is called computing the manipulator kinematics and *inverse kinematics*.

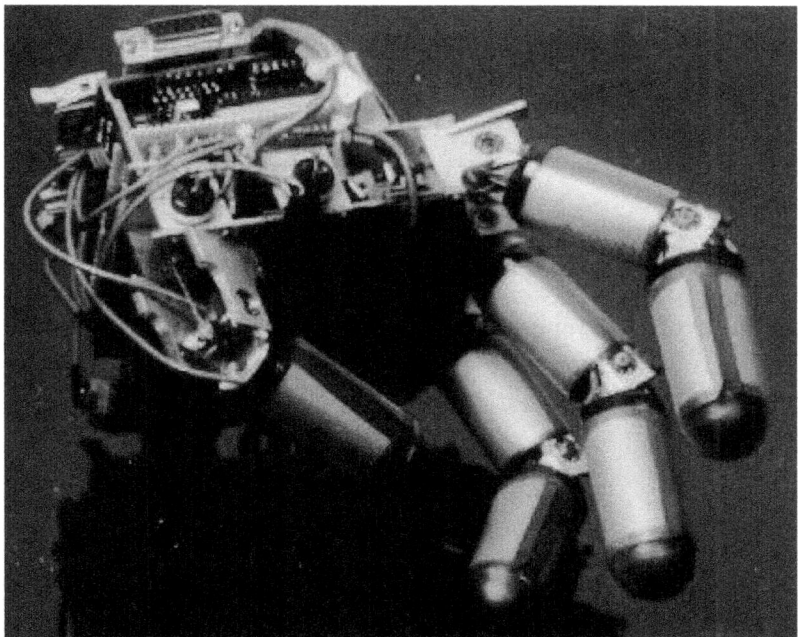

The process of converting the Cartesian (x,y,z) position into a set of joint angles for the arm (thetas) is called inverse kinematics. Kinematics are the rules of what is attached to what, the body structure. Inverse kinematics is computationally intense. And the problem is even harder if the manipulator (the arm) is redundant.

Manipulation involves
- trajectory planning (over time)
- inverse kinematics
- inverse dynamics
- dealing with redundancy

Manipulators are effectors. Joints connect parts of manipulators. The most common joint types are :
- rotary (rotation around a fixed axis)
- prismatic (linear movement)

These joints provide the DOF for an effector, so they are planned carefully.

Robot manipulators can have one or more of each of those joints. Now recall that any free body has 6 DOF; that means in order to get the robot›s end effector to an arbitrary position and orientation, the robot requires a minimum of 6 joints. As it turns out, the human arm has 7 DOF. That›s sufficient for reaching any point with the hand, and it is also redundant, meaning that there are multiple ways in which any point can be reached. This is good news and bad news; the fact that there are multiple solutions means that there is a larger space to search through to find the best solution. Now consider end effectors. They can be simple pointers (*i.e.*, a stick), simple 2D grippers, screwdrivers for attaching tools (like welding guns, sprayer, *etc.*), or can be as complex as the human hand, with variable numbers of fingers and joints in the fingers. Problems like reaching and grasping in manipulation constitute entire subareas of robotics and AI. Issues include : finding grasp-points (COG, friction, *etc.*); force/strength of grasp; compliance (*e.g.*, in sliding, maintaining contact with a surface); dynamic tasks (*e.g.*, juggling, catching). Other types of manipulation, such as carefully controlling force, as in grasping fragile objects and maintaining contact with a surface (so-called *compliant motion*), are also being actively researched. Finally, dynamic manipulation tasks, such as juggling, throwing, catching, *etc.*, are already being demonstrated on robot arms.

Having talked about navigation and manipulation, think about what types of sensors (external and proprioceptive) would be useful for these general robotic tasks. *Proprioceptive* sensors sense the robot›s actuators (*e.g.*, shaft encoders, joint angle sensors, *etc.*); they sense the robot›s own movements. You can think of them as perceiving internal state instead of external state. External sensors are helpful but not necessary or as commonly used.

ACTUATORS

Actuators, also known as drives, are mechanisms for getting robots to move. Most actuators are powered by pneumatics (air pressure), hydraulics (fluid pressure), or motors (electric current). Most actuation uses electromagnetic motors and gears but there have been frequent uses of other forms of actuation including NiTinOL "muscle-wires" and inexpensive Radio Control servos. To get a motor under computer control, different motor types and actuator types are used. Some of the motor types are Synchronous, Stepper, AC servo, Brushless DC servo, and Brushed DC servo. Radio Control servos for model airplanes, cars and other vehicles are light, rugged, cheap and fairly easy to interface. Some of the units can provide very high torque speed. A Radio Control servo can be controlled from a parallel port. With one of the PC's internal timers cranked up, it is possible to control eight servos from a common parallel port with nothing but a simple interrupt service routine and a cable. In fact, power can be pulled from the disk drive power connector and the PC can run all servos directly with no additional hardware. The only down side is that the PC wastes some processing power servicing the interrupt handler.

DC Motors

The most common actuator you will use (and the most common in mobile robotics in general) is the *direct current (DC) motor*. They are simple, cheap, and easy to use. Also, they come in a great variety of sizes, to accommodate different robots and tasks. *DC motors convert electrical into mechanical energy.* They consist of permanent magnets and loops of wire inside. When current is applied, the wire loops generate a magnetic field, which reacts against the outside field of the static magnets. The interaction of the fields produces the movement of the shaft/ armature. Thus, electromagnetic energy becomes motion. As with any physical system, DC motors are not perfectly efficient, meaning that the energy is not converted perfectly, without any waste. Some energy is wasted as heat generated by friction of mechanical parts. *Inefficiencies* are minimized in well-designed (and more expensive) motors, and their performance can be brought up to the 90th percentile, but cheap motors (such as the ones you may use) can be as low as 50%. (In case you think this is very inefficient, remember that other types of effectors, such as miniature electrostatic motors, may have much lower efficiencies still.) A motor requires a power source within its *operating voltage, i.e.,* the recommended voltage range for best efficiency of the motor. Lower voltages will usually turn the motor (but provide less power). Higher voltages are more tricky : in some cases they can increase the power output but almost always at the expense of the operating life of the motor. *E.g.*, the more you rev your car engine, the sooner it will die. When constant voltage is applied, *a DC motor draws current in the amount proportional to the work it is doing*. For example, if a robot is pushing against a wall, it is drawing more current (and draining more of its batteries) than when it is moving freely in open space.

The reason is the resistance to the motor motion introduced by the wall. If the resistance is very high (*i.e.*, the wall just won't move no matter how much the robot pushes against it), the motor draws a maximum amount of power, and stalls. This is defined as the *stall current* of the motor : the most current it can draw at its specified voltage. Within a motor's *operating current* range, the more current is used, the more *torque* or *rotational force* is produced at the shaft. In general, the strengths of the magnetic field generated in the wire loops is directly proportional to the applied current and thus the produced torque at the shaft. Besides stall current, a motor also has its *stall torque*, the amount of rotational force produced when the motor is stalled at its operating voltage. Finally, the amount of *power* a motor generates is the product of its shaft's *rotational velocity* and its *torque*. If there is no load on the shaft, *i.e.*, the motor is spinning freely, then the rotational velocity is the highest, but the torque is 0, since no mechanism is being driven by the motor. The output power, then, is 0 also. In contrast, when the motor is stalled, it is producing maximum torque, but the rotational velocity is 0, so the output power is 0 again.

Between free spinning and stalling, the motor does useful work, and the produced power has a characteristic parabolic relationship demonstrating that the motor produces the most power in the middle of its performance range. Most DC motors have unloaded speeds in the range of 3,000 to 9,000 RPM (revolutions per minute), or 50 to 150 RPS (revolutions per second). That turns out to put them in the high-speed but low-torque category (compared to some other alternatives). For example, how often do you need to drive something very light that rotates very fast (besides a fan)? Yet that is what DC motors are naturally best at. In contrast, robots need to pull loads (*i.e.*, move their bodies and manipulators, all of which have significant mass), thus requiring more torque and less speed. As a result, the performance of a DC motor typically needs to be adjusted in that direction, through the use of *gears*.

Gearing

The force generated at the edge of a gear is equal to the product of the radius of the gear and its torque (F = r t), in the line tangential to its circumference. By combining gears with different radii, we can manipulate the amount of force/torque the mechanism generates. The relationship between the radii and the resulting torque is well defined, as follows : Suppose Gear1 with radius r1 turns with torque t1, generating a force of t1/r1 perpendicular to its circumference. Now if we mesh it with Gear2, with r2, which generates t2/r2, then t1/r1 = t2/r2. To get the torque generated by Gear2, we get : t2 = t1 r2/r1. Intuitively, this means : the torque generated at the output gear is proportional to the torque on the input gear and the ratio of the two gear's radii. If r2 > r1, we get a bigger number, if r1 > r2, we get a smaller number.

If the output gear is larger than the input gear, the torque increases. If the output gear is smaller than the input gear, the torque decreases. Besides the change in torque that takes place when gears are combined, there is also a corresponding change in speed. To measure speed we are interested in the circumference of the gear, C= 2 * pi * r. Simply put, if the circumference of Gear1 is twice that of Gear2, then Gear2 must turn twice for each full rotation of Gear1. If the output gear is larger than the input gear, the speed decreases. If the output gear is smaller than the input gear, the speed increases. In summary, when a small gear drives a large one, torque is increased and speed is decreased. Analogously, when a large gear drives a small one, torque is decreased and speed is increased. Thus, gears are used in DC motors (which we said are fast and low torque) to trade off extra speed for additional torque. Gears are combined using their teeth. The number of teeth is not arbitrary, since it is the key means of proper reduction. Gear teeth require special design so that they mesh properly. If there is any looseness between meshing gears, this is called *backlash*, the ability for a mechanism to move back \& forth within the teeth, without turning the whole gear.

Reducing backlash requires tight meshing between the gear teeth, but that, in turn, increases *friction*. As you can imagine, proper gear design and manufacturing is complicated. To achieve "three to one gear reduction (3 :1)", we apply power to a small gear (say one with 8-teeth) meshed with a large one (with 3 * 8 = 24 teeth). As a result, we have slowed down the large gear by 3 and have tripled its torque. Gears can be organized in series ("ganged"), in order to multiply their effect. For example, 2 3 :1 gears in series result in a 9 :1 reduction. This requires a clever arrangement of gears. Or three 3 :1 gears in series can produce a 27 :1 reduction. This method of multiplying reduction is the underlying mechanism that makes DC motors useful and ubiquitous.

Electronic Control of Motors

It should come as no surprise that motors require more battery power (*i.e.*, more current) than electronics (*e.g.*, 5 milliamps for the 68HC11 processor v. 100 milliamps - 1 amp for a small DC motor). Typically, specialized circuitry is required. You need to learn about H-bridges and *pulse-width modulation* there.

Servo Motors

It is sometimes necessary to be able to move a motor to a specific position. If you consider your basic DC motor, it is not built for this purpose. Motors that can turn to a specific position are called *servo motors* and are in fact constructed out of basic DC motors, by adding :

- some gear reduction
- a position sensor for the motor shaft
- an electronic circuit that controls the motor's operation

Servos are used in toys a great deal, to adjust steering on steering in RC cars and wing position in RC airplanes. Since positioning of the shaft is what servo motors are all about, most have their movement reduced to 180 degrees. The motor is driven with a waveform that specifies the desired angular position of the shaft within that range. The waveform is given as a series of pulses, within a *pulse-width modulated* signal. Thus, the width (*i.e.*, length) of the pulse specifies the control value for the motor, *i.e.*, how the shaft should turn. Therefore, the exact width/length of the pulse is critical, and cannot be sloppy. There are no milliseconds or even microseconds to be wasted here, or the motor will behave very badly, jitter, and go beyond its mechanical limit. This limit should be checked empirically, and avoided. In contrast, the duration between the pulses is not critical at all. It should be consistent, but there can be noise on the order of milliseconds without any problems for the motor. This is intuitive : when no pulse arrives, the motor does not move, so it simply stops. As long as the pulse gives the motor sufficient time to turn to the proper position, additional time does not hurt it.

SolarSpeeder

Continuous Rotation Motors

A regular DC motor can be used for continuous rotation. Furthermore, servo motors can also be retrofitted to provide continuous rotation (remember, they only to 180 otherwise), like this :

- remove mechanical limit (revert back to DC motor shaft)
- remove pot position sensor (no need to tell position)
- apply 2 resistors to fool the servo to think it is fully turning

Related Products for Drives and Actuators

Research into shape memory alloys, polymer gels and micro-mechanism devices is ongoing, and changing often. Nickel-titanium alloys were first discovered by the Naval Ordinance Laboratory decades ago and the material was termed NiTinOL. These materials have the intriguing property that they provide actuation through cycling of current through the materials. It undergoes a 'phase change' exhibited as force and motion in the wire. At room temperature Muscle Wires are easily stretched by a small force. However, when conducting an electric current, the wire heats and changes to a much harder form that returns to the "unstretched" shape -- the wire shortens in length with a usable amount of force. Nitinol can be stretched by up to eight percent of their length and will recover fully, but only for a few cycles. However when used in the three to five percent range, Muscle Wires can run for millions of cycles with very consistent and reliable performance.

CONTROLLERS

The robot connects to a computer, which keeps the pieces of the arm working together. This computer is the controller. The controller functions as the "brain" of the robot. The controller can also network to other systems, so that the robot may work together with other machines, processes, or robots.

Given that the robot arm movement is appropriate to its application, that the arm strength and rigidity meet the payload needs and that servo drives provide the necessary speed of response and resolution, a robot controller is required to manage the arm articulations, its End Effector, and the interface with the workplace. The simplest type of control, still widely used, is "record-playback," or "lead-through". An operator positions arm articulations to desired configurations. At each desired location the articulation encoder positions are recorded in memory. Step by step, an entire work-cycle sequence is recorded. Then in playback mode the sequence is observed and modified.

As applications become more challenging, some jobs require continuous path control of an End Effector. For this action all articulations must be programmed in speeds appropriate to the particular task. This requires programming for the control of the robot. Robots today have controllers run by programs -- sets of instructions written in code. The program sets limits on what the robot can do. These requirements call into play sophisticated computer-based controllers and so-called robot languages. These languages permit a kind of robot control known as hierarchical control, in which decision making by the robot takes place on several levels. These levels are interconnected by feedback mechanisms that inform the next higher level of the status of previous actions. The advantage of a general-purpose robot arm is that it can be programmed to do many jobs. The disadvantage is that the programming tends to be a job for highly paid engineers. Even when a factory robot can perform a task more efficiently than a person, the job of programming it and setting up its workplace can be more trouble than its worth. Commotion Systems, a new California firm, is developing easier ways to program robots using pre-designed software modules. For now though, the job of "training" robots is still one of the main reasons that they are not used more. In the future, controllers with Artificial Intelligence could allow robots to think on their own, even program themselves. This could make robots more self-reliant and independent. Angelus Research has designed an intelligent motion controller for robots that mimics the brain's three-level structure, including instinctive, behavioral, and goal levels. The controller, which can be used in unpredictable circumstances, uses a Motorola 68HC11 microprocessor.

Feedback (Closed Loop) Control

Feedback control is a means of getting a system (in our case a robot) to achieve and maintain a desired state by continuously comparing its current and desired state. The *desired state* is also called the *goal state* of the system. Note that it can be an external or internal state : for example, a thermostat monitors and controls external state (the temperature of the house), while a robot can control its internal state (*e.g.*, battery power, by recharging at proper times) or external state (*e.g.*, distance from a wall). If the current and desired state are the same, the control system does not need to do anything. But if they are not, how does it decide what to do? That is what the design of the controller is all about. A control system must first find the difference between the current and desired states. This difference is called the *error*, and the goal of any control system is to minimize that error. In some systems, the only information available about the error is whether it is 0 or non-0, *i.e.*, whether the current and desired states are the same. This is very little information to work with, but it is still a basis for control and can be exploited in interesting ways. Additional information about the error would be its *magnitude*, *i.e.*, how "far" the current state is from the desired state. Finally, the last part of the error information is its *direction*, *i.e.*, is the current state too close or too far from the desires state (in whatever space it may be). Control is easiest if we have frequent feedback providing error magnitude and direction. Notice that the behavior of a feedback system oscillates around the desired state. In the case of a thermostat, the temperature oscillates around the *set point*, the desired setting. Similarly, the robot's movement will oscillate around the desired state, which is the optimal distance from the wall. How can we decrease this oscillation? We can use a smoother/larger turning angle, and we can also use a *range* instead of a *set point* distance as the goal state. Now what happens when you have sensor error in your system? What if your sensor incorrectly tells you that the robot is far from a wall, but in fact it is not? What about vice versa? How might you address these issues? Feedback control is also called *closed loop control* because it closes the loop between the input and the output, *i.e.*, it provides the system with a measure of "progress."

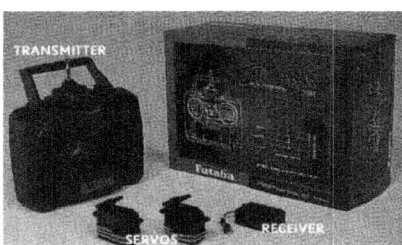

Open Loop Control

The alternative to closed loop control is *open loop control*. This type of control does not require the use of sensors, since state is not fed back into the system. Such systems can operate (perform repetitive, state-independent tasks) only if they are extremely well calibrated and their environment does not change in a way that affects their performance. We have talked about feedback control so far, but there is also an important notion of *feed forward control*. In such a system, the controller determines set points and sub-goals for itself ahead of time, without looking at actual state data.

Next Step Microcontroller Kit

Reactive Control

Reactive control is based on a tight loop connecting the robot's sensors with its effectors. Purely reactive systems do not use any internal representations of the environment, and do not look ahead : they *react* to the current sensory information. Thus, reactive systems use a direct mapping between sensors and effectors, and minimal, if any, state information. They consist of collections of rules that map specific *situations* to specific *actions*. If a reactive system divides its perceptual world into *mutually exclusive* or unique situations, then only one of those situations can be triggered by any sensory input at any one time, and only one action will be activated as a result. This is the simplest form of a reactive control system.

Mini SSC II
Serial Servo Controller

It is often too difficult to split up all possible situations this way, or it may require unnecessary encoding. Consider the case of multiple sensors : to have mutually-exclusive sensory inputs, the controller must encode rules for all possible sensory combinations. There is an exponential number of those. This is, in fact, the robot's entire sensory space (as we defined earlier in the semester). This space then needs to be mapped to all possible actions (the action space), resulting in the complete control space for that robot. Although this mapping is done while the system is being designed, *i.e.*, not at run-time, it can be very tedious, and it results in a large look up table which takes space to encode/store in a robot, and can take time to search, unless some clever parallel look up technique is used. In general, this complete mapping is not used in hand-designed reactive systems. Instead, specific situations trigger appropriate actions, and default actions are used to cover all other cases. Human designers can effectively reduce the sensory space to only the inputs/situations that matter, map those to the appropriate actions, and thus greatly simplify the control system. If the rules are not triggered by mutually-exclusive conditions, more than one rule can be triggered in parallel, resulting in two or more different actions being output by the system. Deciding among multiple actions or behaviors is called *arbitration*, and is in general a difficult problem. Arbitration can be done based on :

- fixed priority hierarchy (processes have pre-assigned priorities)
- a dynamic hierarchy (process priorities change at run-time)
- learning (process priorities may be initialized or not, and are learned at run-time, once or repeatedly/dynamically)

If a reactive system needs to support parallelism, *i.e.*, the ability to execute multiple rules at once, the underlying programming language must have the ability to *multi-task*, *i.e.*, execute several processes/pieces of code in parallel. The ability to multi-task is critical in reactive systems : if a system cannot monitor its sensors in parallel, but must go from one to another in sequence, it may miss some event, or at least the onset of an event, thus failing to react in time. Now that we understand the building blocks of a reactive system (reactive rules coupling sensors and effectors, *i.e.*, situations and actions), we need to consider principled ways of organizing reactive controllers. We will start with the best known reactive control architecture, the Subsumption Architecture, introduced by Rod Brooks at MIT in 1985.

The Subsumption Architecture

The following are the guiding principles of the architecture :
1. Systems are built from the bottom up
2. Components are task-achieving actions/behaviors (not functional modules)
3. Components can be executed in parallel
4. Components are organized in layers, from the bottom up lowest layers handle most basic tasks
5. Newly added components and layers exploit the existing ones
6. Each component provides and does not disrupt a tight coupling between sensing and action
7. There is no need for internal models : "the world is its own best model"

Here is a rough image of how the system works : If we number the layers from 0 up, we can assume that the 0th layer is constructed, debugged, and installed first. As layer 1 is added, layer 0 continues to function, but may be influenced by layer 1, and so on up. If layer 1 fails, layer 0 is unaffected. When layer 1 is designed, layer 0 is taken into consideration and utilized, *i.e.*, its existence is *subsumed*, thus the name of the architecture. Layer 1 can *inhibit the outputs* of layer 0 or *suppress its inputs*. Subsumption systems grow from the bottom up, and layers can keep being added, depending on the tasks of the robot. How exactly layers are split up depends on the specifics of the robot, the environment, and the task. There is no strict recipe, but some solutions are better than others, and most are derived empirically. The inspiration behind the Subsumption Architecture is the *evolutionary process*, which introduces new competencies based on the existing ones. Complete creatures are not thrown out and new ones created from scratch; instead, solid, useful substrates are used to build up to more complex capabilities.

Behavior Based Control

Behavior-based systems (BBS) use behaviors as the underlying module of the system, *i.e.*, they use a behavioral decomposition. Behaviors can vary greatly from one BBS to another, but typically have the following properties :

1. Behaviors are feedback controllers
2. Behaviors achieve specific tasks/goals (*e.g.*, avoid-others, find-friend, go-home)
3. Behaviors are typically executed in parallel/concurrently
4. Behaviors can store state and be used to construct world models/representation
5. Behaviors can directly connect sensors and effectors (*i.e.*, take inputs from sensors and send outputs to effectors)
6. Behaviors can also take inputs from other behaviors and send outputs to other behaviors (this allows for building networks)
7. Behaviors are typically higher-level than actions (go-home rather than turn-left-by-37.5-degrees)
8. Behaviors are typically closed-loop but extended in time (this is often a consequence of #7 above)
9. When assembled into distributed representations, behaviors can be used to look ahead but at a time-scale comparable with the rest of the behavior-based system

Behavior-based systems are not limited in the ways that reactive systems are. As a result, behavior-based systems have the following key properties :

1. The ability to react in real-time

2. The ability to use representations to generate efficient (not only reactive) behavior
3. The ability to use a uniform structure and representation throughout the system (so no intermediate layer)

The key challenge is in how representation (*i.e.*, any form of world model) can be effectively *distributed* over the behavior structure. In order to avoid the pitfalls of deliberative systems, the representation must be able to act on a timescale that is close if not the same as the real-time parts of the system. Similarly, to avoid the pitfalls of the hybrid systems approach, the representation needs to use the same underlying behavior structure as the rest of the system. Note that behavior-based systems can have reactive components to them, *i.e.*, not every part of a behavior-based system needs to be involved with representational computation. In fact, many behavior-based systems did not use complex representations at all. As long as they use behaviors , they are BBS.

ARMS

The robot arm comes in all shapes and sizes and is the single most important part in robotic architecture. The arm is the part of the robot that positions the End Effector and Sensors to do their pre-programmed business. Many (but not all) resemble human arms, and have shoulders, elbows, wrists, even fingers. This gives the robot a lot of ways to position itself in its environment.

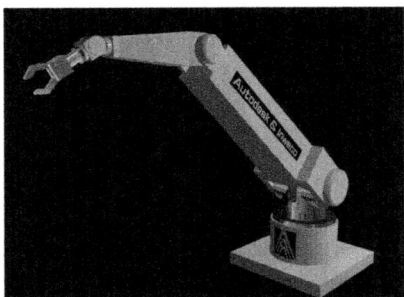

Many robots resemble human arms, and have shoulders, elbows, wrists, even fingers. This gives a robot lots of options for moving, and helps it do things in place of a human arm. In order to reach any possible point in space within its work envelope, a robot uses a total of 7 degrees of freedom. Each direction a joint can go gives an arm 1 degree. So, a simple robot arm with 3 degrees of freedom could move in 3 ways : up and down, left and right, forward and backward. Many robots of today are designed to move with these 7 degrees of freedom.

The human arm is an amazing design. It allows us to place our all-purpose end effector, the hand, where it is needed. Jointed arm robots mimic the ability of human arms to be flexible, precise, and ready for a wide variety of tasks. The jointed-arm robot has six degrees of freedom, which enable it to perform jobs that require versatility and dexterity. The design of a jointed-arm robot is similar to a human arm, but not exactly the same.

Seven Degrees of Freedom

The robotic arm is very similar to the human arm in that it has the capability of having 7 and not 6 degrees of movement, as previously thought. Out of the 7 degrees of movement, your shoulder has 3 of the arm's 7 degrees of freedom. The easiest way to explain the movement of the robotic arm is to use your own arm as an example. Hold out your arm and follow along . . .

First Degree : Shoulder Pitch

To find your arm's first degree of freedom : Point your entire arm straight out in front of you. Move your shoulder up and down. The up and down movement of the shoulder is called the shoulder pitch.

Second Degree : Arm Yaw

To find your arm's second degree of freedom : Point your entire arm straight out in front of you. Move your entire arm from side to side. This side to side movement is called the arm yaw.

Third Degree : Shoulder Roll

To find your arm's third degree of freedom : Point your entire arm straight out in front of you. Now, roll your entire arm from the shoulder, as if you were screwing in a light bulb. This rotating movement is called a shoulder roll.

Fourth Degree : Elbow Pitch

To find your arm's fourth degree of freedom : Point your entire arm straight out in front of you. Hold your arm still, then bend only your elbow. Your elbow can move up and down. This up and down movement of the shoulder is called the shoulder pitch.

Fifth Degree : Wrist Pitch

To find your arm's fifth degree of freedom : Point your entire arm straight out in front of you. Without moving your shoulder or elbow, flex your wrist up and down. This up and down movement of the wrist is called the wrist pitch.

Sixth Degree : Wrist Yaw

To find your arm's sixth degree of freedom : Point your entire arm straight out in front of you. Without moving your shoulder or elbow, flex your wrist from side to side. The side to side movement is called the wrist yaw.

Seventh Degree : Wrist Roll

To find your arm's seventh degree of freedom : Point your entire arm straight out in front of you. Without moving your shoulder or elbow, rotate your wrist, as if you were turning a doorknob. The rotation of the wrist is called the wrist roll.

ARTIFICIAL INTELLIGENCE

The term "artificial intelligence" is defined as systems that combine sophisticated hardware and software with elaborate databases and knowledge-based processing models to demonstrate characteristics of effective human decision making. The criteria for artificial systems include the following : (1) functional : the system must be capable of performing the function for which it has been designed; (2) able to manufacture : the system must be capable of being manufactured by existing manufacturing processes; (3) designable : the design of the system must be imaginable by designers working in their cultural context; and (4) marketable : the system must be perceived to serve some purpose well enough, when compared to competing approaches, to warrant its design and manufacture.

Robotics is one field within artificial intelligence. It involves mechanical, usually computer-controlled, devices to perform tasks that require extreme precision or tedious or hazardous work by people. Traditional Robotics uses Artificial Intelligence planning techniques to program robot behaviors and works toward robots as technical devices that have to be developed and controlled by a human engineer. The Autonomous Robotics approach suggests that robots could develop and control themselves autonomously. These robots are able to adapt to both uncertain and incomplete information in constantly changing environments. This is possible by imitating the learning process of a single natural organism or through

Evolutionary Robotics, which is to apply selective reproduction on populations of robots. It lets a simulated evolution process develop adaptive robots.

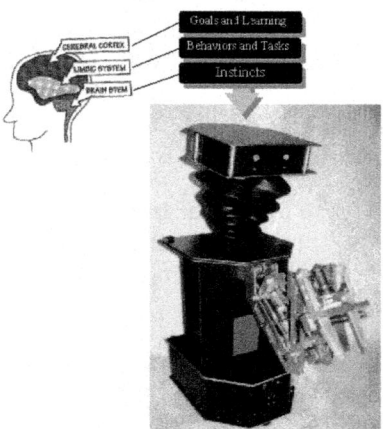

The artificial intelligence concept of the "expert system" is highly developed. This describes robot programmers ability to anticipate situations and provide the robot with a set of "if-then" rules. For example, if encountering a stairwell, stop and retreat. The more sophisticated concept is to give the robot the ability to "learn" from experience. A neural network brain equipped onto a robot will allow the robot to sample its world at random. Basically, the robot would be given some life-style goals, and, as it experimented, the actions resulting in success would be reinforced in the brain. This results in the robot devising its own rules. This is appealing to researchers and the community as it parallels human learning in lots of ways.

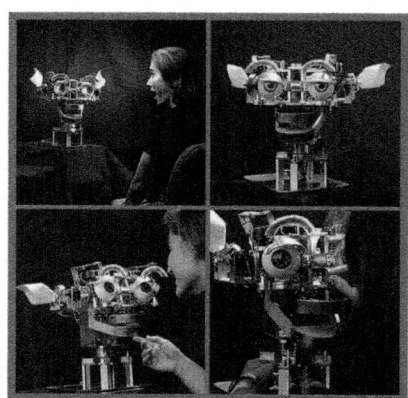

Artificial intelligence dramatically reduces or eliminates the risk to humans in many applications. Powerful artificial intelligence software helps to fully develop the high-precision machine capabilities of robots, often freeing them from direct human control and vastly improving their productivity. When a robot interacts with a richly populated and variable world, it uses it senses to gather data and then compare the sensate inputs with expectations that are imbedded in its world model. Therefore the effectiveness of the robot is limited by the accuracy to which its programming models the real world.

MOBILITY

Industrial robots are rarely mobile. Work is generally brought to the robot. A few industrial robots are mounted on tracks and are mobile within their work station. Service robots are virtually the only kind of robots that travel autonomously. Research on robot mobility is extensive. The goal of the research is usually to have the robot navigate in unstructured environments while encountering unforeseen obstacles. Some projects raise the technical barriers by insisting that the locomotion involve walking, either on two appendages, like humans, or on many, like insects. Most projects, however, use wheels or tractor mechanisms. Many kinds of effectors and actuators can be used to move a robot around. Some categories are :

- legs (for walking/crawling/climbing/jumping/hopping)
- wheels (for rolling)
- arms (for swinging/crawling/climbing)
- flippers

Wheels

Wheels are the locomotion effector of choice. Wheeled robots (as well as almost all wheeled mechanical devices, such as cars) are built to be statically stable. It is important to remember that wheels can be constructed with as much variety and innovative flair as legs : wheels can vary in size and shape, can consist of simple tires, or complex tire patterns, or tracks, or wheels within cylinders within other wheels spinning in different directions to provide different types of locomotion properties. So wheels need not be simple, but typically they are, because even simple wheels are quite efficient. Having wheels does not imply holonomicity. 2 or 4-wheeled robots are not usually holonomic. A popular and efficient design involves 2 differentially-steerable wheels and a passive caster. Differential steering means that the two (or more) wheels can be steered separately (individually) and thus differently. If one wheel can turn in one direction and the other in the opposite direction, the robot can spin in place. This is very helpful for following arbitrary trajectories. Tracks are often used (*e.g.*, tanks).

Legs

While most animals use legs to get around, legged locomotion is a very difficult robotic problem, especially when compared to wheeled locomotion. First, any robot needs to be stable (*i.e.*, not wobble and fall over easily). There are two kinds of stability : static and dynamic. A *statically stable* robot can stand still without falling over. This is a useful feature, but a difficult one to achieve : it requires that there be enough legs/wheels on the robot to provide sufficient static points of support. For example, people are *not* statically stable. In order to stand up, which appears effortless to us, we are actually using active control of our balance, though nerves and muscles and tendons. This balancing is largely unconscious, but must be learned, so that's why it takes babies a while to get it right, and certain injuries can make it difficult or impossible. With more legs, static stability becomes quite simple. In order to remain stable, *the robot's center of gravity (COG) must fall under its polygon of support*. This polygon is basically the projection between all of its support points onto the surface. So in a two-legged robot, the polygon is really a line, and the COG cannot be stably aligned with a point on that line to keep the robot upright. However, a three-legged robot, with its legs in a tripod organization, and its body above, produces a stable polygon of support, and is thus statically stable. But what happens when a statically stable robot lifts a leg and tries to move. Does its COG stay within the polygon of support? It may or may not, depending on the geometry. For certain robot geometries, it is possible (with various numbers of legs) to always stay statically stable while walking. This is very safe, but it is also very slow and energy inefficient. A basic assumption of the static gait (statically stable gait) is that the weight of a leg is negligible compared to that of the body, so that the total center of gravity (COG) of the robot is not affected by the leg swing. Based on this assumption, the conventional static gait is designed so as to maintain the COG of the robot inside of the support polygon, which is outlined by each support leg's tip position. The alternative to static stability is *dynamic stability* which allows a robot (or animal) to be stable while moving. For example, one-legged hopping robots are dynamically stable : they can hop in place or to various destinations, and not fall over. But they cannot stop and stay standing (this is an inverse pendulum balancing problem).

A statically stable robot can use dynamically-stable walking patterns, to be fast, or it can use statically stable walking. A simple way to think about this is by how many legs are up in the air during the robot's movement (*i.e.*, gait). 6 legs is the most popular number as they allow for a very stable walking gait, the tripod gait . If the same three legs move at a time, this is called the alternating tripod gait. if the legs vary, it is called the ripple gait. A rectangular 6-legged robot can lift three legs at a time to move forward, and still retain static stability. How does it do that? It uses the so-called *alternating tripod gait*, a biologically common walking pattern for 6 or more legs. In this gait, one middle leg on one side and two non-adjacent legs on the other side of the body lift and move forward at the same time, while the other 3 legs remain on the ground and keep the robot statically stable. Roaches move this way, and can do so very quickly. Insects with more than 6 legs (*e.g.*, centipedes and millipedes), use the ripple gate. However, when they run really fast, they switch gates to actually become airborne (and thus not statically stable) for brief periods of time.

Statically stable walking is very energy inefficient. As an alternative, dynamic stability enables a robot to stay up while moving. This requires active control (*i.e.*, the inverse pendulum problem). Dynamic stability can allow for greater speed, but requires harder control. Balance and stability are very difficult problems in control and robotics, so that is why when you look at most existing robots, they will have wheels or plenty of legs . Research robotics, of course, is studying single-legged, two legged, and other dynamically-stable robots, for various scientific and applied reasons. Wheels are more efficient than legs. They also do appear in nature, in certain bacteria, so the common myth that biology cannot make wheels is not well founded. However, evolution favors lateral symmetry and legs are much easier to evolve, as is abundantly obvious. However, if you look at population sizes, insects are most populous animals, and they all have many more than 2 legs.

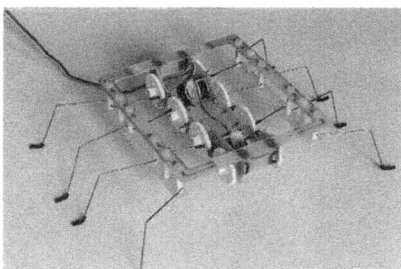

The Spider, A Legged Robot

In solving problems, the Spider is aided by the spring quality of its 1 mm steel wire legs. Hold one of its feet in place relative to the body and the mechanism keeps turning, the obstructed motor consuming less than 40 mA while it bends the leg. Let go and the leg springs back into shape. As I write this, the Spider is scrambling up and over my keyboard. Some of its feet get temporarily stuck between keys, springing loose again as others push down. It has no trouble whatsoever with this obstacle, nor with any of the others on my cluttered desk - even though it is still utterly brainless.

Mobility Limits of the Spider

As the feet rise to a maximum of 2 cm off the floor, a cassette box is about the tallest vertical obstacle that the Spider is able to step onto. Another limitation is slope. When asked to sustain a climb angle of more than about 20 degrees, the Spider rolls over backwards. And even this fairly modest angle (extremely steep for a car, by the way) requires careful gait control, making sure that both rear legs do not lift at the same time. Improvements are certainly possible. Increasing step size would require a longer body (more distance between the legs) and thus a different gear train. A better choice might be more legs, like 10 or 12 on a longer body, but with the same size gear wheels. That would give better traction and climbing ability. And if a third motor is allowed, one might construct a horizontal hinge in the `backbone'. Make a gear shaft the center of a nice, tight hinge joint. Then the drive train will function as before. Using the third motor and a suitable mechanism, the robot could raise its front part to step onto a tall obstacle, somewhat like a caterpillar. But turning on the spot becomes difficult.

Flying and Underwater Robots

Most robots do not fly or swim. Recently, researchers have been exploring the possibilities and problems involved with flying and swimming robots.

Chapter 9

ROBOT SOFTWARE

Robot software is the set of coded commands that tell a mechanical device and electronic system, known together as a robot, what tasks to perform. Robot software is used to perform tasks and automate tasks. Many software systems and frameworks have been proposed to make programming robots easier.

Some robot software aims at developing intelligent mechanical devices. Common tasks include feedback loops, data filtering, control, pathfinding, and locating.

INTRODUCTION

Robot software is the coded commands that tell a mechanical device (known as a robot) what tasks to perform and control its actions. Robot software is used to perform tasks and automate tasks to be performed. Programming robots is a non-trivial task. Many software systems and frameworks have been proposed to make programming robots easier.

Some robot software aim at developing intelligent mechanical devices. Though common in science fiction stories, such programs are yet to become common-place in reality and much development is yet required in the field of artificial intelligence before they even begin to approach the science fiction possibilities. Pre-programmed hardware may include feedback loops such that it can interact with its environment, but does not display actual intelligence.

Data flow programming techniques are used by most robot manufacturers, and is based on the concept that when the value of a variable changes, the values of other variables affected should also change. A programming language that incorporates data flow principles is called a data flow language. In addition to numeric processing, data flow languages also incorporate functional concepts. Unlike other programming languages which use imperative programming, data flow programming is modeled as a sequence of functions.

With any programming software, the state of a program at any given time is an important consideration. The state provides an indication of the various conditions at a particular instant. In order to function properly, most programming languages require a significant amount of state information. This information is invisible to the programmer.

Another key concept – which is associated with any type of robot programming, is the concept of run-time. When a program is running, or executing, it is said to be in run-time. The term run-time is also used as a short form when referring to a run-time library, which is a library of code instructions used by a computer language to manage a program written in the language. The term is also used by software developers to specify when errors in a program can occur. A runtime error is an error that happens while the program is executing. For example, if a robot arm was programmed to turn left, and it turned right, then that would be a runtime error.

The software architecture of a system consists of the various software components used to design and operate the software. All programming methods rely on software architecture as a method of organizing a software system since it not only provides communication support but is also a critical component in hardware and software interfaces.

INDUSTRIAL ROBOT SOFTWARE

Software for industrial robots consists of data objects and lists of instructions, known as program flow (list of instructions). For example

Go to Jig1

is an instruction to the robot to go to positional data named Jig1. Of course programs can also contain implicit data for example

Tell axis 1 move 30 degrees.

Data and program usually reside in separate sections of the robot controller memory. One can change the data without changing the program and vice versa. For example one can write a different program using the same Jig1 or one can adjust the position of Jig1 without changing the programs that use it.

EXAMPLES OF PROGRAMMING LANGUAGES FOR INDUSTRIAL ROBOTS

Due to the highly proprietary nature of robot software, most manufacturers of robot hardware also provide their own software. While this is not unusual in other automated control systems, the lack of standardization of programming methods for robots does pose certain challenges. For example, there are over 30 different manufacturers of industrial robots, so there are also 30 different robot programming languages required. Fortunately, there are enough similarities between the different robots that it is possible to gain a broad-based understanding of robot programming without having to learn each manufacturer's proprietary language.

OTHER ROBOT PROGRAMMING LANGUAGES

Visual Programming Language

The software system for the Lego Mindstorms NXT robots is worthy of mention. It is a graphical user interface (GUI) written with LabVIEW. The approach is to start with the program rather than the data. The program is constructed by dragging icons into the program area and adding or inserting into the sequence. For each icon you then specify the parameters. For example for the motor drive icon you specify which motors and by how much they move. When the program is written it is downloaded into the Lego NXT 'brick' (microcontroller) for test.

Scripting Languages

A scripting language is a high-level programming language that is used to control the software application, and is interpreted in real-time, or "translated on the fly", instead of being compiled in advance. A scripting language may be a general-purpose programming language or it may be limited to specific functions used to augment the running of an application or system program. Some scripting languages, such as RoboLogix, have data objects residing in registers, and the program flow represents the list of instructions, or instruction set, that is used to program the robot.

Programming languages are generally designed for building data structures and algorithms from scratch, while scripting languages are intended more for connecting, or "gluing", components and instructions together. Consequently, the scripting language instruction set is usually a streamlined list of program commands that are used to simplify the programming process and provide rapid application development.

Parallel Languages

Another interesting approach is worthy of mention. All robotic applications need parallelism and event-based programming. Parallelism is where the robot does two or more things at the same time. This requires appropriate hardware and software. Most programming languages rely on threads or complex abstraction classes to handle parallelism and the complexity that comes with it, like concurrent access to shared resources. URBI provides a higher level of abstraction by integrating parallelism and events in the core of the language semantics.

ROBOT APPLICATION SOFTWARE

Regardless which language is used, the end result of robot software is to create robotic applications that help or entertain people. Applications include command-and-control and tasking software. Command-and-control software includes robot control GUIs for tele-operated robots, point-n-click command software for autonomous robots, and scheduling software for mobile robots in factories. Tasking software includes simple drag-n-drop interfaces for setting up delivery routes, security patrols and visitor tours; it also includes custom programs written to deploy specific applications. General purpose robot application software is deployed on widely distributed robotic platforms.

Safety Considerations

Programming errors represent a serious safety consideration, particularly in large industrial robots. The power and size of industrial robots mean they are capable of inflicting severe injury if programmed incorrectly or used in an unsafe manner. Due to the mass and high-speeds of industrial robots, it is always unsafe for a human to remain in the work area of the robot during automatic operation. The system can begin motion at unexpected times and a human will be unable to react quickly enough in many situations, even if prepared to do so. Thus, even if the software is free of programming errors, great care must be taken to make an industrial robot safe for human workers or human interaction, such as loading or unloading parts, clearing a part jam, or performing maintenance. The *ANSI/RIA R15.06-1999 American National Standard for Industrial Robots and Robot Systems - Safety Requirements* book from the Robotic Industries Association is the accepted standard on robot safety. This includes guidelines for both the design of industrial robots, and the implementation or integration and use of industrial robots on the factory floor. Numerous safety concepts such as safety controllers, maximum speed during a teach mode, and use of physical barriers are covered.

ROBOT PROGRAMMING METHODS

There are three basic methods for programming industrial robots but currently over 90% are programmed using the teach method.

Teach Method

The logic for the program can be generated either using a menu based system or simply using a text editor but the main characteristic of this method is the means by which the robot is taught the positional data. A teach pendant with controls to drive the robot in a number of different co-ordinate systems is used to manually drive the robot to the desired locations.

These locations are then stored with names that can be used within the robot program. The co-ordinate systems available on a standard jointed arm robot are :-

Joint Co-ordinates
The robot joints are driven independently in either direction.

Global Co-ordinates
The tool centre point of the robot can be driven along the X, Y or Z axes of the robots global axis system. Rotations of the tool around these axes can also be performed.

Tool Co-ordinates

Similar to the global co-ordinate system but the axes of this one are attached to the tool centre point of the robot and therefore move with it. This system is especially useful when the tool is near to the workpiece.

Workpiece Co-ordinates

With many robots it is possible to set up a co-ordinate system at any point within the working area. These can be especially useful where small adjustments to the program are required as it is easier to make them along a major axis of the co-ordinate system than along a general line. The effect of this is similar to moving the position and orientation of the global co-ordinate system.

This method of programming is very simple to use where simple movements are required. It does have the disadvantage that the robot can be out of production for a long time during reprogramming. While this is not a problem where robots do the same task for their entire life, this is becoming less common and some robotic welding systems are performing tasks only a few times before being reprogrammed.

Lead Through

This system of programming was initially popular but has now almost disappeared. It is still however used by many paint spraying robots. The robot is programmed by being physically moved through the task by an operator. This is exceedingly difficult where large robots are being used and sometimes a smaller version of the robot is used for this purpose. Any hesitations or inaccuracies that are introduced into the program cannot be edited out easily without reprogramming the whole task. The robot controller simply records the joint positions at a fixed time interval and then plays this back.

Off-line Programming

Similar to the way in which CAD systems are being used to generate NC programs for milling machines it is also possible to program robots from CAD data. The CAD models of the components are used along with models of the robots being used and the fixturing required. The program structure is built up in much the same way as for teach programming but intelligent tools are available which allow the CAD data to be used to generate sequences of location and process information. At present there are only a few companies using this technology as it is still in its infancy but its use is increasing each year. The benefits of this form of programming are :-

- Reduced down time for programming.
- Programming tools make programming easier.
- Enables concurrent engineering and reduces product lead time.
- Assists cell design and allows process optimisation.

PROGRAMMABLE DEVICES

Microcontrollers

These days using programmable components is easy. In the early days you had to write code in assembly, burn it in an EPROM, plug the EPROM in its socket and hope you didn't make any mistake. Because if you did, you had to hunt down the error in hundreds of lines of cryptic assembly code and had to use a new EPROM. These days you've got flash memory which can be reprogrammed in circuit. You've got C and Basic compilers to write your code. Most μcontrollers have a whole lot of hardware on chip (UART, Watchdog, Real Time Clock, RAM, ROM, EEPROM,...) and libraries are available for different programming languages which make coding LC-displays or Servos very easy. Not to mention you've got emulators (A special connector plugged into the μcontroller socket which allows a PC to pretend to be the μcontroller) and simulators (simulates the μcontroller on a PC and allows to run its code) to hunt down the bugs.

AVR

A good μcontroller to start with is the AVR ATtiny2313 or ATMEGA8535 (Atmel). It's cheap, has all the essentials on chip and has a whole lot of websites dedicated to it. The 8535 is more expensive, but has more memory and has an on board ADC. An AVR combined with Bascom-AVR is pretty much the easiest way to start with μcontrollers.

PC

The PC has many advantages over μcontrollers when it comes to controlling a robot. However it has 2 flaws that make it nearly impossible to use in small robot-projects :
- First : it's large and heavy.
- Secondly : A modern pc consumes enormous amounts of power.

This means that PCs are limited to tethered or large robots. If power isn't a problem the PC offers enormous amounts of RAM and HD space and plenty of CPU power. Plus a very handy user interface. Also network capabilities can be very useful for some applications. Providing your robot with a wireless connection (IR, radio, WiFi, bluetooth,...) and a similar interface on the PC reduces the wire-problem to a limited-distance (or in the case of IR a line-of-sight) problem.

If you intend to use a PC in your project, you might want to use one of the flavors of linux as the OS, as it allows easy access to any of the parts of the computer. Another choice for an OS is the older MS-DOS, although MS-DOS lacks multithreading (but those can be added in your software) and has ugly memory management (remember the 640Kb limit?).

Laptop

The laptop has only its size against it. A laptop provides the benefits of a PC without the enormous power consumption. The downside to laptops is their limited battery life.

PDA

The **PDA** (Personal Digital Assistant) is nothing more than a hand held **PC**. Most of the time these units used scaled down **RISC** (Reduced Instruction Set Chip) processors to allow for fast execution times and also allow the unit to remain smaller and lighter than your average laptop. Some PDA's even have built-in GPS which makes it very interesting for outdoor robot applications. The downside is their price. The PDA is useful for play, but also serves as a great business tool especially when they are equipped with a stylus.

Gameboy

The Gameboy (the classic, advance, color and DS) can be a very powerful device. These have been used in many projects . The only drawback is constructing the special PCB to act as a connector.

Gameboy Advance (GBA)

The Gameboy Advance is also excellent as a robotics platform. There are a few complete open-source development suites available (GCC based C/C++ plus ARM assembler). There is also a good kit available to interface Lego robotics motors/sensors and give much more control compared to the Lego RCX. With 4MB of FLASH and a few hundred K of RAM the kit has huge potential. The kit is not really for the complete beginner though as setting up the GNU tools can be quite complicated although the development environment is consequently very rich.

Programmable Logic Controllers

Programmable logic controllers, PLCs, are special purpose industrial computers, made to be easy to interface with electrical circuits. Most PLC are expandable but there are some compact "all-in-one" solutions as well. They have excellent bit-manipulating possibilities, and add-on-cards for several special signals as high-speed counters, analog I/O, networks and fieldbuses (RS-232, RS422, RS-485, Ethernet, DeviceNet, Profibus...), pulse output, servo control, *etc.* Increasingly, special inputs are included in the PLC CPU unit. PLCs are traditionally used to control automated factories, possibly several machines per PLC. They normally use special-purpose programming languages, most commonly ladder or SFC.

PLCs are available in sizes from approximately 5x5x5 cm. The smallest ones are very limited though, with only a few I/Os and very limited memory, and also very limited instruction set with regards to data processing and calculations. PLCs are especially suitable if you have a lot of experience of electric systems, but feel intimidated by designing and soldering PCBs and programming computers (*i.e.*, you are an electrician). They also shine in very large system with many special purpose sensors and outputs, as mostly there will be a standard solution from the manufacturer that is possible to use.

The downside with PLCs are their price - They can be quite costly, normally at least an order of magnitude higher than microcontrollers. They also draw more current than a microcontroller, however less than a computer. On the other hand, compared to a computer, they are computationally weak. They aren't suited for heavy signal-processing work.

PLC's are best suited for large robots, but could be usable in medium-sized projects as well.

Combinations

It's common to use both a computer/laptop/PDA and (a) microcontroller(s) as those 2 complement each others limitations. *e.g.* The first has the advantage of large amounts of memory and processing power, however lacks specific outputs that are very handy in robotics, like PWM. Microcontrollers commonly have PWM channels that operate independently of the rest of the microcontroller, but are limited in their processing speed and memory. Linking those 2 devices provides the best of both worlds.

Chapter 10

ROBOT PROGRAMMING LANGUAGES

WHICH LANGUAGE TO PICK?

The choice of language depends on a few points :
- previous experience. If you're already comfortable with a particular programming language, you'll probably want to use that language for programming your robot.
- How much time and effort you intend to invest. Not every language is equally hard/easy to use. Most of the complexity of a language comes from allowing more low-level access. Basically the more control a language gives to the user, the harder it is to use, but also the more powerful it is. In theory Assembly would allow a programmer to write the fastest and smallest code. However this is only true for an experienced programmer. Modern compilers can generate code that comes near to hand written Assembly. In most cases, using Assembly needlessly complicates a program, with the exception of using inline assembly in a C or Basic program. Usually these are a few lines of code which work directly with memory, or have to execute in a known time. For many if not most projects, a language such as Basic or C suffices. Only if you intend to push a microcontroller or PC to the limits of its capabilities is it worth the trouble of writing Assembly code.
- your goals. If you want a simple robot, you wouldn't need Assembly code at all. With modern microcontroller you can easily write sloppy Basic code and still have sufficient speed and memory. If you intend to build a cutting edge robot with a few dozen sensors, image recognition and speech recognition you'll need to write perfect code and will need Assembly for some of the components in order to have enough speed and be able to fit the code in the limited memory of the microcontroller(s). Most projects will fall in between these two extremes and good written Basic or C code

would be more than sufficient. Knowing (some) Assembly can be useful as it'll give some insight on how processors work. Such insights make you a better programmer.

- Availability. Not all languages are available for all microcontrollers. Urbi is open source, C and Basic and Forth are common, often as freeware, other languages might be available commercially, or not at all.

WRITING YOUR SOFTWARE

- Style. "Style" is how you indent your lines, how you pick your variable and function names. Pick one and stick with it throughout your program. Especially for functions and variables you should decide when to use capital letters, underscores and when to stick words together. Doing this the same everywhere makes it easier not to miss-spell names.

- Use informative names. Functions, procedures and variables should have informative names. Their purpose should be clear.

- Plan before coding. In software development there are two important steps before coding : requirement analysis and software architecture. The first is about finding out what the program is expected to do, what inputs to expect, which output it should generate and what the limitations of its environment are going to be. The latter involves working out how the program is going to be structured, which data types to use, which algorithms, how the input is going to enter the program and how to make sure it's valid and how to format the output. Using a lite version of these steps will make it much easier to write a decent program. If you intend to build a more complicated robot, you'll going to need to invest more time in planning out your software. Have all your software requirements written out first. Then pick a few of the most critical requirements and refine them. Implement those first. After you're done with those (this includes throughfully testing them), pick a new set and refine and implement those. By doing it in small steps you avoid having to write out everything at the start (and making many assumptions) and get to use the insights you get by coding parts of it.

- Use PDL (Program Development Language) PDL is a method to write functions. It involves writing down the individual steps of a function in plain English without referring to language-specific things, then refining these steps into smaller steps, until it's easier to write the code than to split the steps further. Afterward leave the PDL lines in as comments.

- Simulation and Debugging Learn to use the simulator. It's easier to find errors there than in the hardware. Also learn to use the debugger. It's your best friend for finding errors.

- If your software runs on a PC a log-file can be useful. But be picky with what you let the software log. If you intend to run it for a long time those log-files can become massive. If you want to log sensor data over time, know that these can be imported in Excel by saving the data with spaces between them to separate columns and newlines to separate rows.

ROBOT PROGRAMMING LANGUAGES

A robot will require a *programming language* for describing the operations that are to be done. Recently, there are plenty of robot programming languages available. Among them, *five* robot languages are commonly and basically used. They are :
- RAIL
- AML
- VAL
- AL
- RPL

RAIL :

RAIL will be a best language for controlling two major tasks such as the *manipulation* and *vision* system. It is a high – level robot language based on Pascal, and it will implement a Motorola–68000 central processor, teach pendant, and terminal. This language was designed by Automatix for arc welding and inspection purposes.

RAIL comes with three different kinds of systems, namely :
- Hitachi Process Robot – Arc Welding
- Cartesian Arm – Assembly functions
- Vision system without arm

AML :

AML (A Manufacturing Language) is a *high level language* based on sub routine, which is mainly implemented to manage RS / 1 Assembly Robot, End Effectors Active Force Feedback, and Cartesian Arm with hydraulic motors. The RS / 1 assembly robot incorporates a mini – computer , 192 KB memory, matrix printer, disk drive, display terminals, and keyboard. AML was developed by IBM Corporation for robot programming. An important reason for creating this language is to offer simple subsets and as well as powerful base language.

VAL :

VAL (Variable Assembly Language) Robot Programming Language is adopted mainly for *Unimation* Robots. As this language is designed with simple syntax, it is capable of illustrating the robot functions very easily. It includes two major tasks such as :

- Program instructions are used to provide VAL programs in order to manage the robot functions.
- Monitor commands are used to execute the user written programs.

AL :

AL robot language was developed in Artificial Intelligence Lab at Stanford University. It is the second generation language based on simultaneous Pascal. The programs are written and executed on PDP – 10. If the program is developed with high level code, then it should be written in Stanford Artificial Intelligence Language (SAIL). The AL system includes a big mainframe computer, and it generally runs on PDP 11/45. The PDP 11/45 implements one terminal, 128 KB RAM memory, and floating point processor. This language has got the capability to control *two Stanford Scheinman* and *two PUMA 600 arms* simultaneously.

RPL :

RPL robot language makes the improvement, checking, and correction of control algorithms very *easy*. It can be done even by an *unskilled programmer* like line foreman, production engineers, *etc*. The RPL programs are translated to interpretable code with the help of a compiler in SRI Robot Programming System. The programs are typically written in BLISS – 11 and run in RT – 11. The DECPDP – 10 is cross compiled into the LSI – 11 or PDP – 11. This robot language was designed in SRI International.

ACTION LANGUAGE

In computer science, an **action language** is a language for specifying state transition systems, and is commonly used to create formal models of the effects of actions on the world. Action languages are commonly used in the artificial intelligence and robotics domains, where they describe how actions affect the states of systems over time, and may be used for automated planning. The best known action language is PDDL.

Action languages fall into two classes : action description languages and action query languages. Examples of the former include STRIPS, PDDL, Language A (a generalization of STRIPS; the propositional part of Pednault's ADL), Language B (an extension of A adding *indirect effects*, distinguishing static and dynamic laws) and Language C (which adds indirect effects also, and does not assume that every fluent is automatically "inertial").

There are also the Action Query Languages P, Q and R. Several different algorithms exist for converting action languages, and in particular, action language C, to answer set programs. Since modern answer-set solvers make use of boolean SAT algorithms to very rapidly ascertain satisfiability, this implies that action languages can also enjoy the progress being made in the domain of boolean SAT solving.

Formal Definition

All action languages supplement the definition of a state transition system with a set F of *fluents*, a set V of values that fluents may take, and a function mapping $S \times F$ to V, where S is the set of states of a state transition system.

ACTION DESCRIPTION LANGUAGE

In artificial intelligence, **Action description language** (**ADL**) is an automated planning and scheduling system in particular for robots. It is considered an advancement of STRIPS. Pednault (a specialist in the field of Data abstraction and modelling who has been an IBM Research Staff Member in the Data Abstraction Research Group since 1996) proposed this language in 1987. It is an example of an action language.

Origins

Pednault observed that the expressive power of STRIPS was susceptible to being improved by allowing the effects of an operator to be conditional. This is the main idea of ADL-A, which is basically the propositional fragment of the ADL proposed by Pednault, with ADL-B an extension of -A. In the -B extension actions can be described with indirect effects by the introduction of a new kind of propositions : "static laws". A third variation of ADL is ADL-C which is similar to -B, in the sense that its propositions can be classified into static and dynamic laws, but with some more particularities.

The sense of a planning language is to represent certain conditions in the environment and, based on these, automatically generate a chain of actions which lead to a desired goal. A goal is a certain partially specified condition. Before an action can be executed its preconditions must be fulfilled; after the execution the action yields effects, by which the environment changes. The environment is described by means of certain predicates, which are either fulfilled or not.

Contrary to STRIPS, the principle of the open world applies with ADL : everything not occurring in the conditions is unknown (Instead of being assumed false). In addition, whereas in STRIPS only positive literals and conjunctions are permitted, ADL allows negative literals and disjunctions as well.

Syntax of ADL

An ADL schema consists of an action name, an optional parameter list and four optional groups of clauses labeled Precond, Add, Delete and Update.

The Precond group is a list of formulae that define the preconditions for the execution of an action. If the set is empty the value "TRUE" is inserted into the group and the preconditions are always evaluated as holding conditions.

The Add and Delete conditions are specified by the Add and Delete groups, respectively. Each group consists of a set of clauses of the forms shown in the left-hand column of the figure 1 :

1. The R represents a relation symbol
2. $\tau_1,...,\tau_n$ represents terms
3. ψ represents a formula
4. The sequence $z_1,...,z_k$ are variable symbols that appear in the terms $\tau_1,...,\tau_n$, but not in the parameter list of the action schema
5. $x_1,...,x_n$ are variable symbols that are different from the variables $z_1,...,z_n$ and do not appear in $\tau_1,...,\tau_n, \psi$, or the parameter list of the action schema

The Update groups are used to specify the update conditions to change the values of function symbols. An Update group consists of a set of clauses of the forms shown in the left column of the figure 2 :

Semantics of ADL

The formal semantic of ADL is defined by 4 constraints. The first constraint is that actions may not change the set of objects that exist in the world; this means that for every action α and every current-state/next-state pair (s, t) ∈ a, it must be the case that the domain of t should be equal to the domain of s.

The second constraint is that actions in ADL must be deterministic. If (s, t1) and (s, t2) are current-state/next-state pairs of action ∃, then it must be the case that t1 = t2.

The third constraint incorporated into ADL is that the functions introduced above must be representable as first-order formulas. For every n-ary relation symbol R, there must exist a formula $\Phi^a_R x_1,...,x_n)$ with free variables $x_2,...,x_n$ such that $f^a_R(s)$ is given by :

$$t(R) = f^a_R(s) = (d_1,..., d_n) \in Dom(s)^n \mid s[d_1/x_1,...,d_n/x_n \vDash \Phi^a_R(x_1,x_n)]$$

Consequently, $F(n_1,...,x_n) = y$ will be true after performing action |= if and only if $\Phi^a_R (x_1,...,x_n,y)$ was true beforehand. Note that this representability requirement relies on the first constraint (Domain of f should be equal to domain of s).

The fourth and final constraint incorporated into ADL is that set of states in which an action is executable must also be representable as a formula. For every action α that can be represented in ADL, there must exist a formula Π^a with the property that s |= Π_a if and only if there is some state t for which (s, t) ∈ α (*i.e.* action α is executable in state s)

Complexity of Planning

In terms of computational efficiency, ADL can be located between STRIPS and the Situation Calculus. Any ADL problem can be translated into a STRIPS instance – however, existing compilation techniques are worst-case exponential.

This worst case cannot be improved if we are willing to preserve the length of plans polynomially, and thus ADL is strictly more brief than STRIPS.

ADL planning is still a PSPACE-complete problem. Most of the algorithms polynomial space even if the preconditions and effects are complex formulae.

Most of the top-performing approaches to classical planning internally utilize a STRIPS like representation. In fact most of the planners (FF, LPG, Fast-Downward, SGPLAN5 and LAMA) first translate the ADL instance into one that is essentially a STRIPS one (without conditional or quantified effects or goals).

Comparison Between STRIPS and ADL

1. The STRIPS language only allows positive literals in the states, while ADL can support both positive and negative literals. For example, a valid sentence in STRIPS could be Rich ∧ Beautiful. The same sentence could be expressed in ADL as ¬Poor ∧ ¬Ugly
2. In STRIPS the unmentioned literals are false. This is called the Closed World Assumption. In ADL the unmentioned literals are unknown. This is known as the Open World Assumption.
3. In STRIPS we only can find ground literals in goals. For instance, Rich ∧ Beautiful. In ADL we can find quantified variables in goals. For example, ∃x At (P1, x) ∧ At(P2, x) is the goal of having P1 and P2 in the same place in the example of the blocks
4. In STRIPS the goals are conjunctions, *e.g.*, (Rich ∧ Beautiful). In ADL, goals may involve conjunctions and disjunctions (Rich ∧ (Beautiful ∨ Smart)).
5. In STRIPS the effects are conjunctions, but in ADL conditional effects are allowed : when P :E means E is an effect only if P is satisfied
6. The STRIPS language does not support equality. In ADL , the equality predicate (x = y) is built in.
7. STRIPS does not have support for types, while in ADL it is supported (for example, the variable p : Person).

The expressiveness of the STRIPS language is constrained by the types of transformations on sets of formulas that can be described in the language. Transformations on sets of formulas using STRIPS operators are accomplished by removing some formulas from the set to be transformed and adding new additional formulas. For a given STRIPS operator the formulas to be added and deleted are fixed for all sets of formulas to be transformed. Consequently, STRIPS operators cannot adequately model actions whose effects depend on the situations in which they are performed. Consider a rocket which is going to be fired for a certain amount of time. The trajectory may vary not only because of the burn duration but also because of the velocity, mass and orientation of the rocket. It cannot be modelled by means of a STRIPS operator because the formulas that would have to be added and deleted would depend on the set of formulas to be transformed.

Although an efficient reasoning is possible when the STRIPS language is being used it is generally recognized that the expressiveness of STRIPS is not suitable for modeling actions in many real world applications. This inadequacy motivated the development of the ADL language. ADL expressiveness and complexity lies between the STRIPS language and the situation calculus. Its expressive power is sufficient to allow the rocket example described above to be represented yet, at the same time, it is restrictive enough to allow efficient reasoning algorithms to be developed.

As an example in a more complex version of the blocks world : It could be that block A is twice as big as blocks B and C, so the action xMoveOnto(B,A) might only have the effect of negating Clear(A) if On(A,C) is already true, or creating the conditional effect depending on the size of the blocks. This kind of conditional effects would be hard to express in STRIPS notation without the conditional effects.

ENCHANTING

Enchanting is a free and open-source cross-platform educational programming language designed to program Lego Mindstorms NXT robots. It is primarily developed by Southern Alberta Robotics Enthusiasts group in the province of Alberta, Canada, and runs on Mac OS X, Windows, and experimentally on Linux devices. Since 2013, the Enchanting version 0.2 has been available.

Technology

Its predecessor, the 1998 Robotics Invention System was developed by Scratch developer team led by Mitch Resnick at MIT Media Lab. Based on BYOB, which is developed by the University of California, Berkley, the current version of Enchanting runs on Windows XP, Windows Vista, Windows 7 and Windows 8 (but not Windows 8 RT); on Mac OS X it runs on version 10.4 and newer; and on Linux it runs on Ubuntu version 10.10.

Educational Resources, Use and Events

It has been used in secondary-to-tertiary computer science program at Monash University in Australia, where an interactive PDF book for use on computer or iPad, titled *Robotics with Enchanting and LEGO® NXT* is available for free download. Most recent *SABRE Games*, organized in 2013 by Southern Alberta Robotics Enthusiasts group, consisted of three disciplines : *Tug Of War*, where two robots are tied together with a string and each tries to pull its opponent over the center line; *Sumo*, where two robots are placed in a sumo ring and each tries to find and push its opponent out without going out of the ring itself; and *Parade*, where robots follow a line trying not to crash into the robot in front.

EUSLISP ROBOT PROGRAMMING LANGUAGE

EusLisp is a Lisp-based programming system. Built on the basis of object orientation, it is designed specifically for developing robotics software. The first version of it ran in 1986 on Unix-System5/Ustation-E20.

LEJOS

leJOS is a firmware replacement for Lego Mindstorms programmable bricks. It currently supports the LEGO RCX brick and leJOS NXJ supports the NXT brick. It includes a Java virtual machine, which allows Lego Mindstorms robots to be programmed in the Java programming language. It is often used for teaching Java to first-year computer science students. The leJOS-based robot *Jitter* flew around on the International Space Station in December 2001.

Pronunciation

According to the official website :

In English, the word is similar to Legos, except there is a J for Java, so the correct pronunciation would be Ley-J-oss. If you are brave and want to pronounce the name in Spanish, there is a word "lejos" which means far, and it is pronounced Lay-hoss.

The name leJOS was conceived by José Solórzano, based on the acronym for Java Operating System (JOS), the name of another operating system for the RCX, legOS, and the Spanish word "lejos."

History

leJOS was originally conceived as TinyVM and developed by José Solórzano in late 1999. It started out as a hobby open source project, which he later forked into what is known today as leJOS. Many contributors joined the project and provided important enhancements. Among them, Brian Bagnall, Jürgen Stuber and Paul Andrews, who later took over the project as José essentially retired from it.

As of August 20, 2006, the original leJOS for the RCX has been discontinued with the 3.0 release. Soon afterwards, iCommand, a library to control the NXT from a Bluetooth-enabled computer via LCP, was released. This library made use of the standard Lego firmware. This library was later superseded by leJOS NXJ 0.8. In January 2007, a full port to the new Lego Mindstorms NXT was released as a firmware replacement. This is far faster (x15 or so) than the RCX version, has more memory available, a menu system, Bluetooth support using the Bluecove library, and allows access to many other NXT features.

In 2008, versions 0.5, 0.6 and 0.7 were released. In addition to numerous improvements to the core classes, the Eclipse plugin was released along with a new version of the tutorial. In 2009, there were 2 more major releases : 0.8 and 0.85. In May 2011 0.9 was released. Broadly speaking, the releases have concentrated on improvements to navigation algorithms, as well as support for numerous 3rd party sensors and the Eclipse plug-in.

In 2013, development began on a port to the Lego Mindstorms EV3 brick. In 2014, the 0.5 and 0.6 alpha versions were released.

Architecture

leJOS NXJ provides support for access to the robot's I^2C ports. This allows access to the standard sensors and motors (ultrasonic distance sensor, touch sensor, sound sensor and light sensor). Other companies, such as MindSensors and HiTechnic have extended this basic set by providing advanced sensors, actuators and multiplexers. leJOS NXJ includes Java APIs for these products.

By taking advantage of the object-oriented structure of Java, the developers of LeJOS NXJ have been able to hide the implementation details of sensors and actuators behind multiple interfaces. This allows the robotics developer to work with high-level abstractions without having to worry about details like the hexadecimal addresses of hardware components. The project includes implementations of the commonly used feedback controller, the PID controller and the Kalman filter noise reduction algorithm. leJOS NXJ also provides libraries that support more abstract functions such as navigation, mapping and behavior based robotics.

Community

Since the first alpha release of leJOS NXJ in 2007, the project has had a consistently active following.

1. Between January 2007 and October 2011 there were over 225,000 downloads
2. In 2011 the downloads averaged between 4000 and 6000 a month
3. In 2011 over 500 topics were discussed in the forums. Each topic often generated several hundred posts.
4. Between May 2012 and March 2013 there were over 36,000 download of release 0.91

The core development team has been a relatively small group. Contributions are accepted from other members of the community. Several of the interfaces to third party sensors and actuators have been contributed by members outside the core team. The platform has been used in university robotics courses, undergraduate research projects and as a platform for robotics research.

NXJ and the Java Platform

As leJOS NXJ is a Java project, it builds on the wealth of functionality inherent in the Java platform. There are leJOS NXJ plugins for the two leading Java IDEs : Eclipse and Netbeans. Robotics developers can take advantage of the standard functionality of an IDE (code completion, refactoring and testing frameworks) as well as point-and-click implementation of NXJ functions : compiling, linking and uploading. A wealth of java open source projects (such as Apache Math) are likewise available to the NXJ robotics developer.

ROBOTML

RobotML ("Robot Markup Language") is an experimental XML-based markup language that is used for communication between autonomous mobile robots and robot components. **RobotML** was developed at the Munich University of Applied Sciences, the first public release was announced for spring 2005.

NEXT BYTE CODES

Next Byte Codes (NBC) is a simple language with an assembly language syntax that can be used to program Lego Mindstorms NXT programmable bricks. The command line compiler outputs NXT compatible machine code, and is supported on Windows, Mac OS and Linux. It is maintained by John Hansen, a Mindstorms Developer Program member.

The NBC compiler is released under the Mozilla Public License. The integrated development environment (IDE) is Bricx Command Center.

The NBC debugger was developed by SorosyDotCom and can be downloaded as freeware.

URBISCRIPT

Urbiscript is a programming language for robotics. It features syntactic support for concurrency and event-based programming. It is a prototype-based object-oriented scripting language. It is dynamic : name resolution is performed during the program execution (late binding); slots (member variables) can be added/removed at runtime, and even prototypes (superclasses) of an object can be changed at runtime.

Memory management is performed by reference counting.

Tightly bound to the Urbi platform it supports seamless integration of C++/Java components.

Syntax and Semantics

Inspiration

From the syntactical point of view, urbiscript belongs to the C-family of programming languages.

Its prototype-based object-oriented design was influenced by the Self and the Io programming languages.

It is designed to program, but also interact with robots; as such, it is influenced by Unix shells and other languages that provide a read-eval-print loop style interactive toplevel. However, contrary to others, there is no prompt for user input but answers from the system are prefixed by a timestamp (in milliseconds) between square brackets :

```
1 + 1; sleep(1s); 1 + 2 * 3;
[00005420] 2
[00006420] 7
```

Sequential Statements and Control Flow

Urbiscript statements include :

- The if statement, which conditionally executes a block of code, along with else.
- The traditional for statement, as in C which iterates over an iterable object, capturing each element to a local variable for use by the attached block.
- Another for statement, which iterates over an iterable object, capturing each element to a local variable for use by the attached block.
- The while statement, which executes a block of code as long as its condition is true.
- The try statement, which allows exceptions thrown in its attached code block to be caught and handled by catch clauses. An optional else clause is run if no exception was thrown. Clean-up code can be guaranteed to be run in every case when given in a finally-clause.
- The assert statement, used during debugging to check for conditions that ought to apply. urbiscript also feature assert blocks, which can be used to factor several assert statements.

Actually, contrary to most C-like languages and despite what the syntax suggests, statements "have a value", and therefore are expressions, provided they are embedded in braces :

```
var status = { if (closed) "closed" else "open" };
var pass = { try { foo } catch { false } else { true } };
```

Concurrent Statements and Control Flow

In urbiscript, some control-flow constructs come in several "flavors" : two types of sequential composition, and two types of concurrent composition. Under the hood, concurrency is implemented using coroutines.

Statement Composition

Like in C, the semicolon denotes sequential composition : a;b stands for "run statement a then run statement b. Other tasks may be run between a and b. Another statement separator, pipe, denotes "tight sequential composition" : no other task can be run between a and b in a | b.

Similarly urbiscript features two means to compose statements concurrently. With a,b, first a is run, and at some point b will be --- possibly while a is still running. This is very similar to the & operator in Unix shells. Alternatively, with a&b, both a and b are started together; in interactive sessions, this means that a won't be run until b is fully entered and properly followed by either a ; or a ,.

Scopes are boundaries for backgrounded jobs, as demonstrated in the following example :

```
{
{ sleep(2s); echo(2) },
{ sleep(1s); echo(1) },
};
echo(3);
[00012451] *** 1
[00013447] *** 2
[00013447] *** 3
```

Concurrent Flavors of Sequential Constructs

Most looping constructs in urbiscript come in several "flavors", which are based on the four statement separators : ;, |, ,, and &.

For instance

```
// This is actually "for;".
for (var i : [0, 1, 2])
{
echo(i);
echo(i ** 2);
};
```
displays
```
[00002919] *** 0
[00002921] *** 0
[00002921] *** 1
[00002922] *** 1
[00002922] *** 2
[00002922] *** 4
```

i.e., the loop bodies are not executed sequentially, while the for& keyword runs the loop bodies concurrently :

```
for& (var i : [0, 1, 2])
```

```
{
  echo(i);
  echo(i ** 2);
};
[00021680] *** 0
[00021680] *** 1
[00021680] *** 2
[00021682] *** 0
[00021682] *** 1
[00021682] *** 4
```

Event-based Programming

Aiming at the development of portable robotic applications, urbiscript relies on specific syntactic constructs to specify reactive behaviors such as "go to the charging dock when the battery is low", "play a friendly sound when a known face is recognized", or "stop when an obstacle is detected".

Explicit Event Handling

Event handling goes into three steps. First, define an event

```
var e = Event.new;
```

Second, specify event handlers

```
at (e?)
  echo("received event e");
```

Third, "emit" this event

```
e!;
[00014333] *** received event e
```

Events can have payloads, and event handlers enjoy pattern matching on the payload :

```
at (e?(1, var x) if x % 2 == 0)
  echo("received event e(1, %s)" % x);
e!(1, 1);
[00014336] *** received event e
e!(1, 2);
[00014336] *** received event e
[00014336] *** received event e(1, 2)
```

Implicit Events

The urbiscript language also allows to monitor expressions :

```
at (batteryLevel <= 0.2)
  robot.goToChargingDock;
```

The following example demonstrates the feature :

```
var x = 0;
[00002165] 0
var y = 0;
[00002166] 0
var z = 0;
[00002167] 0
at (x + y == z)
  echo("%s + %s == %s" % [x, y, z]);
[00002168] *** 0 + 0 == 0
x = 1;
[00002169] 1
z = 1;
[00002170] 1
[00002170] *** 1 + 0 == 1
```

VARIABLE ASSEMBLY LANGUAGE

Variable Assembly Language (VAL) is a computer-based control system and language designed specifically for use with Unimation Inc. industrial robots.

The VAL robot language is permanently stored as a part of the VAL system. This includes the programming language used to direct the system for individual applications. The VAL language has an easy to understand syntax. It uses a clear, concise, and generally self-explanatory instruction set. All commands and communications with the robot consist of easy to understand word and number sequences. Control programs are written on the same computer that controls the robot. As a real-time system, VAL's continuous trajectory computation permits complex motions to be executed quickly, with efficient use of system memory and reduction in overall system complexity. The VAL system continuously generates robot control commands, and can simultaneously interact with a human operator, permitting on-line program generation and modification.

A convenient feature or VAL is the ability to use libraries or manipulation routines. Thus, complex operations may be easily and quickly programmed by combining predefined subtasks.

The VAL language consists of monitor commands and program instructions. The monitor commands are used to prepare the system for execution of user-written programs. Program instructions provide the repertoire necessary to create VAL programs for controlling robot actions.

Terminology

The following terms are frequently used in VAL related operations.

Monitor

The VAL monitor is an administrative computer program that oversees operation of a system. It accepts user input and initiates the appropriate response; follows instructions from user-written programs to direct the robot; and performs the computations necessary to control the robot.

Editor

The VAL editor is an aid for entering information into a computer system, and modifying existing text. It is used to enter and modify robot control programs. It has a list of instructions telling a computer how to do something. VAL programs are written by system users to describe tasks the robot is to perform.

Location

Location is a position of an object in space, and the orientation of the object. Locations are used to define the positions and orientations the robot tool is to assume during program execution.

VAL Programming

Several conventions apply to numerical values to be supplied to VAL commands and instructions. Preceding each monitor-command description are two symbols indicating when the command can be typed by the user. A dot (.) signifies the command can be performed when VAL is in its top-level monitor mode and no user program being executed (that is, when the system prompt is a dot). An asterisk (*) indicates the command can be performed at the same time VAL is executing the program (that is, when the system prompt is an asterisk). If both symbols are present the command can be executed in either case. Most monitor commands and program instructions can be abbreviated. When entering any monitor command or program instruction, the function name can be abbreviated to as many characters as are necessary to make the name unique.

For commands and instructions, angle brackets, < >, are used to enclose an item which describes the actual argument to appear. Thus the programmer can supply the appropriate item in that position when entering the command or instruction. Note that these brackets used here are for clarification, and are never to be included as part of a command or instruction.

Many VAL commands and instructions have optional arguments. For notations, optional arguments are enclosed in square brackets, [ ]. If there is a comma following such an argument, the comma must be retained if the argument is omitted, unless nothing follows. For example, the monitor BASE command has the form :

BASE [<dx>] , [<dy>] , [<dz>] , [<rotation>]

To specify only a 300-millimeter change in the Z direction, the command could be entered in any of the following ways :

- BASE 0,0,300,0
- BASE ,,300,
- BASE ,,300

Note that the commas preceding the number 300 must be present to correctly to relate the number with a Z-direction change. Like angle brackets, square brackets are never entered as part of a command or instruction.

Several types of numerical arguments can appear in commands and instructions. For each type there are restrictions on the values that are accepted by VAL. The following rules should be observed :

1. Distances are entered to define locations to which the robot is to move. The unit of measure for distances is millimeter, although units are never explicitly entered for any value. Values entered for distances can be positive or negative, with their magnitudes limited by a number representative of the maximum reach of the robot (for example, 1024 mm and 700 mm for the PUMA 500 and PUMA 250 robots, respectively). Within the resultant range, distance values can be specified in increments of 0.01 mm. Note, however, that some values cannot be represented internally, and are stored as the nearest representable value.

2. Angles in degrees are entered to define and modify orientations the robot is to assume at named locations, and to describe angular positions of robot joints. Angle values can be positive or negative, with their magnitudes limited by 1800 or 3600 depending on the usage. Within the range, angle values can be specified in increments of 0.01°. Values cannot be represented internally, however they are stored as nearest representable value.

The VAL System

The function of VAL is to regulate and control a robot system by following user commands or instructions. In addition to being a compact stand-alone system, VAL has been designed to be highly interactive to minimize programing time, and to provide as many programming aids as possible.

External Communication

The standard VAL system uses an operator's console terminal and manual control box to input commands and data from the user. The operator console serves as the primary communication device and can be either a direct play terminal or a printing terminal. Interaction with other devices in an automated cell is typically handled by monitoring input channels and switching outputs. By this means the robot can control a modest cell without the need for other programmable devices.

VAL Operating System

The controller has two levels or operation :
- the top level is called the VAL operating system, or monitor, because it administers operations of the system, including interaction with the user;
- the second level is used for diagnostic work on the controller hardware. The system monitor is a computer program stored VAL programmable read-only memory (PROM) in the Computer/Controller.

PROM memory retains its contents finitely, and thus VAL is immediately available when the controller is switched on. The monitor is responsible for control of the robot, and its commands come from the manual control unit, the system terminal, or from programs. To increase its versatility and flexibility, the VAL monitor can perform of its commands even while a user program is being executed. Commands that can be processed in this way include those for controlling the status the system, defining robot locations, storing and retrieving information the floppy disk, and creating and editing robot control programs.

Chapter 11

ROBOT CONTROLLERS

PROGRAMMABLE CONTROLLERS

The programmable controllers can be used in the *robot workcell* when there is a need for a *higher-level controller*. The electromechanical relays are replaced by the introduction of programmable controllers (PC) on the year 1960.

A programmable controller produces the *output signals* according to the logical and other operations carried on the input signals. It is a small-sized *digital operating system* with the programmable memory, which can be highly consistent, and flexible. Most importantly, the human workers with the knowledge of logic diagrams of relay control panels can easily learn its function.

A programmable controller is programmed to find out the *series of operations* and *production of I/O signals*. The program is indicated by the similar logic diagrams known as *ladder diagrams*. The typical functions obtained with the help of a programmable controller are the following :

- Controller relay functions
- Timing functions
- Counting functions
- Arithmetic functions
- Analog control functions

The above functions help a programmable controller to accomplish a higher-level controller for a robot workcell. There are various programmable controllers with *several hundreds of I/O ports*, which is more than a robot controller. As a result, it gets the capacity to control many *complicated activities* on the workplace.

A programmable controller is used as a complete workcell control system in the robots that perform different welding operations with an automobile body *spot welding line*. There are several features in the programmable controllers such as *Safety monitoring, Operator interface,* and *Maintenance & diagnostic* functions, which brings high capabilities to it than the robot controllers.

ROBOT CONTROLLER

Robot controller includes different types of *control technology* such as computer, electronic, and limited sequence controllers. Generally, there is only *limited I/O capability* on complicated types for interfacing with additional equipments. The purpose of I/O interface is to connect the *interlocks* in the workplace. The incoming signals are joined with the work cycle program by the robot controller. As a result, the output signals and robot movements can be accomplished properly. For example, the playback robots may have *10 to 20* numbers of I/O ports.

The *arrangement* of robot controller's I/O ports is described briefly below :

Input Ports

In a robot controller, there may be *10 – 25 input lines*. It is usually used for connecting the incoming signals from exterior equipments to the controller. These signals will be in *binary*, and referenced as *logical conditions* on the robot program. An important use of this connection is to provide the *interlocking* capability. At present, the input ports on the new controllers are incorporated with the ability to read in analog signals.

Input Port

Apart from the above, there may be *five input lines*, which are implemented for *safety interlocking*. The robot controller will *stop the program* and as well as the robot when the signals of outside safety sensors are received in any one of these lines. Sometimes, it may be employed for just *turning OFF* the manipulator.

Output Ports

As like input ports, there may be *10 – 25 output lines* in the controller. The main purpose of these ports is to send the *output interlock signals* to the exterior equipments. The beginning or finishing of signals will be based on the logical conditions in the robot program, which end in *several responses* with the exterior equipments. Recently, the new robot controllers have the ability for analog and binary signal outputs.

The above details describe a robot controller with limited I/O capability for a workplace. As like this type, the recently available controllers are also limited for *sequence control*. Moreover, there will be *advanced robot controllers* in the future for controlling the *superior robots*.

DIFFERENT LEVELS OF ROBOT CONTROLLER

A robot controller is used to decrease the errors of control signal to zero or somewhere close to zero. It can be classified into *six* different types namely :

- ON – OFF control

Robot Controllers

- Proportional control
- Integral control
- Proportional – plus – Integral control (P – I)
- Proportional – plus – Derivative control (P – D)
- Proportional – plus – Integral – plus – Derivative control (P – I – D)

According to the application, anyone of these controllers can be used. The functions of each controller are described briefly below.

ON – OFF Control

The element in the ON – OFF controller offers *two control methods* such as :

- Complete OFF
- Complete ON

$$m(t) = M_1, \text{ if } e(t) > 0$$
$$m(t) = M_2, \text{ if } e(t) < 0$$

Where,

- m(t) denotes the control signal created by the controller.
- e(t) denotes the error on the controller.
- In most of the cases either M1 or M2 will be 0.

The purpose of an ON – OFF control is to protect the controller from swinging with *very high frequency*. This is made possible by moving the error through several ranges before the operation starts. Here, the range is considered as the *differential gap*.

Proportional Control

A control signal produced by this controller is *proportional* to the error. It is basically used as an amplifier by means of a gain (K_p). This is represented as :

$$m(t) = K_p\, e(t)$$

The transfer function will be :

$$M(s)\, /\, E(s) = K_p$$

The proportional controller will be best suited for providing *smooth control action*.

Integral Control

A control signal produced by the integral controller is *altered* at a rate proportional to the error (*i.e.*) the control signal maximizes quickly if the error is big, and the control signal maximizes slowly if the error is small. This can be represented as :

$$m(t) = K_i \int e(t)\, dt$$

Here, the K_i denotes the *integrator gain*.

The transfer function will be :

$$M(s) / E(s) = K_i / s$$

Here, the $1/s$ is used for integration.

Proportional – Plus – Integral Control (PI)

The PI controller is used to *overcome* two major issues. They are :
- The integral control is capable of offering zero errors, but it is set back with its slow response.
- The proportional control provides error while counteracting a load on the system.

This can be represented as K_p.

$$m(t) = K_p e(t) + K_p / T_i \int e(t) \, dt$$

Here,

K_p is used to adjust the proportional and integrator gain.

T_i is used to adjust only the integrator gain.

The transfer function will be :

$$M(s) / E(s) = K_p (1 + 1 / T_i s)$$

Proportional – Plus – Derivative Control (PD)

The control signal produced by the PD controller is proportional to the rate of change of the error. This method is used rarely because of its incapability to provide output without the change of error. An advantage is that it can give changes with *faster responses*. This can be represented as

$$m(t) = K_p e(t) + K_p T_d \, de(t) / dt$$

The transfer function will be :

$$M(s) / E(s) = K_p (1 + T_d s)$$

Proportional – Plus – Integral – Plus – Derivate Control (PID)

The PID controller integrates *three control actions,* and it is the *most frequently* used controller. It is because of its fast response, and low steady – state error. This controller can be represented as

$$m(t) = K_p e(t) + K_p / T_i \int e(t) \, dt + K_p T_d \, de(t) / dt$$

The transfer function will be :

$$M(s) / E(s) = K_p (1 + 1 / T_i s + T_d s)$$

INDUSTRIAL ROBOT CONTROL SYSTEM

A robot must have a *control system* to operate its drive system, which is used to move the arm, wrist, and body of a robot at various paths. When different industrial robots are compared with their control system, they can be divided into four major types. They are :
- Limited Sequence Robots
- Playback Robots with Point – Point Control
- Playback Robots with Continuous Path Control
- Intelligent Robots

Limited Sequence Robots

The limited sequence robots are incorporated with the *mechanical stops* and *limit switches* for determining the finishing points of its joints. These robots do not require any sort of programming, and just uses the *manipulator* to perform the operation. As a result, every joint can only travel to the intense limits. It is considered as the smallest level of controlling, and it will be best for simple operations like pick & place process. This type of robots is generally equipped with the *pneumatic* drive system.

Playback Robots

The playback robots are capable of performing a task by *teaching* the position. These positions are stored in the memory, and done frequently by the robot. Generally, these playback robots are employed with a *complicated* control system. It can be divided into two important types, namely :
- Point to Point control robots
- Continuous Path control robots

Playback Robots with Point to Point Control

The point to point robots are shortly called as *PTP*. It has got the capability to travel from one position to another. The desired paths are taught and stored in the control unit memory. These robots do not move from the desired location for controlling its path. It can be moved in a *small distance* only with the help of programming. This type of robots can be used for spot welding, loading & unloading, and drilling operations.

Playback Robots with Continuous Path Control

The continuous path control is also known as *CP* control. This type of robots can control the path, and can end on any specified position. These robots commonly move in the *straight line*. The initial and final point is first described by the programmer, and the control unit defines the individual joints. This helps the robot to travel in a straight line. Likewise, it can also move in a *curved path* by moving its arm at the desired points. In these robots, the microprocessor is used as a controller. Some of the applications are arc welding, spray painting, and gluing operations.

Intelligent Robots

The intelligent robots can play back the defined motion, and can also work according to their environment. It uses *digital computer* as a controller. The sensor is incorporated in these robots for receiving the information during the process. The programming language will be based on *high level language*. This kind of robots is capable of communicating with the programmers in the work volume. It will be best for arc welding, and assembly purposes.

Chapter 12

SENSORS AND ACTUATORS

USES OF SENSORS IN ROBOTICS

The *sensors* are one of the useful technologies, which play a *vital role* in the robotics field. There are *four important categories* where uses of sensors are highly required in robotics such as :

- Safety monitoring
- Interlocking in work cell control
- Quality control in work part inspection
- Data collection of objects in the robot work cell

Safety Monitoring

The sensors are extremely used in industrial robotics for *monitoring* the hazardous and safety conditions in the robot cell layout. This certainly helps in avoiding the *physical injuries* and other damages caused to the human workers.

Interlocking in Work Cell Control

In robot work cell, the *series of activities* of different equipments are controlled by using *interlocks*. Here, sensors are employed for *verifying* the conclusion of the current work cycle before progressing to the next cycle.

Quality Control in Work Part Inspection

In olden days, the quality control was performed with a *manual inspection system*. Nowadays, sensors are employed in the inspection process for determining the quality features of a work part *automatically*. A major advantage of using sensors in this category provides *high accurate results*. One disadvantage in this automatic inspection is that the sensors are only able to examine a *limited* variety of work part features and faults.

Data Collection of Objects in the Robot Work Cell

Sensors are used in this category to determine the *position or other related data* about the fixtures, work parts, equipment, human workers, and so on. Apart from sensing the position, it is also implemented to find out the other information like work part's color, orientation, size, shape, *etc*. The *key reasons* for determining the above information while executing a robot program includes :

- Recognition of work parts
- Random position and orientation of work parts
- Improved accuracy of robot position using the feedback data

The sensors incorporated in the above four categories will require a *component* in the control system for accomplishing the *specified control tasks*. This component is a part of a larger control system called as a *work cell control system*.

ELECTRIC MOTORS – AC SERVOMOTORS AND STEPPER MOTORS

Apart from DC servomotors, other two commonly used electric-type actuator in the robots are *AC servomotors* and *Stepper motors*.

AC Servomotors

The AC servomotors are incorporated with a *large resistance* and *small-sized rotor*, which is mostly used for obtaining a *perfect* and *quick response*.

The AC servomotors can be categorized into two different types such as :

- Synchronous
- Asynchronous

An AC servomotor is a *reversible induction* and *two – phase motor*, which is transformed for performing the servo process. When there is a possibility of rapid reversals, stops, and starts, this motor use to deliver a *low inertia*.

Most importantly, the performance of an AC servomotor can be increased similar to a DC motor with the help of an *appropriate electronics package*.

Advantages

- When compared with DC motors, the manufacturing cost for AC servomotor is low.
- It is capable of providing high – power output.
- It does not include any sort of brushes.

Stepper Motors

The stepper motors are classified under the brushless DC servomotor category, and it is also known as *stepping motors*. This type of actuator is currently employed in the *computer I/O devices*.

The stepper motors are generally incorporated in the robots, which require a *light – duty function*. Moreover, it is mostly applied in the *open – loop systems* when compared with the closed – loop systems.

A sequence of *discrete electrical pulses* is used to power a stepper motor. The motor shaft is rotated to single step for each electrical impulse. As a result, it produces the *discrete angular motion increments* as the output.

Advantages

- It can support both digital and analog feedback signals.
- It provides high dynamic torque with low pulse rates.

Disadvantages

- Less efficient
- During high load inertia, it might require damping for avoiding oscillation.

ELECTRIC MOTORS – DC SERVOMOTORS

Nowadays, the *electric motors* are playing a major role as actuators in the robots. It is said so because they deliver *high controllability* with *less maintenance*. One of the most commonly used electric motors in the robots is *DC servomotors*.

DC Servomotors

DC servomotors can be classified into *two types* such as :
- Brushed DC servomotors
- Brushless DC servomotors

Brushed DC Servomotors

A brushed DC servomotor consists of two major components such as *stator* and *rotor*. The stator is incorporated with *brush assemblies* and a *permanent magnet*, while the rotor has *commutator assembly* and an *armature*.

At first, the current is passed through the armature windings in which a magnetic field is set opposite to the field placed by the magnets. This process leads to the *generation of torque* in the rotor.

The armature gets current from the commutator and brush assemblies when the rotor starts rotating. As a result, the field will stay opposite to the field placed by the magnets. Similar operation provides a *constant torque* all over the rotation.

Advantages

- Less expensive than brushless DC servomotor.

- Simple variable resistors like rheostat or potentiometer will be sufficient for regulating the motor.

Disadvantages

- It is not highly efficient than brushless DC servomotor.

Brushless DC Servomotors

A brushless DC servomotor includes a rotor with *permanent magnet* and an *electromagnetic* stator. Moreover, it has *electronically controlled commutator* instead of brushes. It is assembled similar to an *inside – out* DC servomotor.

A *DC current* is used to power the system. The poles in the permanent magnet are attracted to the opposing magnetic rotational poles, which is available in the stator. This operation produces a *torque*.

This type of electric motor also includes *linearly* connected *torque, rpm, voltage,* and *current*.

Advantages

- Electromagnetic interference is decreased.
- Improved reliability
- Highly efficient
- Long lasting life-time
- Less noise

Disadvantages

- Complex systems are required for controlling the speed.
- Very costly

ROBOTIC ACTUATORS – HYDRAULIC AND PNEUMATIC

The actuators are used in the robots for providing the *power* to the *robot joints*. It can be powered by anyone of the following sources :

- Hydraulic – pressurized fluid
- Pneumatic – compressed air
- Electric – electricity

Hydraulic and Pneumatic Actuators

Both hydraulic and pneumatic actuators use moving fluids for powering a device. Here, the fluid represents *pressurized oil* for hydraulic and *compressed air* for pneumatic actuators.

SENSORS AND ACTUATORS

When these actuators are compared with each other, two things can be observed such as :
- Both actuators are similar in terms of their operational functions.
- These actuators differ only in their capability to pressure the fluid.

The hydraulic actuator is capable of providing the pressure at $1000 - 3000\ lb/in^2$, while the pneumatic actuator can only deliver around $100\ lb/in^2$.

Let us consider a *cylinder*, which is one of the simplest devices for powering the fluids. In this device, a moving piston is available for actuating the *linear joint*.

This type of device is known as *single – ended cylinder*, because it has only one end in which the piston rod comes out. There are also some types of cylinders like :
- Rodless cylinder
- Double – ended cylinder

A set of relationships describe the *particular interest* of the actuators :
- The actuator velocity relating to the input power.
- The actuator force relating to the input power.

These relationships for a cylinder-type actuator can be given by :

$$V(t) = f(t) / A$$
$$F(t) = P(t) A$$

Here,

P (t) represents the *fluid pressure*.

V (t) represents the *piston velocity*.

F (t) represents the *force*.

A represents the *piston area*.

f (t) represents the *flow rate of the fluid*.

The relations illustrated to select a suitable robot can be according to the *requirements of a robot* that is necessary for carrying a load at a specified speed.

POSITION SENSORS

Position sensors are used in robotics for sensing and controlling the *arm position*. These types of devices are available in plenty and the *three most widely* used devices are :
- Encoders
- Potentiometers
- Resolvers

Encoders

Encoders are used for converting the *angular or linear displacement* into the *digital signals*. Some types of encoders are :

- Linear encoders : It is used to calculate the directions and positions that are in the linear form. Example : Linear motors.
- Rotary encoders : It helps to calculate the directions and positions that are in the angular form. Examples : Motor.
- Incremental encoders : It is used in robots for sensing the position from its last position.
- Absolute encoders : It brings a definite position that is proportional to a fixed reference position.

The incremental encoders have a *glass disk,* which is noticeable with discontinuous stripes. The disk includes a *phototransmitter* on one surface and a *photoreceiver* on the other surface. As soon as the disk starts rotating, the light beams are finished alternately and broken down. The photoreceiver gets the output as a *pulse train,* and its frequency is proportional to the rotational speed of the disk.

The construction of absolute encoders is almost same as the incremental encoders. The difference is that it adds *many stripes* and as well as photo transmitters and receivers. It is mostly used to determine the *absolute position* of a part. The stripes are set to give a binary number that is relative to the shaft angle. The resolution of this type of encoder can be given by :

$$\text{Resolution} = 2^n$$

The above equation describes n as the *number of the tracks* on the glass disk.

Potentiometers

Potentiometers produce an output voltage that is proportional to the position of the wiper. They are the *analog devices* used for calculating the *linear or rotary movements* based on the design. It consists of a rotating wiper, which is in contact with a resistive element. This wiper is connected with an object that is motion. An AC or DC voltage is supplied to the resistive element. The wiper and ground voltage is proportional to the ratio of the wiper's one side resistance to the resistive element's total resistance.

Potentiometers are basically a *voltage divider system.* The wiper separates the voltage of the resistive element into two parts. By measuring the voltage obtained, it is possible to find out the *position of the wiper.*

Resolvers

A resolver is also an analog device as like potentiometers, and it is a *rotary electrical transformer* basically implemented for calculating the *degrees of rotation.* It requires only AC signal for excitation because the use of DC current will not produce any output signal. The output signal of a resolver is proportional to the angle of rotating element with respect to the fixed element.

The resolvers have a stator with a pair of windings and a rotor with one winding. The *sinusoidal electric current* excites the rotor winding, and hence the electromagnetic induction makes the current to pass through the stator windings. By this process, it generates a cosine and sine feedback current with the stator windings, which are fixed at 90°. The magnitudes are measured to find out the *angle of the rotor* that is proportional to the stator windings.

VELOCITY SENSORS

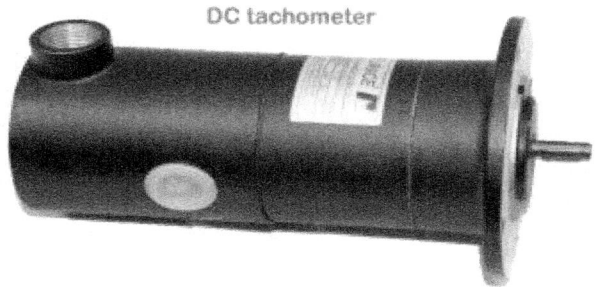

A most important device that is used to provide velocity feedback is the *tachometer*. It is also known as *rpm gauge,* and *revolution counter*. A tachometer is employed in a motor to calculate the rotational speed of a shaft. The output is displayed as RPM (revolution per minute) in an analog device. Nowadays, the use of digital displays has also been increased to a larger extent.

The two common types of a tachometer are :
- Permanent magnet AC tachometer
- DC tachometer

Permanent Magnet AC Tachometer

The construction of an AC tachometer is almost same as the brushless DC motor and stepper motor. It possesses primary and secondary stators with fixed windings, and a rotor with *permanent magnet*.
- If the rotor is stationary, a constant output voltage will be obtained.
- If the rotor is moving, proportional to the speed of a rotor is induced.

This type of tachometer cannot provide information of direction with only one output winding. However, it can be done by having additional winding in the quadrature.

DC Tachometer

DC tachometer is the *most commonly* used instrument in the robotics. It is a *DC generator* implemented to provide an output voltage that is proportional to the angular velocity of the armature. In this mechanism, the rotor and rotational part will be attached directly. It has a stationary device called as *commutator*, which is connected with the split slip rings. It is used for picking the induced output signal from the rotating coil.

Sometimes, the use of commutator will lead to the production of *small ripples* in the output voltage. Most importantly, it is impossible to filter out as well. In this condition, the *AC tachometer* will be the best one because it does not have such problems.

COMPONENTS USED FOR ROBOT ACTUATION AND FEEDBACK

In a robot, some of the devices are used as the *components* of a robot control system for providing *feedback* and *control actions* in order to control the *manipulator*. Such devices are :

Feedback Devices

- Position Sensors
- Velocity Sensors

Actuation Devices

- Actuators
- Power Transmission Systems

Position and Velocity Sensors

Position sensors are the feedback devices used to find out the *accurate movement of joints* for accomplishing the required direction and position of the end effector. A robot requires velocity sensors for *controlling the speed* of the manipulator movement. Position sensors use several devices such as resolvers, potentiometers, and encoders for sensing the position, while the velocity sensors include dc tachometer for providing appropriate speed.

Every single robot requires feedback devices for delivering accurate speed control. Even the recently developed complicated control system also hardly relies on these devices for gaining *better performance* of the robot manipulator in some tasks like deceleration and acceleration.

Actuators and Power Transmission Systems

A controller uses the actuators and power transmission systems for achieving the *control actions*. As a result, these devices help in generating *power* for moving the robot arm.

The actuators are used to provide power to the *robot joints* with the help of the power transmission systems. In some cases, the power transmission systems will not be required if the actuators are attached with the robot joints itself. Some of the devices that are used as the actuators are *pneumatic, hydraulic,* and *electric*.

The power transmission systems make use of *gears, screws,* and *pulleys* for performing this mechanism.

SENSORS USED IN ROBOTICS

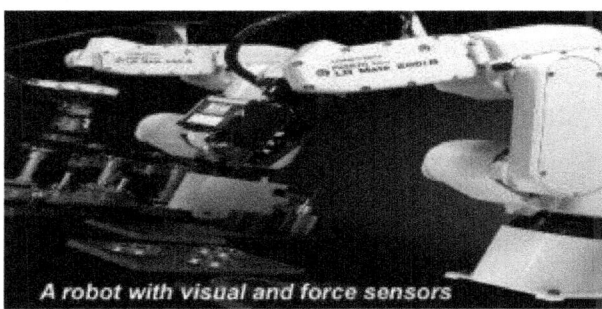

A robot with visual and force sensors

The use of *sensors* in robots has taken them into the next level of creativity. Most importantly, the sensors have increased the performance of robots to a large extent. It also allows the robots to perform several functions like a human being. The robots are even made *intelligent* with the help of Visual Sensors (generally called as machine vision or computer vision), which helps them to respond according to the situation. The Machine Vision system is classified into six sub-divisions such as Pre-processing, Sensing, Recognition, Description, Interpretation, and Segmentation.

Different Types of Sensors

There are plenty of sensors used in the robots, and some of the important types are listed below :

- Proximity Sensor,
- Range Sensor, and
- Tactile Sensor.

Proximity Sensor

This type of sensor is capable of pointing out the availability of a component. Generally, the *proximity sensor* will be placed in the robot moving part such as end effector. This sensor will be turned ON at a specified distance, which will be measured by means of feet or millimeters. It is also used to find the presence of a human being in the work volume so that the accidents can be reduced.

Range Sensor

Range Sensor is implemented in the end effector of a robot to calculate the distance between the sensor and a work part. The values for the distance can be given by the workers on visual data. It can evaluate the size of images and analysis of common objects. The range is measured using the Sonar receivers & transmitters or two TV cameras.

Tactile Sensors

A sensing device that specifies the contact between an object, and sensor is considered as the *Tactile Sensor*. This sensor can be sorted into two key types namely :

- Touch Sensor, and
- Force Sensor.

Touch Sensor

The *touch sensor* has got the ability to sense and detect the touching of a sensor and object. Some of the commonly used simple devices as touch sensors are micro – switches, limit switches, *etc*. If the end effector gets some contact with any solid part, then this sensor will be handy one to stop the movement of the robot. In addition, it can be used as an inspection device, which has a probe to measure the size of a component.

Force Sensors

The *force sensor* is included for calculating the forces of several functions like the machine loading & unloading, material handling, and so on that are performed by a robot. This sensor will also be a better one in the assembly process for checking the problems. There are several techniques used in this sensor like Joint Sensing, Robot – Wrist Force Sensing, and Tactile Array Sensing.

Chapter 13

GRIPPERS

CONSIDERATIONS IN ROBOT GRIPPER SELECTION AND DESIGN

The industrial robots use *grippers* as an end effector for picking up the raw and finished work parts. A robot can perform good grasping of objects only when it obtains a proper gripper selection and design. Therefore, *Joseph F. Engelberger*, who is referred as *Father of Robotics* has described several factors that are required to be considered in gripper selection and design.

- The gripper must have the ability to reach the surface of a work part.
- The change in work part size must be accounted for providing accurate positioning.
- During machining operations, there will be a change in the work part size. As a result, the gripper must be designed to hold a work part even when the size is varied.
- The gripper must not create any sort of distort and scratch in the fragile work parts.
- The gripper must hold the larger area of a work part if it has various dimensions, which will certainly increase stability and control in positioning.
- The gripper can be designed with resilient pads to provide more grasping contacts in the work part. The replaceable fingers can also be employed for holding different work part sizes by its interchangeability facility.

Moreover, it is difficult to find out the *magnitude of gripping force* that a gripper must apply to pick up a work part. The *following significant factors* must be considered to determine the necessary gripping force.

- Consideration must be taken to the weight of a work part.
- It must be capable of grasping the work parts constantly at its centre of mass.

- The speed of robot arm movement and the connection between the direction of movement and gripper position on the work part should be considered.
- It must determine either friction or physical constriction helps to grip the work part.
- It must consider the co-efficient of friction between the gripper and work part.

MISCELLANEOUS ROBOT GRIPPER TYPES

The most commonly used three types of robot grippers in industrial applications are mechanical, vacuum, and magnetic grippers. Apart from these devices, there are also some other gripper types available such as *adhesive, scoops, hooks,* and *inflatable bladder*.

Adhesive Grippers

In this gripper, the *adhesive* body is used to grasp the *fabric objects*. It is also very much suitable for carrying the *light-weight materials*. As like magnetic and vacuum grippers, it uses only *one surface* for gripping the objects.

One disadvantage in this gripper design is that it will lose its *strong* adhesive substances when it is used *continuously*. As a result, the consistency of the gripper is reduced and leads to the improper grasping of objects.

This disadvantage can be solved by constantly applying the adhesive substances on the gripper. It can be done by using a continuous ribbon connected to a feeding system, which is placed in the robot wrist. This process looks very much related to the operation of a typewriter ribbon system.

Scoops

Scoop is used as an end effector for grasping the material in the form of *powder or liquid*. It handles some objects like *molten metal, food, chemicals, and granular*. There are two problems in this gripper such as :
- The control of the material is very difficult.
- It may also spill out the materials, while performing the process.

Hooks

A hook gripper is incorporated in an operation for picking up the *containers of parts*. A hook that grips an object will require a handle for permitting the gripper to grasp it. Moreover, this type of grippers can be used in the *part loading and unloading* process in which the work parts hang from the overhead conveyors.

Inflatable Bladder Type Gripper

The inflatable bladder type gripper is made of *rubber* or some sort of *elastic materials*. It can be used to pick up the *fragile objects* by expanding the inflatable bladder. The mechanical gripper applies concentrated power for grasping a material, while this type gripper applies a *constant handling pressure* in opposition to the object surface

MECHANICAL GRIPPER MECHANISMS

Robot mechanical grippers and its actuating mechanisms can be classified into several methods. The first method is based on *the type of finger movement*. During this arrangement, the opening and closing of the fingers can be actuated by either *pivoting*, or *linear or translational movement*.

Pivoting Movement

In this motion, the rotation of fingers is concerned with the *fixed pivot points* of the gripper for providing open and close actions. It uses the *linkage mechanism* for achieving this movement.

Linear or Translational Movement

During linear motion, the *guide rails* are used to move the fingers parallel to each other for accomplishing closing and opening. In translational movement, the fingers are maintained in a *parallel orientation* to each other. As like pivoting movement, it also uses the linkage mechanism for actuation.

The second method of classifying the mechanical grippers is based on *the type of kinematic device* used for the actuation of finger motions. It can be accomplished by anyone of these types : *linkage, screw, gear and rack, rope and pulley, or cam actuation*.

Linkage Actuation

The design of linkage actuation helps in finding out the conversion of gripper's input force into the gripping force, the time taken to actuate the gripper, and the maximum capability to open the finger. It has *plenty of designs* for opening and closing the finger, and some of its types are shown below.

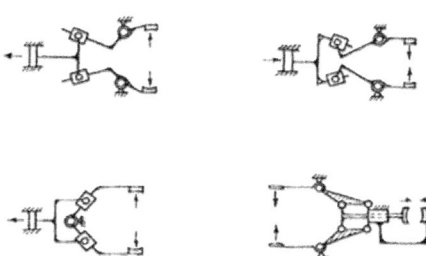

Screw Actuation

The screw-type actuated gripper consists of a *screw* connected with a *threaded block*. To rotate the screw, a motor is used along with a speed reduction device. If the screw is turned in one direction, the threaded block is moved in one direction. Similarly, the threaded block moves in the opposite direction if the screw is turned on the other direction. As the threaded block is attached with gripper, it makes the fingers to open and close.

Gear and Rack Actuation

For this actuation, the gear and rack are connected with a *piston*, which provides *linear-type movement*. The two partial pinion gears are driven when the rack is moved. As it is linked with gripper, the opening and closing of fingers are accomplished.

Rope and Pulley Actuation

In this actuation, a *tension device* is required to go up against the rope movement in the pulley. Suppose, if the pulley is activated in one direction for opening the gripper, the tension device will provide *slack* in the rope. Similarly, the gripper is closed by activating the pulley on the other direction.

Cam Actuation

As like linkage actuated gripper, it also has a wide range of designs for opening and closing the gripper fingers. One of its types is described shortly here. A cam actuated gripper with *spring loaded follower* can be used to provide open and close actions of fingers. The spring is incorporated for forcing the gripper to close if the cam is moved in one direction, while the movement of cam on the other direction causes the gripper to open. This type can be useful for holding *various sizes of work parts*.

Additionally, some mechanisms come under miscellaneous types. An important one is *diaphragm or expandable bladder*, which is inflated or deflated to open and close the fingers.

VACUUM GRIPPERS

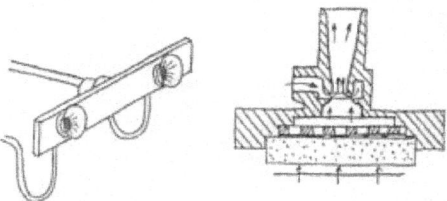

Suction cups vacuum gripper   Venturi type vacuum gripper

Vacuum grippers are used in the robots for grasping the *non – ferrous objects*. It uses *vacuum cups* as the gripping device, which is also commonly known as *suction cups*. This type of grippers will provide good handling if the objects are *smooth, flat, and clean*. It has only one surface for gripping the objects. Most importantly, it is not best suitable for handling the objects with holes.

Vacuum Cups

Generally, the vacuum cups (suction cups) will be in the round shape. These cups will be developed by means of *rubber* or other elastic materials. Sometimes, it is also made of *soft plastics*. Moreover, the vacuum cups are prepared of hard materials for handling the soft material objects.

Two different devices are used in the suction cups for creating the vacuum. They are :

- Venturi
- Vacuum pump

Venturi device is operated with the help of *shop air pressure*, while the vacuum pump is driven either by means of *vane or piston* device. The vacuum pump has the ability to create the *high vacuum*. As the venturi is a simple device, it is more *reliable* and *inexpensive*. Both these devices are very well capable of providing high vacuum if there is a sufficient supply of air pressure.

Types of Vacuum Grippers

- The ball joint type vacuum gripper is capable of changing into various contact angles automatically. Moreover, the bending moments in the vacuum cups are also decreased. It is used for carrying irregular materials, heavy objects, *etc*.
- A vacuum gripper with level compensator can be very helpful in balancing the objects with different levels. It also has the capability to absorb the shocks.

Applications of Vacuum Grippers

- Vacuum grippers are highly useful in the heavy industries, automobiles, compact disc manufacturing, and more for material handling purposes.
- It is also used in the tray & box manufacturing, labeling, sealing, bottling, and so on for packaging purposes.

ROBOT MAGNETIC GRIPPERS

Magnetic grippers are most commonly used in a robot as an end effector for grasping the *ferrous* materials. It is another type of handling the work parts other than the mechanical grippers and vacuum grippers.

Robot with magnetic gripper

Types of Magnetic Grippers

The magnetic grippers can be classified into *two common types*, namely :
Magnetic grippers with
- Electromagnets
- Permanent magnets

Electromagnets

Electromagnetic grippers include a *controller unit* and a *DC power* for handling the materials. This type of grippers is easy to control, and very effective in releasing the part at the end of the operation than the permanent magnets. If the work part gripped is to be released, the polarity level is minimized by the controller unit before the electromagnet is turned off. This process will certainly help in *removing the magnetism* on the work parts. As a result, a best way of releasing the materials is possible in this gripper.

Permanent Magnets

The permanent magnets do not require any sort of external power as like the electromagnets for handling the materials. After this gripper grasps a work part, an additional device called as *stripper push – off pin* will be required to separate the work part from the magnet. This device is incorporated at the sides of the gripper.

The advantage of this permanent magnet gripper is that it can be used in hazardous applications like *explosion-proof apparatus* because of no electrical circuit. Moreover, there is no possibility of *spark production* as well.

Benefits

- This gripper only requires one surface to grasp the materials.
- The grasping of materials is done very quickly.
- It does not require separate designs for handling different size of materials.
- It is capable of grasping materials with holes, which is unfeasible in the vacuum grippers.

Drawbacks

- The gripped work part has the chance of slipping out when it is moving quickly.
- Sometimes oil in the surface can reduce the strength of the gripper.
- The machining chips may stick to the gripper during unloading.

MECHANICAL GRIPPERS

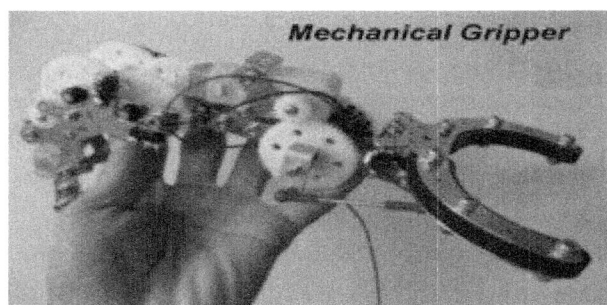

A mechanical gripper is used as an *end effector* in a robot for grasping the objects with its *mechanically* operated fingers. In industries, two fingers are enough for holding purposes. More than three fingers can also be used based on the application. As most of the fingers are of *replaceable* type, it can be easily removed and replaced.

A robot requires either hydraulic, electric, or pneumatic drive system to create the input power. The power produced is sent to the gripper for making the fingers react. It also allows the fingers to perform open and close actions. Most importantly, a *sufficient force* must be given to hold the object.

In a mechanical gripper, the holding of an object can be done by *two different methods* such as:

- Using the finger pads as like the shape of the work part.
- Using soft material finger pads.

In the first method, the contact surfaces of the fingers are designed according to the work part for achieving the *estimated shape*. It will help the fingers to hold the work part for some extent.

In the second method, the fingers must be capable of supplying sufficient force to hold the work part. To avoid scratches on the work part, *soft type pads* are fabricated on the fingers. As a result, the contact surface of the finger and co – efficient of friction are improved. This method is very simple and as well as *less expensive*. It may cause slippage if the force applied against the work part is in the parallel direction. The slippage can be avoided by designing the gripper based on the force exerted.

$$\mu \, n_f F_g = w \qquad \qquad \ldots 1$$

μ => co-efficient of friction between the work part and fingers

n_f => no. of fingers contacting

F_g => Force of the gripper

w => weight of the grasped object

The equation 1 must be *changed* if the weight of a work part is more than the force applied to cause the slippage.

$$\mu\, n_f F_g = w\, g \qquad ...2$$

g => g factor

During rapid grasping operation, the work part will get *twice* the weight. To get rid out of it, the modified equation 1 is put forward by *Engelberger*. The g factor in the equation 2 is used to calculate the acceleration and gravity.

The *values of g factor* for several operations are given below :

- g = 1 - acceleration supplied in the opposite direction.
- g = 2 - acceleration supplied in the horizontal direction.
- g = 3 - acceleration and gravity supplied in the same direction.

END EFFECTOR - ROBOT'S HAND

Robot Grippers

An *End Effector* is considered as the "*hand of a robot.*" It is one of the important devices in a robot. It integrates an arm and a wrist, which helps it to perform several functions like material handling, pick and place, machine loading and unloading, *etc*. It is also known as *EOA* , Robotic Peripherals, and more.

Based on the performance of work tasks, the robot end effector can be described into two important categories such as :

- Grippers, and
- Tools.

Grippers

The *grippers* are mainly used in a robot for grasping / holding an object. The grasped objects will be moved to the preferred place with the help of a robot. The grippers are capable of carrying work parts, bottles, tools, and so on. It can work by means of magnetic, mechanical, vacuum cups, *etc.*

The grippers can also be sorted as :
- Single gripper,
- Double grippers, and
- Multiple grippers.

If a robot wrist contains one grasping device, then it is described as *Single Gripper*. Similarly, a robot wrist with two grasping devices will be termed as *Double Grippers*. It will be very much suitable for machine loading and unloading operations. Moreover, it is better than single gripper, because it finishes plenty of works in a quick time. The *Multiple Grippers* incorporates more than two grasping devices in a robot wrist. The double grippers seem to be the best one than the multiple grippers. The reason is that the multiple grasping mechanisms are pricing very high. The grippers can be further classified into internal gripper and external gripper.

Other types of grippers are :
- Magnetic Grippers,
- Vacuum Cups, and
- Adhesive Grippers.

Tools

The *tools* are implemented in a robot wrist for performing several operations on a part instead of carrying it. Some of the applications of this end effector type are Continuous Arc Welding, Spot Welding, and Spray Painting. The tools can be equipped in a gripper as well. As a result, the change of a tool can be done in a quick succession of time, especially where it requires a lot of tools to be altered.

TOOLS AS END EFFECTORS

A *tool* is equipped in the robot for carrying out *several operations* on the work parts instead of grasping it. A tool acts as an *end effector* when it is attached directly to the *robot's wrist*. In some applications, there will be a need for multi-tool task, and changing the tool all the time from the robot wrist will be *highly difficult*. As a result, a *gripper* is used in this process to grasp and manipulate the tool. It certainly helps the robot to handle several tools in an operation, and thus makes the *multi-tooling function* possible. Moreover, the time taken to change the tool is *very low*.

For example : In a deburring operation, various sizes of deburring tool will be required to hold for reaching every surface of the work part. Here, the tool is equipped in the gripper for quick exchange from one tool to another.

Examples for Tools Used as end Effectors

In robot applications, the most commonly used *three tools as end effectors* are listed below :

- Spot welding tools
- Spray painting nozzle
- Arc welding torch

The *rotating spindle* for routing, drilling, grinding, and wire brushing operations also comes under this category. Some other examples of tools used as end effectors are liquid cement applicators, water jet cutting tool, and heating torches.

In all the above examples, the actuation of a tool must be *coordinated* by an industrial robot. An example of this case includes the spot welding operation in which the robot must control the actuation as a part of its process. The control of this process is very similar to the opening and closing of a *mechanical gripper*.

Chapter 14

SAFETY PRINCIPLES FOR INDUSTRIAL ROBOTS

Industrial robots are found throughout industry wherever high productivity demands must be met. The use of robots, however, requires design, application and implementation of the appropriate safety controls in order to avoid creating hazards to production personnel, programmers, maintenance specialists and system engineers.

WHY ARE INDUSTRIAL ROBOTS DANGEROUS?

One definition of robots is "moving automatic machines that are freely programmable and are able to operate with little or no human interface". These types of machines are currently used in a wide variety of applications throughout industry and medicine, including training. Industrial robots are being increasingly used for key functions, such as new manufacturing strategies (CIM, JIT, lean production and so on) in complex installations. Their number and breadth of applications and the complexity of the equipment and installations result in hazards such as the following :

- movements and sequences of movements that are almost impossible to follow, as the robot's high-speed movements within its radius of action often overlap with those of other machines and equipment
- release of energy caused by flying parts or beams of energy such as those emitted by lasers or by water jets
- free programmability in terms of direction and speed
- susceptibility to influence by external errors (*e.g.*, electromagnetic compatibility).
- human factors.

Investigations in Japan indicate that more than 50% of working accidents with robots can be attributed to faults in the electronic circuits of the control system. In the same investigations, "human error" was responsible for less than 20%. The logical conclusion of this finding is that hazards which are caused by system faults cannot be avoided by behavioural measures taken by human beings. Designers and operators therefore need to provide and implement technical safety measures.

Special operating control system for the setting up of a mobile welding robot.

Accidents and Operating Modes

Fatal accidents involving industrial robots began to occur in the early 1980s. Statistics and investigations indicate that the majority of incidents and accidents do not take place in normal operation (automatic fulfilment of the assignment concerned). When working with industrial robot machines and installations, there is an emphasis on special operation modes such as commissioning, setting up, programming, test runs, checks, troubleshooting or maintenance. In these operating modes, persons are usually in a danger zone. The safety concept must protect personnel from negative events in these types of situations.

International Safety Requirements

The 1989 EEC Machinery Directive establishes the principal safety and health requirements for machines. A machine is considered to be the sum total of interlinked parts or devices, of which at least one part or device can move and correspondingly has a function. Where industrial robots are concerned, it must be noted that the entire system, not just one single piece of equipment on the machine, must meet the safety requirements and be fitted with the appropriate safety devices. Hazard analysis and risk assessment are suitable methods of determining whether these requirements have been satisfied.

Safety Principles for Industrial Robots

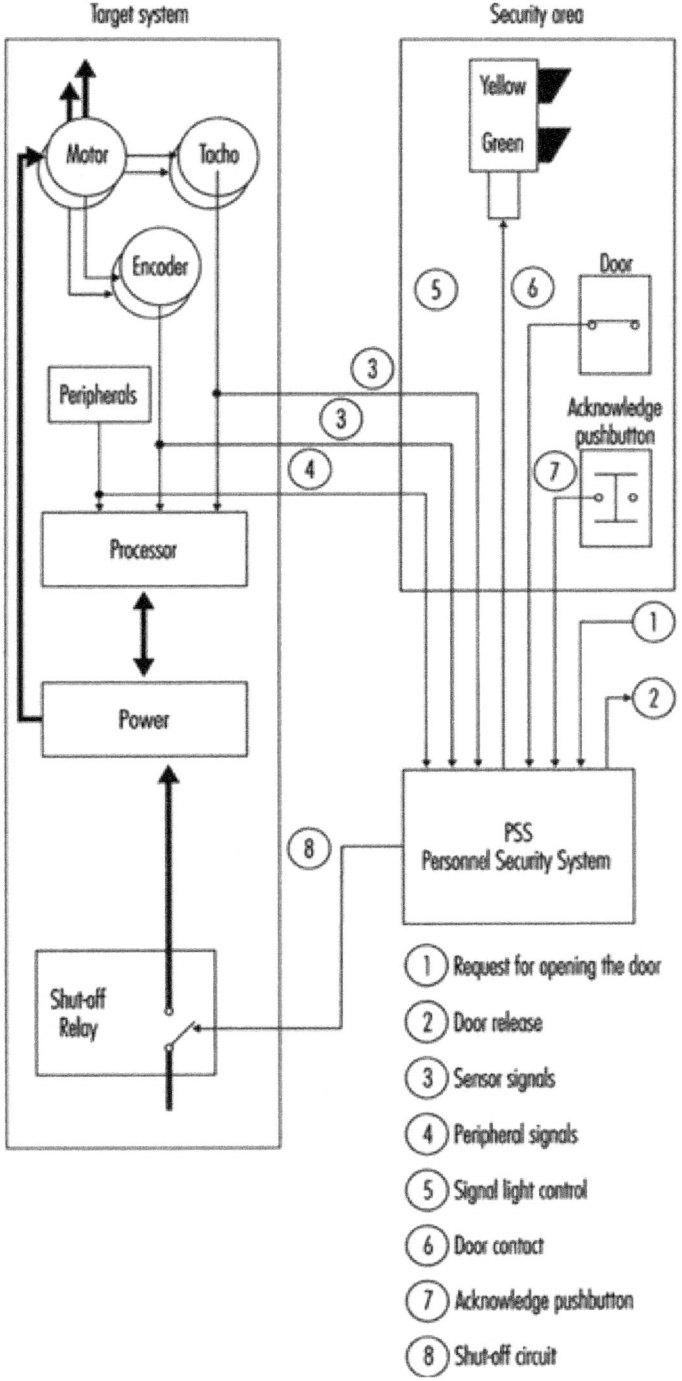

Block diagram for a personnel security system

Requirements and Safety Measures in Normal Operation

The use of robot technology places maximum demands on hazard analysis, risk assessment and safety concepts. For this reason, the following examples and suggestions can serve only as guidelines :

1. Given the safety goal that manual or physical access to hazardous areas involving automatic movements must be prevented, suggested solutions include the following :
   - Prevent manual or physical access into danger zones by means of mechanical barriers.
   - Use safety devices of the sort which respond when approached (light barriers, safety mats), and take care to switch off machinery safely when accessed or entered.
   - Permit manual or physical access only when the entire system is in a safe state. For example, this can be achieved by the use of interlocking devices with closure mechanisms on the access doors.
2. Given the safety goal that no person may be injured as a result of the release of energy (flying parts or beams of energy), suggested solutions include :
   - Design should prevent any release of energy (*e.g.*, correspondingly dimensioned connections, passive gripper interlocking devices for gripper change mechanisms, *etc.*).
   - Prevent the release of energy from the danger zone, for example, by a correspondingly dimensioned safety hood.
3. The interfaces between normal operation and special operation (*e.g.*, door interlocking devices, light barriers, safety mats) are necessary to enable the safety control system to automatically recognize the presence of personnel.

Demands and Safety Measures in Special Operation Modes

Certain special operation modes (*e.g.*, setting up, programming) on an industrial robot require movements which must be assessed directly at the site of operation. The relevant safety goal is that no movements may endanger the persons involved. The movements should be

- only of the scheduled style and speed
- prolonged only as long as instructed
- those which may be performed only if it can be guaranteed that no parts of the human body are in the danger zone.

A suggested solution to this goal could involve the use of special operating control systems which permit only controllable and manageable movements using acknowledgeable controls. The speed of movements is thus safely reduced (energy reduction by the connection of an isolation transformer or the use of fail-safe state monitoring equipment) and the safe condition is acknowledged before the control is allowed to activate.

Six-axis industrial robot in a safety cage with material gates.

Demands on Safety Control Systems

One of the features of a safety control system must be that the required safety function is guaranteed to work whenever any faults arise. Industrial robot machines should be almost instantaneously directed from a hazardous state to a safe state. Safety control measures needed to achieve this include the following safety goals :

- A fault in the safety control system may not trigger off a hazardous state.
- A fault in the safety control system must be identified (immediately or at intervals).

Suggested solutions to providing reliable safety control systems would be :

- redundant and diverse layout of electro-mechanical control systems including test circuits
- redundant and diverse set-up of microprocessor control systems developed by different teams. This modern approach is considered to be state-of-the-art; for example, those complete with safety light barriers.

SAFETY GOALS FOR THE CONSTRUCTION AND USE OF INDUSTRIAL ROBOTS

When industrial robots are built and used, both manufacturers as well as users are required to install state-of-the-art safety controls. Apart from the aspect of legal responsibility, there may also be a moral obligation to ensure that robot technology is also a safe technology.

Normal Operation Mode

The following safety conditions should be provided when robot machines are operating in the normal mode :

- The field of movement of the robot and the processing areas used by peripheral equipment must be secured in such a way as to prevent manual or physical access by persons to areas which are hazardous as a result of automatic movements.
- Protection should be provided so that flying workpieces or tools are not allowed to cause damage.
- No persons must be injured by parts, tools or workpieces ejected by the robot or by the release of energy, due to faulty gripper(s), gripper power failure, inadmissible speed, collision(s) or faulty workpiece(s).
- No persons may be injured by the release of energy or by parts ejected by peripheral equipment.
- Feed and removal apertures must be designed to prevent manual or physical access to areas which are hazardous as a result of automatic movements. This condition must also be fulfilled when production material is removed. If production material is fed to the robot automatically, no hazardous areas may be created by feed and removal apertures and the moving production material.

Special Operation Modes

The following safety conditions should be provided when robot machines are operating in special modes :

The following must be prevented during rectification of a breakdown in the production process :

- manual or physical access to areas which are hazardous due to automatic movements by the robot or by peripheral equipment
- hazards which arise from faulty behaviour on the part of the system or from inadmissible command input if persons or parts of the body are in the area exposed to hazardous movements
- hazardous movements or conditions initiated by the movement or removal of production material or waste products
- injuries caused by peripheral equipment
- movements that have to be carried out with the safety guard(s) for normal operation removed, to be carried out only within the operational scope and speed, and only as long as instructed. Additionally, no person(s) or parts of the body may be present in the area at risk.

The following safe conditions should be assured during set up :

No hazardous movements may be initiated as a result of a faulty command or incorrect command input.

Safety Principles for Industrial Robots

- The replacement of robot machine or peripheral parts must not initiate any hazardous movements or conditions.
- If movements have to be carried out with the safety guard(s) for normal operation removed when conducting setting-up operations, such movements may be carried out only within the directed scope and speed and only as long as instructed. Additionally, no person(s) or parts of the body may be present in the area at risk.
- During setting-up operations, the peripheral equipment must not make any hazardous movements or initiate any hazardous conditions.

During programming, the following safety conditions are applicable :

- Manual or physical access to areas which are hazardous due to automatic movements must be prevented.
- If movements are carried out with the safety guard(s) for normal operation removed, the following conditions must be fulfilled :
- (a) Only the command to move may be carried out, and only for as long as it is issued.
- (b) Only controllable movements may be carried out (*i.e.*, they must be clearly visible, low-speed movements).
- (c) Movements may be initiated only if they do not constitute a hazard to the programmer or other persons.
- Peripheral equipment must not represent a hazard to the programmer or other persons.

Safe test operations require the following precautions :

Prevent manual or physical access to areas which are hazardous due to automatic movements.

- Peripheral equipment must not be a source of danger.

When inspecting robot machines, safe procedures include the following :

- If it is necessary to enter the robot's field of movement for inspection purposes, this is permissible only if the system is in a safe state.
- Hazards caused by faulty behaviour on the part of the system or by inadmissible command input must be prevented.
- Peripheral equipment must not be a source of danger to inspection personnel.

Troubleshooting often requires starting the robot machine while it is in a potentially hazardous condition, and special safe work procedures such as the following should be implemented :

- Access to areas which are hazardous as a result of automatic movements must be prevented.
- The starting up of a drive unit as a result of a faulty command or false command input must be prevented.

- In handling a defective part, all movements on the part of the robot must be prevented.
- Injuries caused by machine parts which are ejected or fall off must be prevented.
- If, during troubleshooting, movements have to be carried out with the safety guard(s) for normal operation removed, such movements may be carried out only within the scope and speed laid down and only as long as instructed. Additionally, no person(s) or parts of the body may be present in the area at risk.
- Injuries caused by peripheral equipment must be prevented.

Remedying a fault and maintenance work also may require start-up while the machine is in an unsafe condition, and therefore require the following precautions :

- The robot must not be able to start up.
- The handling of various machine parts, either manually or with ancillary equipment, must be possible without risk of exposure to hazards.
- It must not be possible to touch parts that are "live".
- Injuries caused by the escape of liquid or gaseous media must be prevented.
- Injuries caused by peripheral equipment must be prevented.

Chapter 15

ROBOTS : RESHAPING MANUFACTURING

INDUSTRIAL ROBOTS ARE REDEFINING INDUSTRY, PRODUCTS AND WORKING PRACTICES

Industrial robots have been around for over 50 years. The first one, an automated die-casting machine that took over hazardous tasks from workers, was installed at a General Motors plant in the US in 1961. Industrial robots have since become much more complex and are used in a variety of other industries throughout the world. A number of IEC TCs and SCs (Subcommittees) prepare International Standards for countless components and systems that are used in robots and are fundamental to their safe operation.

"Anything that is Manufactured is Manipulated"

Robots were initially met with scepticism by managers and distrust by workers when they were first introduced to US car plants in the early 1960s, but then were gradually adopted by a number of industrialized countries to replace workers for repetitive and often hazardous tasks.

George C. Devol, the man credited with inventing the first industrial robot, gave an obvious reason for the introduction of robots in the industry : "Anything that is manufactured is manipulated. Every part is manipulated while it is made. Every part is manipulated while it is assembled. A part is manipulated when it is delivered from a plant. Everything is manipulated," he said in a 1983 interview.

In a logical move, robots graduated from their original assignments in die-casting and welding to lifting and moving car parts for assembly. Initially the US and Japanese car industries were the main outlets for industrial robots, accounting for around 40% of the total number used in the early 1980s. The potential of robots to carry out relatively simple tasks accurately, without interruption and at a quick pace, led to their adoption in many other industrial sectors such as electronics, the food industry and handling some products.

Relentless March

Industrial robots gained in popularity rapidly as they allowed high productivity as well as accuracy and quality. According to the IFR (International Federation of Robotics), "total accumulated sales, measured since the introduction of industrial robots at the end of the 1960s, amounted to more than 2 310 000 units by the end of 2011". Including early robots which are no longer in service, the IFR estimates "that the total worldwide stock of operational industrial robots was at the end of 2011 in the range of 1 153 000 and 1 400 000 units".

In 2011, the sales of industrial robots increased by 38% to 166 028 units and the worldwide market value for robot systems (including the cost of software, peripherals and system engineering) for that year was estimated at USD 25,5 billion. The systems therefore represent a major industrial sector, which has the added benefit of increasing industrial productivity.

Contrary to widely-held assumptions robots do not destroy but create many jobs both directly and indirectly, according to a Metra Martech report for the IFR. Examining the correlation between increased robotization and declining unemployment rates in 6 countries, the report states that robots carry out work in areas that would be unsafe for humans, that would not be economically viable in a high wage economy and that would be impossible for humans. Robotization should create between 700 000 and 1 million jobs in the countries concerned between 2011 and 2016, Metra Martech says.

More than a Fixed One-armed Machine

The first generation of industrial robots could best be described as one-armed manipulators that were installed in a permanent position and carried out simple tasks and routines. Safety represents a major issue; it is best to exclude human workers from the vicinity and to place robots into protected enclosures that cannot be entered by workers until the machines have been disabled, either actively or automatically.

However, advances in robotics have enabled new characteristics to be introduced to industrial robots. They include so-called "cooperative working" : the skills of human workers are combined with the precision and force that robots can provide, allowing both to work side by side without compromising workers' safety. This, and the small-batch assembly that is characteristic of many small- and medium-sized enterprises, is now possible using the mechanism of "guiding and teaching by example" (rather than by inflexible and uneconomic programming) and major advances in various kinds of tactile (*e.g.* pressure), optical or proximity sensors.

The Powers that Drive

Industrial robots may be powered by electric, pneumatic and hydraulic systems, according to the intended purpose. For instance, hydraulic machines are able to perform some heavier duty tasks. Electric robots are efficient and present benefits including easy and direct access to an energy source, using uniform and simple components and presenting no delay in transferring signals. Furthermore, they can move around when powered by batteries.

Pneumatic and hydraulic robots require another source of energy (electricity or hydrocarbons) to provide compressed air or move fluids through their components. Hydraulic fluids must be recycled and may pollute if they leak.

IEC TC 2 : Rotating machinery, prepares International Standards for rotating electrical machines such as drives and motors used in industrial robots. International Standards prepared by TC 22 : Power electronic systems and equipment, and its SCs, are also central to components used in robot drives and other systems.

Changing Industrial Landscape

As noted by the IFR, businesses are investing heavily in industrial robots. The benefits are obvious and the impact on the global industrial landscape and international trade will be significant.

Ever since their introduction, industrial robots have carried out difficult and hazardous tasks. While they will continue to be irreplaceable in this role, they are also able to :

- carry out work that would otherwise not be economically viable
- enhance manufacturing jobs by increasing productivity, flexibility and competitiveness
- improve process quality
- reduce operation costs and material waste
- improve quality of work for workers by carrying out repetitive tasks
- improve health and safety for workers
- reduce labour turnover and recruitment difficulties

In countries where labour costs are traditionally high, a benefit of introducing more industrial robots is inshoring : the repatriation to the local country of activities – and jobs – previously outsourced to low-wage countries.

The latter countries are also introducing industrial robots to improve product quality and move workers to other tasks. For example, in June 2011, Foxconn, a China-based company manufacturing computers and consumer electronics goods, decided to eliminate "monotonous, repetitive tasks" by replacing thousands of its workers with robots, each costing USD 20 000-25 000. Foxconn had installed 30 000 such robots by the end of 2012 and plans to have fully automated plants in 5-10 years.

All signs from the industry point to a healthy growth in years to come as traditional markets in North America, Europe and Asia increase or renew their assets and emerging industrialized countries equip their factories. IEC International Standards will contribute significantly to this global growth of the robotics industry.

ROBOT WORK VOLUME AND ITS COMPARISONS

A space on which a robot can move and operate its *wrist end* is called as a work volume. It is also referred as the *work envelope* and *work space*. For developing a better work volume, some of the physical characteristics of a robot should be considered such as :

- The anatomy of various robots
- The maximum value for moving a robot joint
- The size of the robot components like wrist, arm, and body

Comparison of Work Volumes with Different Robot Configurations

Cylindrical Configuration

A cylindrical configuration robot has an arm that has got the ability to reach horizontal and vertical directions. Moreover, it can make a rotary motion by placing the arm at the centre of the robot. As a result, this robot requires a *cylindrical* type of work volume for performing an operation. It is mostly used in the *material handling* process. One setback of this robot is that it can't pick the tools from the floor.

Polar Configuration

The arm of a polar configuration robot does not move in up and down position on the vertical axis as done in the cylindrical configuration. It just makes the robot to rotate during the horizontal stroke by expanding the inner and outer circles. It results in the formation of an arc movement. As soon as this process finishes, the robot arm swings and gets *spherical work* volume. This type of robot is incorporated in the *Flexible Manufacturing System* (FMS) for picking up the tools from the floor.

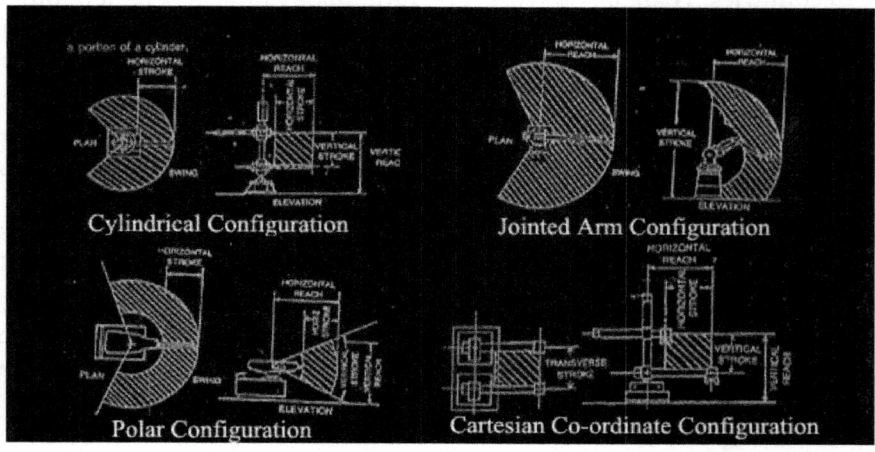

Jointed Arm Configuration

The work volume of a jointed arm configuration robot is a *complex* one. The wrist and elbow of a manipulator are jointly swept in the horizontal and vertical position. It works almost like a human arm. The result achieved from the jointed arm and cylindrical configuration robots are same. The major advantage of the jointed arm configuration robot is that it can move in all directions *very easily and flexibly*. It is used for performing machine loading and unloading operations in the CNC machines.

Cartesian Co-ordinate Configuration

The Cartesian co-ordinate configuration robot arm will move in up & down directions on the horizontal axis, and in & out motions in the vertical axis. The collective movement of traverse and horizontal stroke results in the *rectangular work volume*. It is inflexible, and best suitable for *pick and place* processes.

MACHINE LOADING AND UNLOADING BY INDUSTRIAL ROBOTS

In machine loading and unloading process, a *robot* will be used to move the work parts to or/and from the *production machine*. This application comes under the category of *material handling* operations.

The machine loading and unloading application includes the following *three processes* :

- Machine loading
- Machine unloading
- Machine load and unload

Machine Loading

In this operation, the robot *loads raw work parts* in the machine, and some other systems are used to unload the finished work parts from the machine.

Example : In a *press working process*, a robot is used to load the sheet metal in the press, and the finished work parts are removed from the press with the help of gravity.

Machine Unloading

In machine unloading, the finished work parts are *unloaded* from the machine by a robot, while the loading of raw materials are done *without any robot* support.

Example : *Plastic modeling* and *die casting*.

Machine Load and Unload

In this process, a robot performs both loading and unloading of work parts in and from the machine.

Example : *Machining operation*.

The machine loading and unloading process performed by *industrial robots* are very well characterized by the robot-centered workcell. This cell includes a robot, production machine, and other devices like part delivery system. It helps in *increasing* the usage of a robot by making it to service more than a production machine. As a result, the *productivity* in the cell is also increased to a larger extent. This robot cell can be preferred when a robot is in the *idle state* for a long time.

Moreover, the robots are largely used to carry out loading and unloading process in some *production operations* like forging, die casting, plastic modeling, stamping press, and machining operations.

TYPES OF COSTS REQUIRED FOR DEVELOPING A ROBOT

In a robot project, there are two types of *cost statistics* available for performing the economic analysis such as :

- Investment costs
- Operating costs

Types of Costs Included in the Investment Costs

Several types of costs that come under the investment costs are *robot purchase cost, installation costs, engineering costs, special tooling,* and *miscellaneous costs*.

- Robot purchase cost refers to the primary cost of a robot. In this condition, the robot will have suitable abilities for performing an operation. Moreover, it does not include the end effector.
- Installation costs possess the human workers and equipments required for assembling the installation site.
- Engineering costs include the price of design prepared by the engineering staff of the industry for installing the robot.
- Special tooling comprises the price of part positioners, end effector, and other equipments that are necessary for processing the workstation.
- Miscellaneous costs include other investment costs that do not come under the above types.

Types of Costs Included in the Operating Costs

The operating costs cover several costs like *direct labor cost, indirect labor cost, maintenance cost, utilities,* and *training*.

- The direct labor cost adds the price given to a human worker for operating the robot workcell. It includes the fringe advantages to evaluate the direct labor price, and excludes other overhead costs.
- The indirect labor cost includes the price of management, programming, arrangement, and miscellaneous costs that are not related with anyone of the above six types.
- Maintenance cost involves the expected price for maintaining and servicing a robot in a workcell.
- Utilities possess the costs of electricity, gas, air pressure, and other utilities, which are used for operating a robot workcell.
- As the training cost comes under the basic installation cost, it may be considered as an investment cost. The reason for including this category in operating cost is that it is a continuing activity.

THREE METHODS TO DEVELOP A ROBOT WITH PROFIT

Before starting the development of a robot, some of the data must be collected to carry out *economic analysis* effectively. They are :

- Type of robot to be installed.
- Cost to install a robot.
- Time taken to produce a robot.
- Savings and benefits in the development.

In an industry, the investment put on the development of a robot can be compared and analyzed by *three common methods* such as :

- Payback method
- EUAC (Equivalent Uniform Annual Cost) method
- ROI (Return on Investment) method

Payback Method

The duration taken to equal the *initial investment* and *net accumulated cash flow* in the development of a robot is called as payback period or payback method. If the net annual cash flows are identical to every year, then it can be stated by a formula given below.

Payback period = Investment Cost / Net Annual Cash Flow

EUAC Method

The EUAC is the short form of *Equivalent Uniform Annual Cost method*. It is used to alter the *total cash flows* and *investments* into the equivalent uniform costs over the expected time of developing a robot. It is done by employing different interest features that are connected with the calculations of engineering economy.

ROI Method

The *Return on Investment* is the expansion of ROI method. It is used to determine the *return ratio* of the current project, which is related to the anticipated expenditures and profits. If the rate of return is *low* to the expected cost of a company, then the investment made is *accepted*.

FACTORS FOR SELECTING A ROBOT BASED ON THE APPLICATION

In an industry, the engineers will be having several robots under their control. According to the job given, the engineers must be able to select the perfect robot. This is only possible, if the engineers are well known about the *technical features* of every robot.

The important technical features that are to be noted before selecting a robot for an application are :

- Type of job
- Configuration
- Number of axes
- Drive system
- Programming method
- Control system

If the above features of a robot are well suited for the specified application, then the selection will be the perfect for the operation.

For more details, here are some of the applications and technical features required for selecting a robot.

Spot Welding Robot

- A robot must be of jointed arm or polar configuration for performing this application.

- It must have five or six axes.
- It can have either electric or hydraulic drive system.
- It must be programmed in powered lead through method.
- It requires a point to point control system.

Spray Painting Robot

- For this application, the jointed arm configuration robot will be the appropriate selection.
- The number of axes must be more than or equal to six.
- It must have a hydraulic drive system.
- Manual lead through programming method will be required.
- It must possess a continuous path control system.

Arc Welding Robot

- This type of robot can be of Cartesian, jointed arm, or polar configuration.
- It must be incorporated with five or six axes.
- It must feature either electric or hydraulic drive system.
- It is capable of programming with Powered or Manual Lead through method.
- It must have a continuous path control system.

Machine Loading

- To perform machine loading, a robot should be of either cylindrical, jointed arm, or polar configuration.
- It requires only four or five axes.
- The drive system must be hydraulic for loading heavy weights. Otherwise electric type is enough.
- The programming method needed for this operation will be powered lead through.
- As this is a simple process, it can have either point to point or limited sequence control system.

Assembly Operation

- The configuration of the robot must be Jointed Arm or Cartesian.
- The number of axes can be from three to six.
- It requires an electric drive system for this process.
- It must be capable of programming in textual languages or powered lead through method.
- It must have either continuous path or point to point control system.

Chapter 16

ROBOT OPERATION

ROBOT OPERATING TECHNIQUES

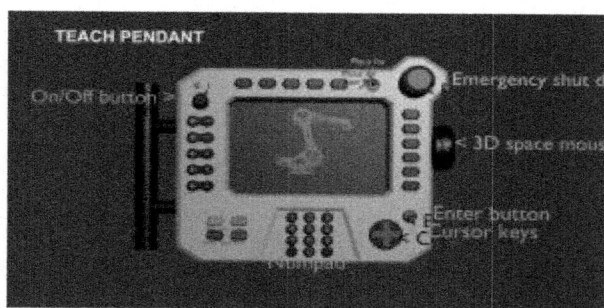

A robot certainly requires a system for operating their various *external mechanisms* in the programming stage. Some important robot operating techniques are:
- Manual Data Input Panel
- Teach Pendant
- Computer Control

Manual Data Input Panel

The manual data input panel is shortly called as *MDI*. It is mainly used to perform three processes such as to enter the data in the program and program control registers, and to edit the data in the program.

There are *three components* used in the MDI panel namely:
- CRT (Cathode – Ray Tube) Display
- Power ON / OFF buttons
- Alphanumeric Keyboard

The CRT *displays* the input given to the program. As a result, the errors in the program can be easily identified and rectified. It also shows the feed rate codes, geometric move codes, positional data, service request codes, and program addresses.

The alphanumeric keyboard is used in the MDI panel instead of the teach pendant. It helps to *input* the data like letters and numbers to the program. This keyboard contains read / write keys for reading or writing the data in the memory storage. It also has an input key to enter the data in the different paths, and a start key to begin the program operation.

The MDI panel has mode key, cursor keys, page down keys, make up keys, and programming keys. An added advantage in this system is that it can control many controllers by using a single MDI panel.

Teach Pendant

A *teach pendant* is very much similar to the MDI panel. It is used to input the location data to the controller memory. It has several indicators to show coordination of axes, operation mode, and positional information. Other than that, it has a dead man button, programming keys, emergency stop, and axis jogging keys. It is very compact in size and therefore, it is *easy to carry* everywhere in the work volume. It can be incorporated in middle and high level controllers.

Computer Control

Nowadays, every operation has become *automated* with the help of a computer. A computer that is controlling every operation in the industries is known as *Flexible Manufacturing System*. A single computer is enough to control various machining processes. The programming is done by either engineering workstations or *CAD* system. It can program, teach, and edit different registers, service codes, feed rate codes, geometric move codes, and control paths.

ROBOT LINKS AND JOINTS

In a robot, the connection of different manipulator joints is known as *Robot Links*, and the integration of two or more link is called as *Robot Joints*. A robot link will be in the form of *solid material*, and it can be classified into two key types – *input link* and *output link*. The movement of the input link allows the output link to move at various motions. An input link will be located nearer to the base.

Different Types of Robot Joints

The Robot Joints is the important element in a robot which helps the links to travel in different kind of movements. There are *five major types* of joints such as :
- Rotational joint
- Linear joint

- Twisting joint
- Orthogonal joint
- Revolving joint

Rotational Joint

Rotational joint can also be represented as R – Joint. This type will allow the joints to move in a *rotary motion* along the axis, which is vertical to the arm axes.

Linear Joint

Linear joint can be indicated by the letter L – Joint. This type of joints can perform both translational and sliding movements. These motions will be attained by several ways such as telescoping mechanism and piston. The two links should be in *parallel axes* for achieving the linear movement.

Twisting Joint

Twisting joint will be referred as V – Joint. This joint makes *twisting motion* among the output and input link. During this process, the output link axis will be vertical to the rotational axis. The output link rotates in relation to the input link.

Orthogonal Joint

The *O – joint* is a symbol that is denoted for the orthogonal joint. This joint is somewhat similar to the linear joint. The only difference is that the output and input links will be moving at the right angles.

Revolving Joint

Revolving joint is generally known as V – Joint. Here, the output link axis is *perpendicular* to the rotational axis, and the input link is *parallel* to the rotational axes. As like twisting joint, the output link spins about the input link.

ROBOT DRIVE SYSTEMS

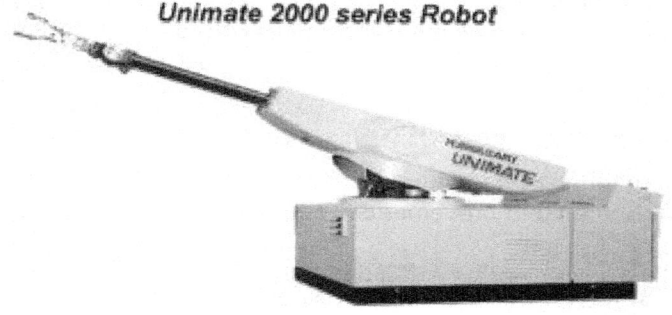

Unimate 2000 series Robot

A robot will require a *drive system* for moving their arm, wrist, and body. A drive system is usually used to determine the capacity of a robot. For actuating the robot joints, there are *three different types* of drive systems available such as :

- Electric drive system,
- Hydraulic drive system, and
- Pneumatic drive system.

The most importantly used two types of drive systems are electric and hydraulic.

Electric Drive System

The electric drive systems are capable of moving robots with *high power* or speed. The actuation of this type of robot can be done by either DC servo motors or DC stepping motors. It can be well – suited for rotational joints and as well as linear joints. The electric drive system will be perfect for *small robots* and precise applications. Most importantly, it has got greater accuracy and repeatability. The one disadvantage of this system is that it is slightly costlier. An example for this type of drive system is *Maker 110 robot*.

Hydraulic Drive System

The hydraulic drive systems are completely meant for the *large – sized robots*. It can deliver high power or speed than the electric drive systems. This drive system can be used for both linear and rotational joints. The rotary motions are provided by the rotary vane actuators, while the linear motions are produced by hydraulic pistons. The *leakage* of hydraulic oils is considered as the major disadvantage of this drive. An example for the hydraulic drive system is *Unimate 2000 series robot*.

Pneumatic Drive System

The pneumatic drive systems are especially used for the *small type robots*, which have less than five degrees of freedom. It has the ability to offer fine accuracy and speed. This drive system can produce rotary movements by actuating the rotary actuators. The translational movements of sliding joints can also be provided by operating the piston. The price of this system is *less* when compared to the hydraulic drive. The drawback of this system is that it will not be a perfect selection for the *faster operations*.

ROBOT MECHANICAL TRANSMISSION SYSTEMS

The links and joints of a robot will require anyone of the *drive systems* (electric, hydraulic, and pneumatic) to perform several motions. However, the necessary speed or torque is not gained even with the use of an actuator. As a result, the mechanical transmission system is applied. There are many methods for transferring mechanical power such as :

- Gears
- Belts
- Harmonic Drives

GEARS

The *rotational* motion of one shaft is transmitted to another with the help of gears. The transferring of power can be either in intersecting or parallel shafts. It is the most frequently used mechanical transmission system in the robots. The added advantage of the gears is that the speed can be *decreased or increased* according to the operation.

Types of Gears

- Spur gears
- Bevel gears
- Helical gears
- Worm and Worm Wheel

a) Spur Gears

The arrangement of *two parallel* and coplanar shafts is done by the spur gears. These gears can be fitted externally and internally as well. The spur gear will rotate in the same way in the internal gearing, and it revolves in the opposite way in the external gearing. It is used in the clocks, gear box, and lathe back gear.

b) Bevel Gears

The bevel gears are implemented for connecting *two coplanar intersecting shafts*. It is capable of transferring power up to 900. It is mostly used in drill chuck key, differential of automobile, *etc.*

c) Helical Gears

The connection of intersecting shafts and non-parallel or parallel shafts is performed by the helical gears. The major benefit of this gear is that it transfers *high power*. The operation is done smoothly and creates less vibration too. It is used in IC engines shaft & timing gears, and automobile gear box.

d) Worm and Worm Wheel

It can integrate shafts that are perpendicular to each other. Here, the driver is worm and driven is worm wheel. The advantage of using worm and worm wheel is that it can transmit high power and efficiency along with the *reduced velocity ratio*. Some of the applications are wind shield wiper, index head, and steering gear box.

Benefits

- Better efficiency
- High Durability
- Compact

Drawbacks

- High manufacturing cost
- Improper arrangement of tooth causes high noise.

BELTS

The belts are another type of power transmission system. It is used in the robots to transmit the power. It is not so physically powerful, but it can deliver *smooth and quiet process* with the help of the shock absorber.

Types of Belts

- V – belt
- Flat belt
- Circular belt

V – Belt

The V – belt is highly used to transmit the *maximum amount of power* to the pulleys. It has the cross-section of trapezoidal. It is the best type to connect the pulleys that are close to each other.

Advantage : (sssh)

- Less floor space
- No possibility of slip
- Easy to maintain

Disadvantage : (sssh)

- Difficult to construct
- High cost

Flat Belt

The flat belt combines the pulleys *up to 10 meters* distance, and transmits a *medium amount of power*. It is of rectangular cross section type. It can be classified into three key types namely cross belt, open belt, and compound belt.

Advantage : (sssh)
- Simple in construction
- Smooth process
- Low cost

Disadvantage : (sssh)
- Huge in size
- Possibility of slip

Circular Belt

The circular belt is also known as the *rope drive* in which the cross section is circular. It is largely included in the robots for gaining *maximum amount of power* between the pulleys. If the center distance of one pulley and the other is more than 5 meters, then this type of belt is mostly used.

HARMONIC DRIVES

The harmonic drives are incorporated for *increasing and decreasing* the speed. It is related to the mechanism of non – rigid material. The decrease ratio of harmonic drives is 1 :1 to Infinity : 1 and it is commonly set as 100 :1. An important advantage is that it can work for life time with *less maintenance*. When this drives is compared to the gears, it is less efficient.

ROBOTICS/FEEDBACK SENSORS/ENCODERS

It's not at all uncommon to require information concerning rotation on a robot, whether from arm or manipulator axis motion to drive wheel speed. Rotary encoders can be found in several varieties, most commonly mechanical or using photodetectors. In mechanical encoders, electrical contacts periodically connect, usually to drive a terminal high or low. These electrical pulses are then used for information. Photodetector encoders can be made smaller and operate much quicker. Typically, a reflective sensor will shine on a marked reflective surface, or a photointerruptor will shine through a disc with transparent and opaque sections, such that the amplitude of received light causes high and low electrical pulses in the receiver. These pulses are then decoded by external circuitry.

Encoding

Depending on the type and reliability of information needed, the patterning on the rotary encoder will be increasingly complex. If only rotational speed is required, a simple alternating pattern will do, but if both speed and direction are needed, more encoded information is required. Encoding position will inherently yield information regarding velocity and direction, as comparing the direction of change in position will give direction of rotation, where the change in position with respect to time will (by definition) indicate speed. The specific type of position encoding will depend on the precision required, as well as the reliability : how much of an error in measured rotational position due to an error in reading the encoded information, is acceptable in the application.

Measuring Rotational Speed

For some applications, only rotational speed needs to be considered. If a toothed wheel which is mechanically limited to spinning a single direction is being measured, the direction is already known, and only the velocity remains. Perhaps a feedback system is being developed for drive speed on a robot, and an encoder on the axle will enable the controller to calibrate speed by varying PWM values; if the controller enables the motor to spin in a specific direction, all it requires is the rotational speed. All that is necessary to measure speed is a single alternating pattern causing the detector to emit a pulsetrain. The pulse rate will be directly proportional to the rotational speed; measuring the time between consecutive rising (or falling) edges and dividing that into the angle of rotation represented by a pulse will yield the rate of rotation.

$$\omega \approx \frac{\frac{2\pi}{n}}{(T_1 - T_0)}, \quad where\ n = pulses\ per\ revolution$$

If the rate is changing during the time of the measurement, the calculated rate will be an average approximating the instantaneous rate. This is where precision comes into play – the more pulses per rotation, the shorter the time between measurements, and the closer the approximation is to correct. Anyone familiar with calculus will recognize that the rate of change of position with respect to time is a derivative, and the definition of the derivative states that as the timestep across which a measurement for approximation is taken approaches zero, the result is equal to the derivative at that time.

$$lim\ \Delta t \to 0\ \frac{\theta(T_0 + \Delta t) - \theta(T_0)}{\Delta t} = \frac{d\theta(t)}{dt} = \omega(t)$$

$$\Delta t = T_1 - T_0$$

$$lim\ (T_1 - T_0) \to 0\ \frac{\theta(T_0 + (T_1 - T_0)) - \theta(T_0)}{T_1 - T_0} = \frac{d\theta(t)}{dt} = \omega(t)$$

$$lim\ (T_1 - T_0) \to 0\ \frac{\theta(T_1) - \theta(T_0)}{T_1 - T_0} = \frac{d\theta(t)}{dt} = \omega(t)$$

$$\theta(T_m) = m\frac{2\pi}{n}$$

$$\theta(T_1) - \theta(T_0) = \frac{2\pi}{n}(1-0) = \frac{2\pi}{n}$$

$$lim\ (T_1 - T_0) \to 0 \quad \frac{\theta(T_1) - \theta(T_0)}{T_1 - T_0} = \frac{\frac{2\pi}{n}}{T_1 - T_0} = \frac{d\theta(t)}{dt} = \omega(t)$$

Measuring the difference in the calculated velocities of successive pulses can give an approximation of rotational acceleration as well.

Measuring Direction

Assume a machine is tuning its PWM routines for powering the drive wheels. The controller obviously needs to know the rotational speed of the wheels so it can determine the speed outputs for given duty cycles. But what if the machine is placed on a hill? It starts to roll backward and the controller sees a substantial rotational speed for no input, which messes up the entire calibration process. If the controller were able to determine that the wheel was spinning backward, it would know to ignore the speed information. In order to determine direction with a rotary encoder, there must be at least two patterns on the disk, so there is a reference for the direction of change. The simplest way to achieve this is with two alternating patters that are half a pulse out of phase with each other.

If the disc is rotating in a particular direction (in this case, such that the pattern slides to the left), the rising (or falling) edge of channel A will be detected before that of channel B; if rotation is in the opposite direction, the edge of channel B will be detected first. By testing for which pattern's edges are detected first, the direction of rotation can be determined. The time between pulses for a single channel can be measured to approximate rotational speed. However, combining detections from the two channels gives an effective pulse rate double that of a single channel, which halves the measuring time and doubles precision. A means of measuring speed and direction with a single channel exists as well. Two receivers are placed on a single channel spaced half a pulse width apart, so that as the disc rotates, one sensor will be in state as the other transitions. The order in which the two sensors transition into a state is dependent on the direction of rotation of the disc.

Measuring Position

The simplest way to determine an angular position on a rotary encoder is to divide the disc into segments and imprint each segment with a unique pattern. The more segments the disc is divided into, the more precision is gained in knowing the location at a given time (each segment will cover a span of angles, so there will be uncertainty, but more segments will give fewer degrees per segment). More segments, however, require more unique identifiers, which means a more complex scheme. This discussion is limited to digital-signal encoders, where the possible outputs for any given channel are limited to 0 or 1. This is conducive to a binary encoding scheme, where each segment is imprinted with a unique binary number. The number of unique IDs required depends on the number of segments; the number of channels is equal to log2(n) where n is the number of segments. A disc with 8 segments has a resolution of 360° / 8 = 45° per segment, and requires log2(8) = 3 separate channels. For a doubling of the number of segments, precision is doubled with an increase of only one pattern channel.

Reliability

One of the problems with using a binary encoding scheme is detector error. There's no guarantee that every channel is going to transition at the same time, which can cause a misread of a segment's ID. If the section of the disc below the detector is transitioning from one segment to the next, briefly the decoder might read only part of the bits correctly. Assume a standard eight-segment binary encoding scheme, where segments are numbered from 0 to 8 sequentially : if the transition from segment 000 to segment 001 doesn't read the rising edge correctly, your read is off by one space, which for the moment is negligible. If, however, it's transitioning from segment 011 to segment 100, and the least significant bit is read in first, briefly your controller will think it's transitioned to 010, in the complete opposite direction. If the most significant bit is read in first, it'll think it jumped from 011 to 111, on the complete opposite side of the wheel. Transition errors when a single bit changes can often be considered negligible, as the error in calculated position is 1. Errors from multi-bit transitions can be catastrophic.

Gray Code

4-bit Gray code
0000
0001
0011
0010
0110
0111
0101
0100
1100
1101
1111
1110
1010
1011
1001
1000

Gray code is a specific type of binary pattern in which there is only a single bit changing during a transition from one element to the next. Being a binary pattern, Gray codes use the same number of channels for the same number of specific patterns; they are just ordered differently. Binary uses patterns in a numerically sequential order; Gray code does not. Encoding positions using this method will require additional software or hardware to decode the ID (as a simple binary conversion will not yield a sequentially-ordered set of segments), but the benefits for the reliability of hardware are substantial – as only a single bit changes during each transition, then the maximum error in detected position for an error in pattern reception is 1 segment. Gray code can be written out bitwise in much the same way binary can. In binary, the least significant bit alternates from 0 to 1 every step; the next bit alternates every two steps, the next every four and so on. In Gray code, each bit holds a pattern of 0-1-1-0; the least bit holds each element for one step; the next bit holds each two steps, the next four and so on. It can be noticed that two-channel Gray code is identical to the two-channel pattern used to find rotational direction.

Chapter 17

HUMAN–ROBOT INTERACTION

Human–robot interaction is the study of interactions between humans and robots. It is often referred as HRI by researchers. Human–robot interaction is a multidisciplinary field with contributions from human–computer interaction, artificial intelligence, robotics, natural language understanding, design, and social sciences.

ORIGINS

Human–robot interaction has been a topic of both science fiction and academic speculation even before any robots existed. Because HRI depends on a knowledge of (sometimes natural) human communication, many aspects of HRI are continuations of human communications topics that are much older than robotics per se.

The origin of HRI as a discrete problem was stated by 20th-century author Isaac Asimov in 1941, in his novel *I, Robot*. He states the Three Laws of Robotics as,

"
1. A robot may not injure a human being or, through inaction, allow a human being to come to harm.

2. A robot must obey any orders given to it by human beings, except where such orders would conflict with the First Law.

3. A robot must protect its own existence as long as such protection does not conflict with the First or Second Law.
"

These three laws of robotics determine the idea of safe interaction. The closer the human and the robot get and the more intricate the relationship becomes, the more the risk of a human being injured rises. Nowadays in advanced societies, manufacturers employing robots solve this issue by not letting humans and robots share the workspace at any time. This is achieved by defining safe zones using liar sensors or physical cages. Thus the presence of humans is completely forbidden in the robot workspace while it is working.

With the advances of artificial intelligence, the autonomous robots could eventually have more proactive behaviors, planning their motion in complex unknown environments. These new capabilities keep safety as the primary issue and efficiency as secondary. To allow this new generation of robot, research is being conducted on human detection, motion planning, scene reconstruction, intelligent behavior through task planning and compliant behavior using force control (impedance or admittance control schemes).

The goal of HRI research is to define models of humans' expectations regarding robot interaction to guide robot design and algorithmic development that would allow more natural and effective interaction between humans and robots. Research ranges from how humans work with remote, tele-operated unmanned vehicles to peer-to-peer collaboration with anthropomorphic robots.

Many in the field of HRI study how humans collaborate and interact and use those studies to motivate how robots should interact with humans.

THE GOAL OF FRIENDLY HUMAN-ROBOT INTERACTIONS

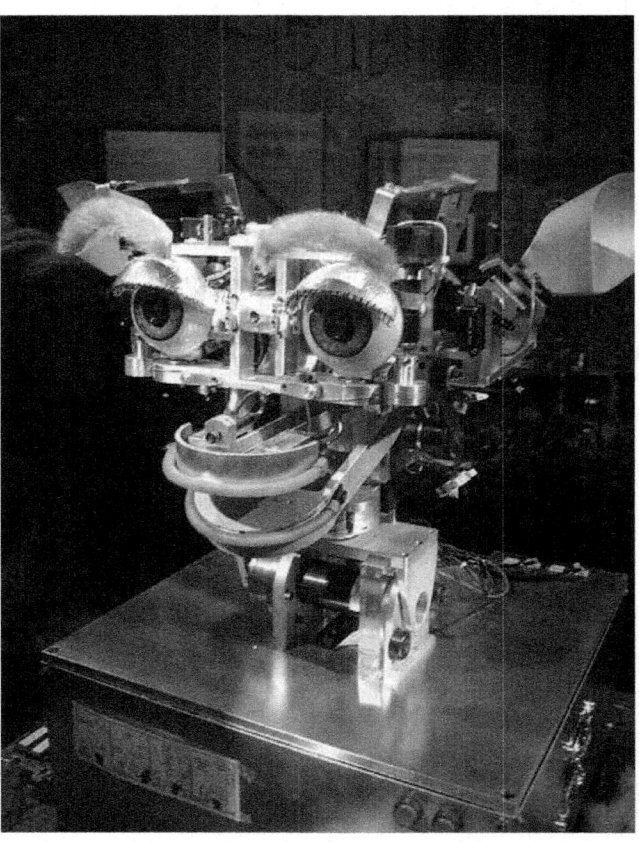

Kismet can produce a range of facial expressions.

Robots are artificial agents with capacities of perception and action in the physical world often referred by researchers as workspace. Their use has been generalized in factories but nowadays they tend to be found in the most technologically advanced societies in such critical domains as search and rescue, military battle, mine and bomb detection, scientific exploration, law enforcement, entertainment and hospital care.

These new domains of applications imply a closer interaction with the user. The concept of closeness is to be taken in its full meaning, robots and humans share the workspace but also share goals in terms of task achievement. This close interaction needs new theoretical models, on one hand for the robotics scientists who work to improve the robots utility and on the other hand to evaluate the risks and benefits of this new "friend" for our modern society.

With the advance in AI, the research is focusing on one part towards the safest physical interaction but also on a socially correct interaction, dependent on cultural criteria. The goal is to build an intuitive, and easy communication with the robot through speech, gestures, and facial expressions.

Dautenhan refers to friendly Human–robot interaction as "Robotiquette" defining it as the "social rules for robot behaviour (a 'robotiquette') that is comfortable and acceptable to humans" The robot has to adapt itself to our way of expressing desires and orders and not the contrary. But every day environments such as homes have much more complex social rules than those implied by factories or even military environments. Thus, the robot needs perceiving and understanding capacities to build dynamic models of its surroundings. It needs to categorize objects, recognize and locate humans and further their emotions. The need for dynamic capacities pushes forward every sub-field of robotics.

On the other end of HRI research the cognitive modelling of the "relationship" between human and the robots benefits the psychologists and robotic researchers the user study are often of interests on both sides. This research endeavours part of human society.

GENERAL HRI RESEARCH

HRI research spans a wide range of field, some general to the nature of HRI.

Methods for Perceiving Humans

Most methods intend to build a 3D model through vision of the environment. The proprioception sensors permit the robot to have information over its own state.

Methods for perceiving humans in the environment are based on sensor information. Research on sensing components and software lead by Microsoft provide useful results for extracting the human kinematics. An example of older technique is to use colour information for example the fact that for light skinned people the hands are lighter than the clothes worn. In any case a human modelled a priori can then be fitted to the sensor data. The robot builds or has (depending on the level of autonomy the robot has) a 3D mapping of its surroundings to which is assigned the humans locations.

A speech recognition system is used to interpret human desires or commands. By combining the information inferred by proprioception, sensor and speech the human position and state .

Methods for Motion Planning

Motion planning in dynamic environment is a challenge that is for the moment only achieved for 3 to 10 degrees of freedom robots. Humanoid robots or even 2 armed robots that can have up to 40 degrees of freedom are unsuited for dynamic environments with today's technology. However lower-dimensional robots can use potential field method to compute trajectories avoiding collisions with human.

Cognitive Models and Theory of Mind

A lot of data has been gathered with regards to user studies. For example, when users encounter proactive behaviour on the part of the robot and the robot does not respect a safety distance, penetrating the user space, he or she might express fear. This is dependent on one person to another. Only intensive experiment can permit a more precise model.

It has been shown that when a robot has no particular use, negative feelings are often expressed. The robot is perceived as useless and its presence becomes annoying.

In another experiment, it has occurred that people tend to attribute to the robot personality characteristics that were not implemented.

APPLICATION-ORIENTED HRI RESEARCH

In addition to general HRI research, researchers are currently exploring application areas for human-robot interaction systems. Application-oriented research is used to help bring current robotics technologies to bear against problems that exist in today's society. While human-robot interaction is still a rather young area of interest, there is active development and research in many areas.

Search and Rescue

First responders face great risks in search and rescue (SAR) settings, which typically involve environments that are unsafe for a human to travel. In addition, technology offers tools for observation that can greatly speed-up and improve the accuracy of human perception. Robots can be used to address these concerns . Research in this area includes efforts to address robot sensing, mobility, navigation, planning, integration, and tele-operated control.

SAR robots have already been deployed to environments such as the Collapse of the World Trade Center.

Other application areas include :
- Entertainment
- Education
- Field robotics
- Home and companion robotics
- Hospitality
- Rehabilitation and Elder Care
- Robot Assisted Therapy (RAT)

Chapter 18

DECENTRALIZED SENSOR FUSION FOR UBIQUITOUS NETWORKING ROBOTICS IN URBAN AREAS

Alberto Sanfeliu[1,*], Juan Andrade-Cetto[1], Marco Barbosa[2], Richard Bowden[3], Jesus Capitan[5], Andreu Corominas[1], Andrew Gilbert[3], John Illingworth[3], Luis Merino[4], Josep M. Mirats[1], Plínio Moreno[2], Aníbal Ollero[5,6], João Sequeira[2] and Matthijs T.J. Spaan[2]

[1] Institut de Robotica i Informatica Industrial, CSIC-UPC, Barcelona, Spain; E-Mails: cetto@iri.upc.edu (J.A.-C); acoromin@iri.upc.edu (A.C.); jmirats@iri.upc.edu (J.M.M.)
[2] Instituto Superior Tecnico & Institute for Systems and Robotics, Lisbon, Portugal; E-Mails: mafb@isr.ist.utl.pt (M.B.); plinio@irs.ist.utl.pt (P.M.); jseq@isr.ist.utl.pt (J.S.); mtjspaan@isr.ist.utl.pt (M.TJ.S.)
[3] Centre for Vision Speech and Signal Processing, University of Surrey, Guildford, UK; E-Mails: r.bowden@surrey.ac.uk (R.B.); a.gilbert@surrey.ac.uk (A.G.); j.illingworth@surrey.ac.uk (J.I.)
[4] Pablo de Olavide University, Seville, Spain; E-Mail: lmercab@upo.es
[5] Robotics, Vision and Control Group, University of Seville, Seville, Spain; E-Mail: aollero@cartuja.us.es
[6] Center for Advanced Aerospace Technology, Seville, Spain

[*] Author to whom correspondence should be addressed; E-Mail: sanfeliu@iri.upc.edu; Tel.: +34-934015751; Fax: +34-934015750.

ABSTRACT

In this article we explain the architecture for the environment and sensors that has been built for the European project URUS (Ubiquitous Networking Robotics in Urban Sites), a project whose objective is to develop an adaptable network robot architecture for cooperation between network robots and human beings and/or the environment in urban areas. The project goal is to deploy a team of robots in an urban area to give a set of services to a user community. This paper addresses the sensor architecture devised for URUS and the type of robots and sensors used, including environment sensors and sensors onboard the robots. Furthermore, we also explain how sensor fusion takes place to achieve urban outdoor execution of robotic services. Finally some results of the project related to the sensor network are highlighted.

Keywords

Network robot systems; distributed sensors; robot sensors; camera network.

1. INTRODUCTION

In the URUS project [1] we needed to interconnect and manage multiple services that enable urban robots to perform social urban tasks. To this end, we have designed an architecture that takes advantage of all information available within a large set of ubiquitous sensors, including fixed cameras, wireless sensors, Mica2 sensors and various sensing devices onboard the robots, such as cameras and laser range finders. Two characteristics of this architecture are of particular relevance: it is a highly distributed architecture, and it is scalable. Contrary to other architectures for multiple robot interoperability, our system does not make use of a central server that receives and combines all the information available to obtain, for instance, a fused estimation of a given variable, say a tracked person position [2]. Such kind of systems are dependent on a central node, thus not robust to communication failures, latencies or drop outs, and they do not scale well with the number of nodes. In the URUS architecture, we opted for a decentralized system in which each subsystem only manages local information and exchanges with its peers local estimates of any given variable. Decentralized data fusion takes place by means of an information filter (IF), dual to the more common Kalman filter typically used for data fusion [3]. The IF has very nice characteristics for decentralization, with applications in other robotics contexts such as aerial robot decentralized perception [4]. One of the applications of decentralized data fusion in URUS is person tracking and guiding, employing the local sensors of the robot and the environment camera network, efficiently coping with occlusions, and single module tracking failures.

The article is organized as follows. We first present the objectives of the URUS project, the partners and the robot sites where the experiments take place. Then, we explain the distributed URUS architecture and the sensors in a network that combines cameras and Mica2 sensors. Next, we describe the sensors that are included in some of the robots used for the experiments. The robotic systems described in this paper include Tibi and Dabo from the Institut de Robótica i Informática Industrial (CSIC-UPC), Romeo from the Asociación de Investigacióny Cooperación Industrial de Andalucía (AICIA) and a fleet of Pioneer robots from the Instituto Superior Técnico (IST). Next we describe how information fusion from several sensors takes place, and the software architecture that we have developed to manage such heterogeneous set of sensors and systems. Finally, we explain how we use these sensors for several robot services as required by some of the URUS experiments, including for instance robot localization and people tracking.

2. THE URUS PROJECT

2.1. Objectives of the URUS Project

The general objective of the URUS project is the development of new ways for the cooperation between network robots and human beings and/or the environment in urban areas, in order to achieve efficiently tasks that for single systems would be too complex, time consuming or costly. For instance, the cooperation between robots and video cameras can solve surveillance problems for blind spots in urban areas, or the cooperation between robots and wireless communications devices can improve efficiency in people assistance services. The focus of the project is in urban pedestrian areas, for which there exists a growing interest in reducing the number of cars and improving the quality of life. Network robots become an important instrument towards these goals.

Network robots is a new concept that integrates robots, sensors, communications and mobile devices in a cooperative way. Meaning that, not only a physical connection between these elements exists, but also, that there is a need for the development of novel intelligent cooperation methods for task oriented purposes, new ways of communication between the different elements, and new robot mobility methods using the ubiquity of sensors and robots.

The URUS project is focused on designing and developing a network of robots that in a cooperative way interact with human beings and the environment for tasks of guidance and assistance, transportation of goods, and surveillance in urban areas. Specifically, our objective has been the design and development of a networked robot architecture that integrates cooperating urban robots, intelligent sensors (video cameras, acoustic sensors, *etc.*), intelligent devices (PDA, mobile telephones, *etc.*) and communications. The main scientific and technological challenges that have been addressed during the course of the project are: navigation and motion coordination among robots; cooperative environment perception; cooperative map building and updating; task negotiation within cooperative systems; human robot interaction; and wireless communication strategies between users (through mobile phones, PDAs), the environment (cameras, acoustic sensors, *etc.*), and the robots.

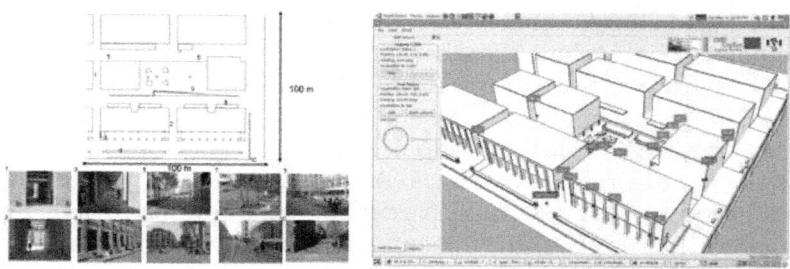

Figure 1. Barcelona Robot Lab.

Moreover, we have devised two demonstration scenarios in the Barcelona Robot Lab, a 10,000 m² area devoted to urban robotics experimentation (Figure 1). Scenario 1 involves transporting a person or goods to a destination; and Scenario 2 is devoted to drive people slowly and orderly towards the main exit in public spaces at closing hour. In the first case, a person calls, by means of a mobile phone, for a robots in order to receive the service. The robot that has the transport functionality, is available and closest to that person, approaches the person, identifies the person, and guides him or her to the requested final destination. All this, with the aid of the distributed sensor network for the tasks of localization, identification, guidance, and robot navigation. In the second case, the trigger signal for the surveillance service of public space is the closing time or a human gesture. Then, the appropriate robots available for this service approach the area where people is gathered and directs them to leave the space.

2.2. Project Participants

The project participants are:

- Institut de Robótica i Informática Industrial, CSIC-UPC, Barcelona, Spain
- Laboratoire d'Analyse et d'Architecture des Systémes, CNRS, Toulouse, France
- Swiss Federal Institute of Technology Zurich, Switzerland
- Asociación de Investigaciony Cooperación Industrial de Andalucia, Seville, Spain
- Scuola Superiore di Studi Universitari e di Perfezionamento Sant'Anna, Pisa, Italy
- Universidad de Zaragoza, Spain
- Institute for Systems and Robotics, Instituto Superior Técnico, Lisbon, Portugal
- University of Surrey, Guildford, UK
- Urban Ecology Agency of Barcelona, Spain
- Telefónica I + D, Spain
- RoboTech, Italy

2.3. Barcelona Robot Lab

The Barcelona Robot Lab, built within the context of this project, is an outdoor urban experimental robotics site located at the Universitat Politécnica de Catalunya (UPC) campus, which includes 6 university buildings over a 10,000 m² area. In this area we have placed 21 fixed color video cameras connected through a Gigabit Ethernet connection to a computer rack. The position of the cameras is shown in Figure 2, with varying density in the positioning of the cameras, according to the various objectives of the project. Specifically, there are five cameras directed towards the Computer Science School (FIB) square, to have enough resolution for the detection of human gestures. There is also increased density in the positioning of the cameras in front of building A6, for the same reason. The rest of the area is covered with sparser density only to guarantee almost complete coverage of the area during people and robot tracking.

This outdoor facility is also equipped with 9 WLAN antennas with complete area coverage. Specifically for this project, we have installed the IEEE 802.11a protocol, contrary to the more common b/g/ or n networks. The reason was to limit as much as possible interferences with the university WLAN also covering the same area. The laboratory is also equipped with 802.15.4 wireless sensors for location purposes. Further information on the Barcelona Robot Lab experimental site can be seen at (http://www.urus.upc.es).

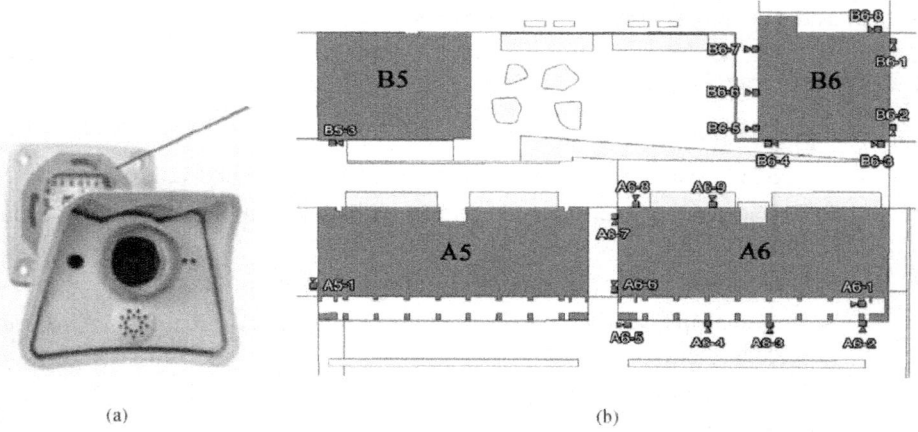

Figure 2. Camera network. (a) Colour video camera deployed in the URUS project. (b) The position of the cameras within the Barcelona Robot Lab.

2.4. ISRobotNet

The Institute for Systems and Robotics at Instituto Superior Técnico, in Lisbon, has developed also a testbed for NRS, the Intelligent Sensor and Robot Network (ISRobotNet), that enables testing a wide range of perception and robot navigation techniques, as well as human robot-interaction, distributed decision making, and task and resource allocation. The ISRobotNet testbed, also part of the URUS project, is composed of a 160 m^2 indoor area with 10 webcams placed at the ceiling such that some of their fields of view do not overlap. The cameras are distributed in 4 groups, each of which is managed by its own computer for image acquisition. The managing computers are connected to the global ISR/IST network and can be accessed by duly authorized external agents. Ongoing work will extend the number of cameras and the usable indoor space to include multiple floors. Robots will use the same elevators as ordinary people to move between floors. Besides the camera sensors, four Pioneer ATs and one ATRV-Jr robots are available. Each of the robots is equipped with sonar, onboard cameras, laser range finder and is Wi-Fi connected to the network. Figure 3 shows one of the floors at ISR/IST where the testbed is implemented and a map with the camera fields of view.

ISRobotNet is built around the kernel of a service-oriented architecture (MeR-MaID) [5], and extended to use the YARP networking software [6]. The testbed is being used to develop different components, namely, cooperative perception, people and robot detection and tracking using the camera network, and decision making. Arbitrary computational resources can be distributed over the network for fully decentralized use. Basic services such as robot teleoperation and direct image acquisition and recording from the camera network, as well as event logging, are also available.

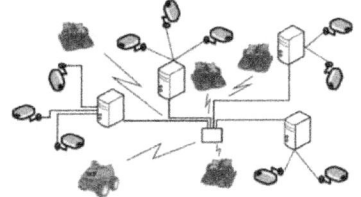

Figure 3. ISRobotNet.

3. The URUS Architecture

The sensor architecture of URUS has being designed to manage distributed information coming from different sensors and systems, and they have to be used for flexible and dynamic robot services. The architecture is distributed among subsystems which are interconnected through various means, Ethernet cable, WLAN and GSM/3G (see Figure 4).

The architecture is divided in three layers:

- Environment Layer:
  - The networked cameras oversee the environment and are connected through a Gigabit connection to a rack of servers.
  - The wireless Zigbee sensors all communicate to a single subsystem which is also connected to the system through one computer.
  - The WLAN environment antennas are connected through the Gigabit connection to the rack servers.
  - People use devices to connect to the robots and the environment sensors. For instance, a mobile phone with PDA features is connected through GSM/3G to the system.
- Robot Sensor Layer:
  - The robots have their own sensors connected through proprietary networks (usually Ethernet) which are connected through WLAN and GSM/3G to the system. A proprietary communications service has been

developed to transparently switch between WLAN and 3G depending on network availability.

- Server Layer:
    - The server rack (8 servers with 4 cores each) are connected through Ethernet to the Environment Layer and the Robot Sensor Layer.

The network of cameras is used to detect, track and identify people gestures in the urban area, as well as to detect and identify the robots. They can also be used for other services, such as surveillance, obstacle detection, *etc*. The wireless sensors, in this case Mica 2 sensor motes, are used mainly to enhance person and robot localization, based on radio signals. The Mica2 localization estimates are robustly fused with map-based localization methods in a decentralized way. Robot localization using GPS signals is not possible in the Barcelona Robot Lab, as in many urban areas due to satellite occlusion. The localization of the robots is so important, that we need to use as many methods as possible in order to minimize localization uncertainty.

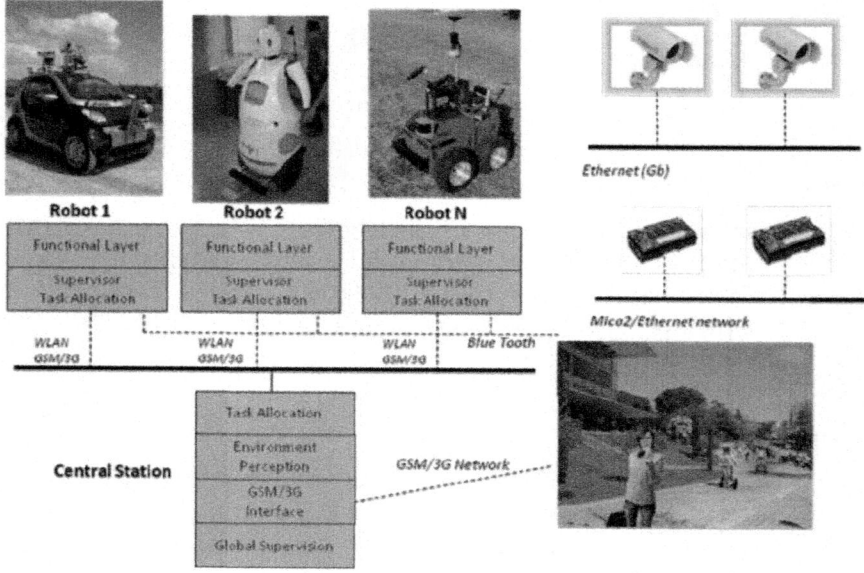

Figure 4. URUS global architecture.

The sensors onboard the robots are used mainly for navigation, localization and security; for robot and person identification, including the identification of human gestures and actions; and for human robot interaction.

Robots share information with other robots and with the server layer, through WLAN and GSM/3G. If WLAN is lost, for example when a robot crosses an area where the signal strength of the assigned AP (Access Point) is lost, then the robot either connects to another AP, or the robot connects to the system through GSM/3G.

The fully distributed software architecture is developed over YARP [6]. It can be used on any type of robots, sensors and human wearable devices. Specifically, we have developed standard procedures for connecting software modules allowing the share of information among the robots, sensors and human wearable devices.

At the level of sensor fusion for estimating system variables such as robot or people localization, data was fused in a decentralized form, with each component implementation independent from each other. However, at the experiment level, a central station was devised to handle task allocation and human operator commands through a GSM interface. This station had a finite state automata that took care of processing the experiment scripts and handling contingencies.

4. SENSORS IN THE URBAN SITE

4.1. Camera Network

The camera network consists of 21 IP color video cameras distributed around the Barcelona Robot Lab. The cameras are connected to each building switch with a Gigabit link. From each building, there is a Gigabit connection to the computer rack with servers dedicated to process URUS software. These servers are in a rack located in another building 200m away from the NRS (Network Robot System) area. Each camera has a dedicated IP and is only accessible from the proprietary network that we have built. The parameters of the cameras can be modified independently and they include a time stamp that is used in the tracking functions.

The camera network serves as a mean to detect, localize and map environmental information in a globally coherent frame of reference. Persons and robots must be localized in a unique coordinate system even though they are observed by distant cameras. We perform image-feature registration for camera calibration. In contrast to other approaches that use a calibration pattern or stereo geometry over architectural features to infer depth, depth information is obtained from the range map. The reason to use the range map to geometrically relate the cameras is because no overlapping fields of view are available in the camera network. Additionally, being an outdoor system, it is constantly susceptible to weather conditions, such as rain and wind, and thus it is expected to have slight but visible positioning and orientation changes from time to time. The calibration methodology must therefore encompass simple self-adjusting mechanisms.

The development of powerful laser sensors combined with Simultaneous Location and Mapping (SLAM) methodologies [7, 8] allow the possibility to have available high precision Laser Range Finder (LRF) data registered over large areas. These large outdoor LRF datasets have started recently to be acquired also for the purpose of creating robot navigation systems in urban areas. The LRF map is acquired over the complete area of the network and, in particular, contains the areas corresponding to the fields of view of the cameras. We exploit these novel technologies proposing a methodology for calibrating an outdoor distributed camera network using LRF data.

The laser based map, whose construction is discussed further in this section, is used as external information to aid the calibration procedure of the distributed camera network. For our case, in which the cameras have non-overlapping fields of view, it is impossible to estimate the relative position between them unless external data is used to refer the camera calibration parameters to a global reference frame. Since calibration inevitably requires some user intervention, in large camera networks this can be a very tedious procedure if one does not develop practical and semiautomated methods that facilitate and speed up user input. See Figure 5.

Figure 5. Camera calibration. (a) Lines are automatically extracted by intersecting planes on the LRF map, (b) The optimization method registers these 3D features with 2D lines on the images, and (c) Final reprojection of the segmented laser data after calibration.

The idea of the approach is the following: in a first stage, the LRF map is registered to an aerial view of the site and the user sets up the position and nominal calibration parameters of the cameras in the network. This allows user selection of an initial camera field of view onto the LRF area of interest likely to be observed by each camera. In a second stage, lines extracted from the LRF area of interest are represented in the nominally calibrated camera coordinate system and are reprojected to the real-time cameras' acquired images. This allows the user to perceive the calibration errors and input information to a non-linear optimization procedure that refines both intrinsic and extrinsic calibration parameters. The optimization process matches 3D lines to image lines. The 3D lines are extracted by intersecting planes on the segmented LRF set. A novel approach to 3D range segmentation based on local variation is used [9]. To show the applicability of the calibration results, homographies of the walking areas are computed [10].

4.2. Mica2 Network

Latest advances in low-power electronics and wireless communication systems have made possible a new generation of devices able to sense environmental variables and process this information. Moreover, they have wireless communication capabilities and are able to form ad-hoc networks and relay the information they gather to a gateway. This kind of system is usually called a Wireless Sensor Network (WSN).

In URUS, a network of wireless Crossbow's Mica2 sensor nodes is deployed in the urban scenario. Each Mica2 node (see Figure 6, a) has a suite of sensorial devices, including accelerometers, light and sound sensors, humidity sensors, *etc.*, a tiny processor, based on the ATmega128L microcontroller, and a communication module on the 868/916 MHz band. These sensor nodes can be used for monitoring applications. In addition, the signal strength received by the set of static nodes of the WSN (Received Signal Strength Indicator, RSSI) can be used to infer the position of a mobile object or a person carrying one of the nodes. And thus, one of the applications developed in the project is an algorithm for mobile sensor tracking, which can be used for tracking persons in urban scenarios.

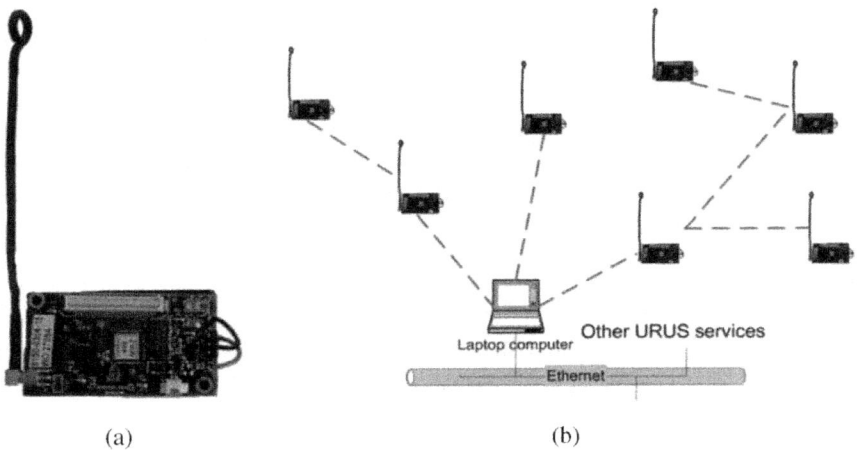

Figure 6. Mica2 network. (a) One of the sensor nodes employed in the system. (b) Scheme of the integration of the nodes into the architecture.

These sensors are integrated within the URUS system through a gateway which receives the information from the sensor nodes (see Figure 6b). The sensor nodes build themselves an ad-hoc network to send all the information to this gateway. In the gateway, a service that provides estimations on the position of mobile nodes is placed.

4.3. Site Map

We need to build the required maps for our heterogeneous fleet of service robots in urban settings [1]. With these maps the robots should be able to perform path planning and navigate to accomplish their tasks, such as guidance, assistance, transportation of goods, and surveillance. The experimental area has several levels and underpasses, poor GPS coverage, moderate vegetation, several points with aliasing (different points in the environment which cannot be disambiguated form local sensory data), large amounts of regularity from building structures, and sunlight exposure severely subject to shadows.

Our technique to Simultaneous Localization and Mapping (SLAM) is described in detail in [7, 11, 12], and is summarized here. We introduce a principled on-line approach for Pose SLAM, the variant of SLAM in which the state vector contains a history of visited robot locations, which only keeps non-redundant poses and highly informative links (see Figure 7). This is achieved by computing two measures; the distance between a given pair of poses and the mutual information gain when linking two poses. In [11], we show that, in Pose SLAM, the exact form of these two measures can be computed in constant time. When compared to other existing approaches [13], the proposed system produces a more compact map that translates into a significant reduction of the computational cost and a delay of the filter inconsistency, maintaining the quality of the estimation for longer mapping sequences.

By applying the proposed strategy, the robot closes only few loops and operates in open loop for long periods, which is feasible using recent odometric techniques [14, 15]. Our Pose SLAM approach includes a novel state recovery procedure at loop closure that takes advantage of the inherent sparsity of the information matrix scaling linearly both in time and memory. This computational cost is further amortized over the period where the robot operates in open loop, for which we introduce a factorization of the cross- covariance that allows the state to be updated in constant time. Thus, the proposed state recovery strategy outperforms state of the art approaches that take linear time for very sparse matrices (*i.e.*, when the robot operates in open loop), but are worst case quadratic when many loops are closed [13]. After this, the bottleneck for real time execution is not state recovery, but data association. That is, detecting poses close to the current one for which feature matching is likely. Exploiting the factorization of the cross-covariance, we introduced also in [11] a tree-based method for data association using interval arithmetic to encode the internal nodes of the tree. The main advantage of that method is that it improves the search up to logarithmic time. Moreover, by taking into account the cross-covariances from the very beginning it also avoids false positives, typically present in existing tree-based data association techniques[16]. This offers the possibility to use Pose SLAM for mapping large scale environments, such as the Barcelona Robot Lab.

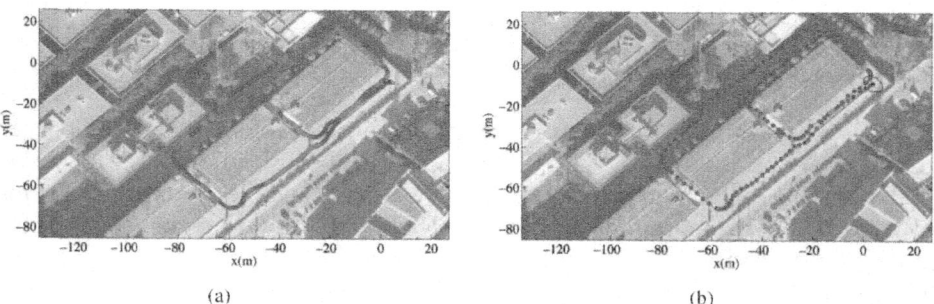

Figure 7. Filtered trajectory (in red) using encoder and visual odometry on a dataset collected at the Barcelona Robot Lab. Loop closure links are displayed in green, the black ellipse indicates the final robot location and its associated covariance at a 95% confidence level. (a) standard approach: incorporating all poses and all links to the filter; and (b) proposed approach: incorporating only relevant poses and links.

Aerial views of the site are shown in Figures 7 and 8. In order to build the maps described in this paper, we built our proprietary 3D scanning system, using a Leuze RS4 scanner and controlling its pitch with a DC motor and a computer. The system was installed atop an Activmedia Pioneer 2AT robotic platform. The system yields 3D point clouds with ranges up to 30 m, and sizes of about 76,000 points. The sensor noise level is 5 cm in depth estimation for each laser beam. Figure 9 portrays the complete device. The robot was tele-operated through the site along a path of over 600 m (see Figure 9b).

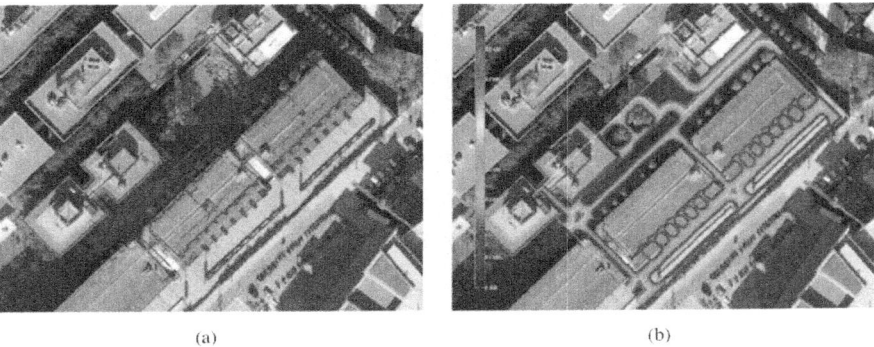

Figure 8. Traversability map built from 2D layers of three-dimensional aligned point clouds. (a) 2D layer superimposed on an aerial image. (b) Corresponding traversability map. Velocity varies from 0 m/s (blue) to 1 m/s (red).

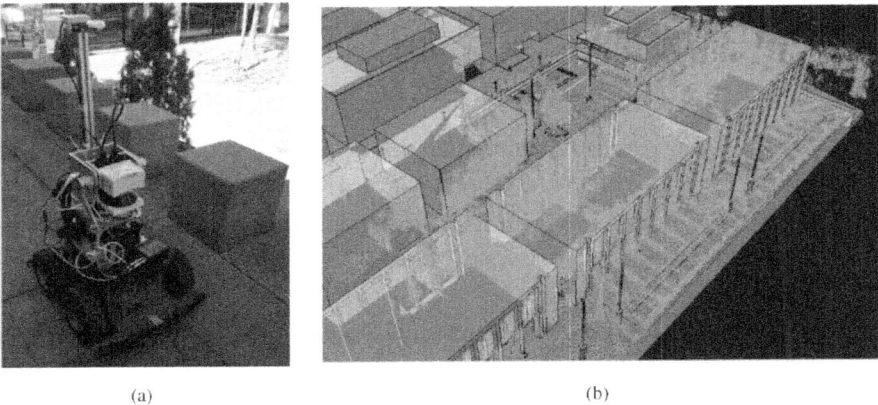

Figure 9. Site map. (a) Custom built 3D laser range scanner mounted on a Pioneer robotic platform. (b) Top view of the 3D map.

The figure contains results from state augmentation by concatenation of ICP computed motion constraints through the EIF Pose SLAM algorithm that closes 19 loops in the trajectory [7, 11, 12]. The figure also shows a comparison of the mapping results to a manually built CAD model of the experimental site. This model is made using geo-referenced information.

5. SENSORS INCLUDED IN URBAN ROBOTS

5.1. Sensors in the Robots Tibi and Dabo - Architecture and Functionalities

Tibi and Dabo (Figure 10) are two robots built at IRI with functionalities to navigate in urban areas, to assist and guide people from one origin to a destination. Traction is based on Segway RMP200 robotic platforms, and they have been customized with a variety of sensors:

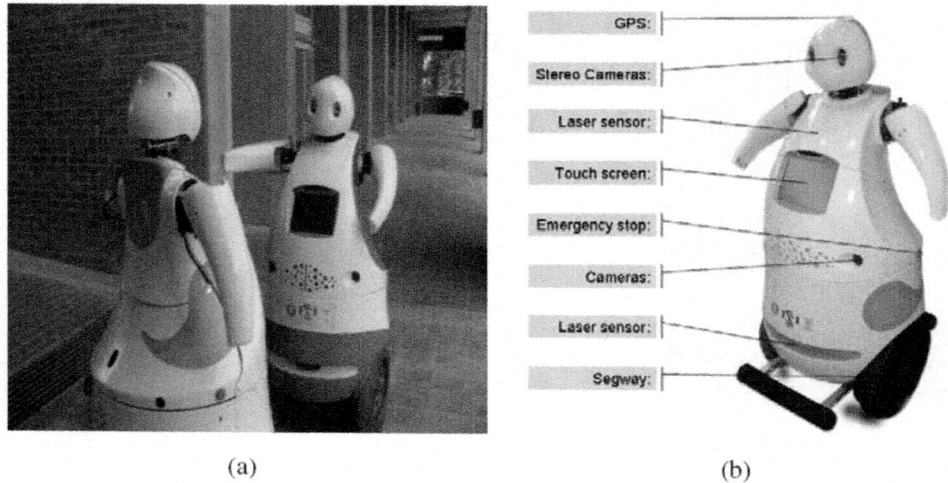

Figure 10. (a) Tibi and Dabo; (b) Tibi sensors.

- Sensors for navigation: Two Leuze RS4 and one Hokuyo laser rangefinders, as well as Segway's own odometric sensors (encoders and IMU).
  - The first Leuze rangefinder is located in the bottom front with a 180° horizontal view. This sensor is used for localization, security and navigation.
  - The second Leuze rangefinder is located in the bottom back, also with a 180° horizontal view. This sensor is used for localization, security and navigation.
  - The Hokuyo rangefinder is located in the front, but placed vertically about the robot chest, also with a 180° field of view. It is used for navigation and security.
- Sensors for global localization: GPS and compass.
  - The GPS is used for low resolution global localization, and can only be used in open areas where several satellites are visible. In particular, for

the URUS scenario, this type of sensor has very limited functionality due to loss of line of sight to satellites from building structures.
- The compass is used also used for recovering robot orientation. Also, in the URUS scenario, this sensor has proven of limited functionality due to its large uncertainty in the presence of metallic structures.
- Sensors for map building: One custom built 3D range scanner and two cameras.
  - The two cameras are located to the sides of the robots and facing front to ensure a good baseline for stereo triangulation, and they are used for map building in conjunction with the laser sensors. These cameras can also be used for localization and navigation.
  - A custom built 3D laser range finder unit has been developed in the context of the project. This unit, placed on top of a Pioneer platform has been used to register finely detailed three dimensional maps of the Barcelona Robot Lab that allow localization, map building, traversability computation, and calibration of the camera sensor network.
  - Vision sensors: One Bumblebee camera sensor.
  - The Bumblebee camera sensor is a stereo-vision system that is used for detection, tracking and identification of robots and human beings. Moreover we use this camera as image supplier for robot teleoperation.
- Tactile display:
  - The tactile display is used for Human Robot Interaction (HRI), to assist people and to display information about the status of the robot, as well as task specific information, such as destination information during a guidance service.

The internal architecture of these robots can be seen in Figure 11. We can see all hardware and sensors on the left side and their connection to the Acquisition/Actuation system. Colors in the figure indicate development state at an early stage of the project. Green indicates software modules that were completed, yellow indicates software modules that were in development at that time, and red indicates software modules that were still not developed. The second Acquisition/Actuation subsystem, second column in the figure, includes all drivers that connect hardware and sensors to the localization, motion controller, visual tracking, obstacle avoidance, gesture server and TS interaction modules. The safety subsystem in the third column uses the process monitor to control the robot platform, using the information from the robot sensors and the heart beats of the robot systems. The Estimation and Planning subsystem use the localization and path planning to control the robot motion. The different operation modes for these robots are: path execution, guiding person, tele-operation and human robot interaction. Each one of these operation modes are used in the URUS project to give specific services.

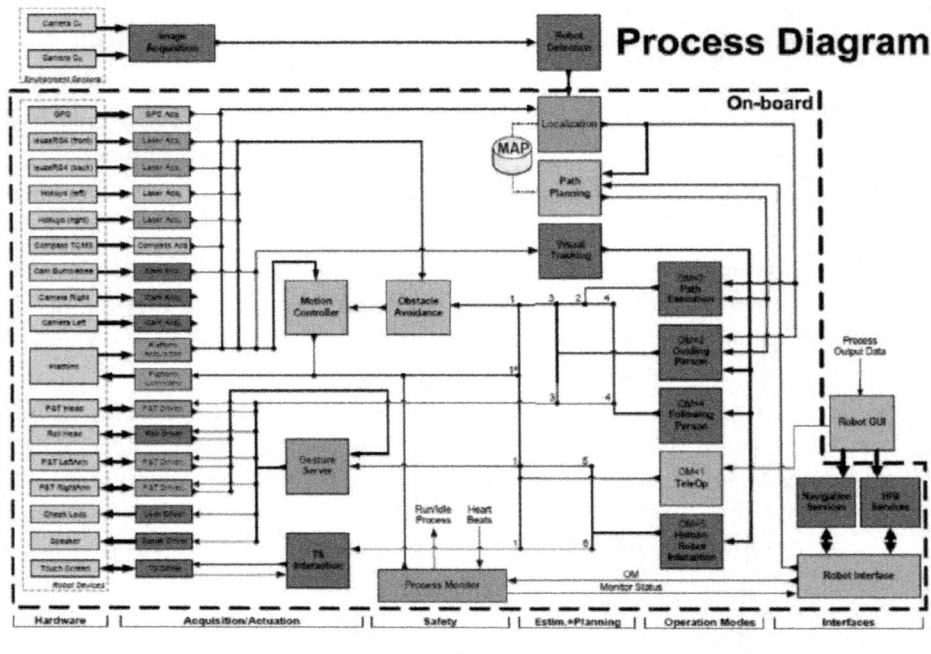

Figure 11. Internal architecture of Tibi & Dabo.

5.2. Romeo Sensors—Architecture and Functionalities

Romeo is an electric car modified with sensors and actuators to navigate autonomously in outdoor scenarios, including urban environments. Figure 12 shows the robot and its main sensors:

- Odometric sensors: Romeo has wheel encoders for velocity estimation, and a KVH Industries' gyroscope and an Inertial Measurement Unit (IMU) for angular velocity estimation.
- Rangefinders: Romeo has one SICK's LMS 220-30106 laser rangefinder located in the frontal part of the robot, at a height of 95 cm, for obstacle avoidance and localization. Moreover, it has 2 Hokuyo's URG-04LX (low range, up to 4 meters) in the back for backwards perception, and 1 Hokuyo's UTM-30LX (up to 30 meters) at the top of Romeo's roof and tilted for 3D perception.
- Novatel's OEM differential GPS receiver.
- Firewire color camera, which can be used for person tracking and guiding.
- Tactile screen, which is used for robot control and for human-robot interaction.

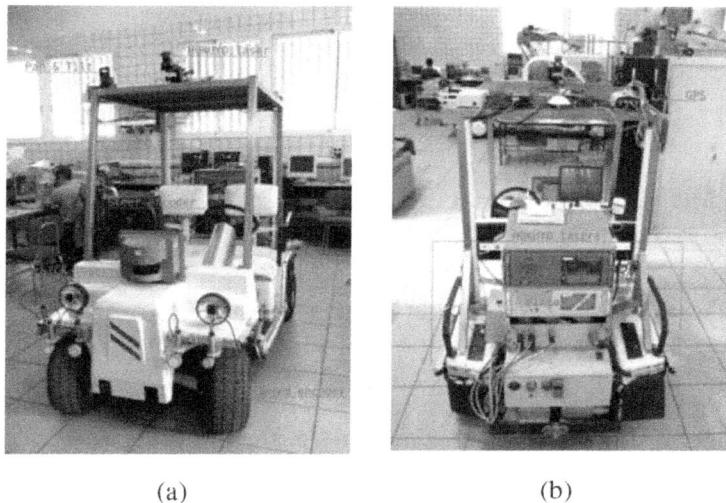

(a)   (b)

Figure 12. Sensors on board Romeo. Sensors for localization (GPS, gyro, encoders, Sick laser), map building and navigation (laser rangefinders). (a) front view; (b) rear view.

Figure 13, a, shows the basic software architecture for robot navigation in pedestrian environments. For localization, Romeo fuses all its odometric sensors (encoders, gyroscope, IMU) with an Extended Kalman Filter (EKF-LOC) to estimate its 6D pose. In order to compensate the drift associated to odometry, Romeo carries a differential GPS receiver. However, as stated before, in urban environments, often the GPS measurements are not available, or are affected by multi-path effects, which can produce erroneous estimates, thus localization is map-based and only assisted by GPS. All robots, Tibi, Dabo, and Romeo use the same map-based localization mechanism developed by Corominas *et al.* [17, 18], and described further in this paper, allowing the cancelation of drift error at a rate of 1 Hz.

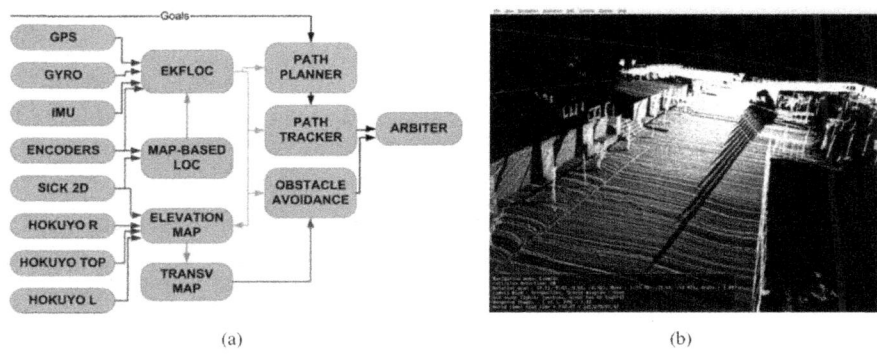

(a)   (b)

Figure 13. (a) basic architecture of Romeo. (b) 3D information obtained by Romeo from the Romeo lasers. Different obstacles can be identified, like stairs, trees or small steps.

For navigation, Romeo combines the pose information from the localization algorithm with the data from its lasers to build a 3D representation of the environment (see Figure 13b). From this representation, it is possible to obtain a traversability map; that is, a grid in which each cell indicates if it can be drivable by Romeo or not, such as the one shown in Figure 8b. The analysis to decide if a place is drivable or not depends mainly on the slope at each point of the map. At the same time, mobile obstacles can be also identified. This is used by the obstacle avoidance module to command the robot.

Romeo carries on board cameras that can be used for person tracking and guiding. The algorithms employed for this are based on a combination of person detection and tracking. The tracking algorithm is based on the mean shift technique [19]. In parallel, a face detection algorithm is applied to the image [20], the results from the tracking and the detection applications are combined, so that the robot employs the face detector when the tracker is lost to recover the track. As a result, the robots can obtain estimates of the pose of a person face on the image plane (see Figure 14).

Figure 14. People tracking from Romeo for guidance.

5.3. ISTRobotNet Architecture and Functionalities

IST is using Pioneer 3AT robots equipped with odometry, ultrasound, and laser range finder sensors (Figure 15). Basic navigation capabilities, namely path planning and obstacle avoidance, are available for each robot. Localization for these robots is also is based on the fusion of odometry and range information and the a priori defined world map. Contrary to the robots that navigate in the Barcelona Robot Lab, the CARMEN package is used for map building and localization at the ISRobotNet site. A grid is superimposed on the map obtained with CARMEN, and the centers of each cell constitute the set of waypoints used in the robot navigation.

An auction based high level supervision is implemented by a distributed component [21]. This approach is flexible enough to allow fast prototyping in behavior development. The robot tasks consist in the use of navigation behaviors to reach a given cell in the grid map. The robot supervisor is synthesized by formulating POMDP's for each possible task [22]. Thus, the robots task fitness is the value function of the POMDP for each task. The benefit of using the POMDP approach is that it is possible to compare the fitness of unrelated tasks, such as navigation and the use of onboard sensors for tracking.

The camera detection events are sent to the speculators, see Figure 15, which process this information and determine if a robot should approach the cell where the detection occurred. Since in general the number of detection events and mobile robots is different, an auction protocol is implemented to compute the optimal task assignment solution given the robots individual fitness values.

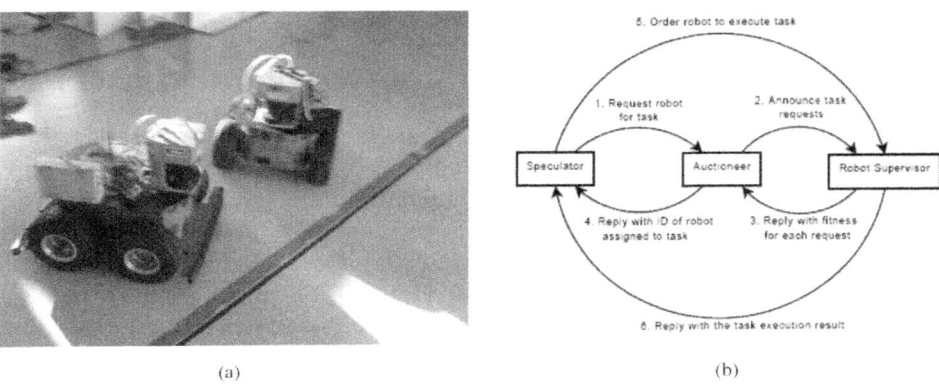

Figure 15. (a) IST robots and, (b) the auction based distributed approach to high level supervision.

Experiments integrating all the components required for the URUS project have been carried out at the ISRobotNet. A typical experiment can be identified with a rendezvous scenario where a service robot meets a person detected by the camera network. Each camera detecting persons and robots associates an uncertainty measure (covariance matrix) to each localization hypothesis in world coordinates. When targets are observed by more than one camera, a Bayesian data fusion process joins the information from the several cameras and provides a posteriori distribution of the data, assuming Gaussian uncertainties.

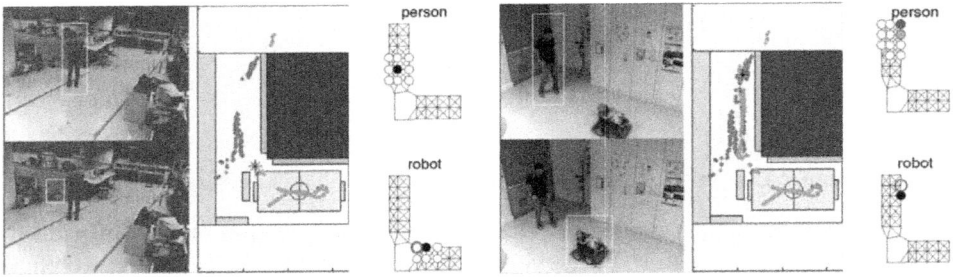

Figure 16. Rendezvous experiment. Integrated experiment at ISRobotNet.

Figure 16 shows (i) the detections provided by the camera network; (ii) the person and robot positions displayed in the 2D world map and (iii) the environment discrete grid cells. The maps show trajectories of persons and robots obtained by gathering information from all the cameras. False positive detections are due mainly to occlusions in the field of view of the cameras, but the false detection rate is very low and does not influence the probabilistic localization methods. The green and red dots indicate estimated robot and people trajectories, respectively. The box and circles map represents the likelihood of person and robot positions (the darker the cell, the higher the likelihood) and the blue circle shows the selected cell for commanding the next robot position, determined by the decision process. This information is then used to plan minimal cost paths accounting, for instance, for the relevance of visiting extra locations [23, 24].

6. DECENTRALIZED SENSOR FUSION FOR ROBOTIC SERVICES

From the perception point of view, in the URUS system (Figure 17), the information obtained by the fixed camera network or the wireless sensor network is fused with the information each robot obtains about the environment to improve perception estimates. In a guiding task for instance, the targets identified by the camera network (see Figure 17, right) can be combined with the information from the other systems (robots, and WSN, see Figure 17) to improve the tracking of the person being guided. Combining information from several sources, allows to cope with occlusions, to obtain better tracking capabilities as information of different modalities is employed, and to obtain predictions for uncovered zones, as the robots can move to cover these zones.

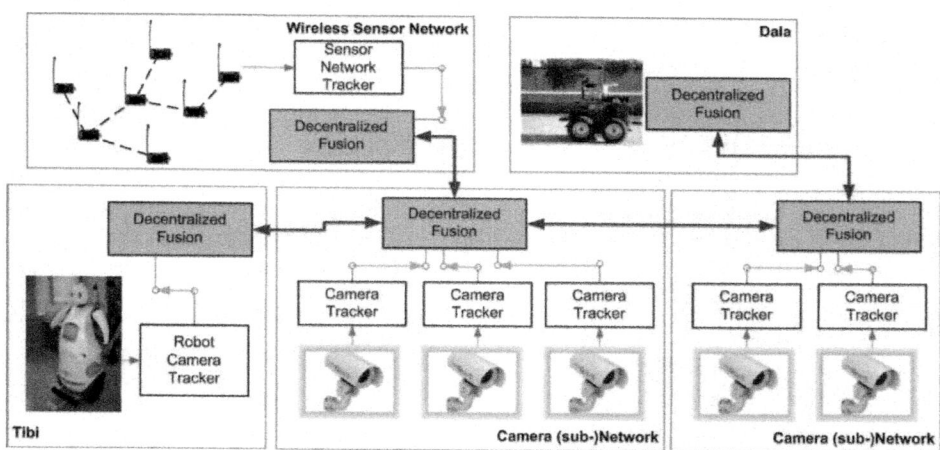

Figure 17. A block description of the sensor fusion architecture in URUS.

One option is to have a centralized system that receives all the information and performs data fusion. However, such approach presents some drawbacks, as it does not scale well with the number of robots and other systems; the central element can be a bottleneck; and communication failures are critical. Therefore, in URUS we have implemented a decentralized estimation system (as shown in Figure 17). In a decentralized data fusion system [25, 26], there are a set of fusion nodes which only employ local information (data from local sensors; for instance, a camera subnet, or the sensors on board the robot) to obtain a local estimate on the quantity of interest (for instance, the position of the person being guided).

Then, these nodes share their local estimates among themselves if they are within communication range. The main idea is that, as the nodes only use local communications, the system is scalable. Also, as each node accumulates information from its local sensors, temporal communication failures can be coped without losing information. Finally, fusing information from neighbor nodes allows for consistent global estimation in a decentralized form.

Figure 18. (a-b) Tracks obtained by the camera on board Romeo. (c-d) Tracks obtained by the camera network in the same experiment.

Using the trackers developed in the project, the camera network and the robots are able to obtain local estimates of the variable at hand *(i.e.,* the position of people on the image plane, see Figure 18). These estimates, characterized as Gaussian distributions (mean and covariance matrix) and the ones provided by the WSN, can be fused in order to obtain an accurate estimate of such variable. The Information Filter (IF), which corresponds to the dual implementation of the Kalman Filter (KF), is very suitable for decentralized data fusion using Gaussian estimates. While the KF represents the distribution using its first and second order moments (mean μ and covariance Σ), the IF employs the so-called canonical representation, encoded by the information vector $\eta = \Sigma^{-1}\mu$ and the information matrix $\Omega = \Sigma^{-1}$. Prediction and updates for the linear IF are easily derived from the linear KF [3]. In the case of nonlinear motion models *(i.e.,* robot kinematics) and nonlinear measurement functions *(i.e.,* perspective projection on cameras), first order linearization leads to the Extended Information Filter (EIF).

The main interest of the EIF, compared to the EKF, is that it can be easily decentralized. In a decentralized approach, each node of the network employs only local measurements z_i, of the event at hand (as said, for instance, the position of the person as observed by a local sensor), to obtain a local estimate of such event (the person trajectory), represented by η_i and Ω_i, and then, shares this estimate with its neighbors. The information coming from neighboring nodes is locally fused in order to improve such local estimate. The decentralized fusion rule produces locally, the same result that would be obtained using a centralized EIF. This is because for the EIF, data fusion is an additive process [26]:

$$\eta_i \leftarrow \eta_i + \eta_j - \eta_{ij}$$
$$\Omega_i \leftarrow \Omega_i + \Omega_j - \Omega_{ij}$$

The update rules state that each node should sum up the information received from other nodes. The additional terms η_{ij} and Ω_{ij} represent the common information between nodes. This common information is due to previous communications between nodes, and should be removed to avoid double counting of information, known as rumor propagation. This common information is maintained by a separated EIF called channel filter [27]. This common information is locally estimated assuming a tree-shaped network topology, *i.e.*, there exist no cycles or duplicated paths of information.

The decentralized system has as advantages that the system is scalable, as each fusion node employs only local communications. Moreover, communications dropouts do not compromise the system (although the performance can be degraded during the dropout). Another advantage of using delayed states is that the belief states can be received asynchronously. Each node in the network can accumulate evidence, and send it whenever it is possible. However, as the dimension of the state trajectories grow with time, the size of the message needed to communicate the estimate of a node also does. For the normal operation of the system, only the state trajectory over a limited time interval is needed, so these trajectories can be bounded. Note that the trajectories should be longer than the maximum expected delay in the network in order not to miss any measurements information.

7. SOFTWARE ARCHITECTURE TO MANAGE SENSORS NETWORKS

The ever-increasing level of autonomy demanded by networked robotic systems is naturally leading to an increase in the complexity in their components. In robotics parlance these components and strategies are known as middleware, following similar designations in computer engineering, referring to computer software that connects software components or applications. MeRMaID (Multiple-Robot Middleware for Intelligent Decision-making) is a robot programming framework, used in URUS, whose goal is to provide a simplified and systematic high-level behavior programming environment for multi-robot teams, supported on a middleware layer [5]. This middleware layer uses the Service concept as its basic building block.

The middleware layer used in URUS extends that developed for RoboCup competitions [28]. The functional architecture, guides the development of the embedded components so as to improve overall dependability. This way, MeRMaID ensures that separately developed components will be more easily assembled together at integration time. The use of this architectural layer is not mandatory. The developer can build its application using only the support layer, without any constraints on how to connect subsystems. Figure 19 depicts an overview of the structure of both levels of MeRMaID. For a thorough description of MeRMaID and its functionalities, refer to [5].

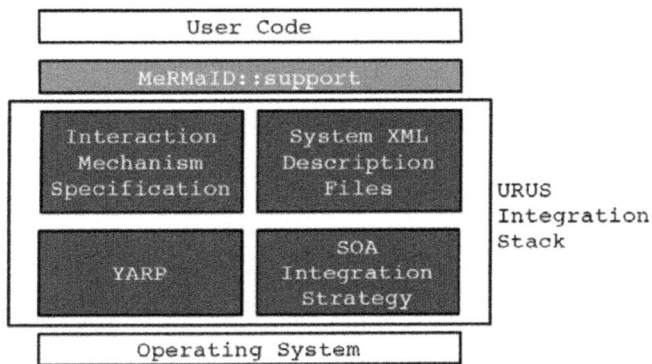

Figure 19. General structure overview of the middleware layer MeRMaID::support.

Among the advantages in using the MeRMaID framework are (i) reduced time to have third part components fully integrated in to the system, and (ii) the automatic validation of the interactions among services, assuring the developer that all protocols and specifications are correctly followed. The integration of functionality provided by image processing software connected to the camera network in this setup is an example of quick integration with software not developed using MeRMaID. The software to detect people and robots was developed by an independent team without any constraint on how they would later integrate it with the rest of the system at ISRobotNet. Given that the software was able to detect events visible by the camera network that occur asynchronously, the Data Feed mechanism was chosen as appropriate to make this information available. In fact, the whole camera network can be seen as just a producer of a stream of asynchronous events. It was quite simple to insert a valid data feed mechanism in the camera network code. Developers only needed to write data to a YARP Port (which is YARP's fundamental building block for communications) and populate the system's description file with a declaration of what data they were exporting. This approach can be applied for any data producing software component in the system. Services developed using MeRMaID Support also benefit from simplified access and full syntactic validation of the sent data.

8. SOME RESULTS IN THE URUS PROJECT

In this section we only explain briefly three examples developed in the URUS project, where a combination of sensors is used for specific robotic functionalities:

localization using robot sensors and environment sensors; tracking people and detection of gestures; and tracking of persons and robots using Mica2 sensors.

8.1. Localization

One of the main functions that any robot operating in an urban environment must have is localization. Without this functionality the robots cannot perform missions like, "Go to destination X to pick up a person and guide him to destination Y". The robot must first know its position and pose with respect to the global reference frame, and then know where is the X and Y positions are within that reference frame, to be able to navigate and guide the person. One could think this is a trivial task in an urban setting that could be solved for instance, with a GPS sensor. However, GPS signals are often unreliable for autonomous robot navigation in urban pedestrian areas: there is lack of satellite visibility for accurate position estimation. For this reason, we have developed a map-based position estimation method for precise and fast localization using the information sensed by the laser rangefinders installed in the robot, a GIS-augmented map of the environment, and the camera network. For the robots Tibi and Dabo, we use the front and rear laser sensors. Usually, the aggregation of evidence from laser data is sufficient for accurate map-based localization. Nonetheless, in some cases of aliasing, estimates on robot location coming from the camera network help disambiguate between similar localization hypotheses, and also as strong localization evidence in densely populated areas [17, 29]. The method in [30] is based on a particle filter approach [31], and is summarized here. Note that the localization mechanism described next accounts for a local node of the decentralized position estimation mechanism detailed in Section 6. Other nodes include for instance, each robot localization estimate coming from each of the cameras in the network.

Let $X^t_r = (x^t_r, y^t_r, \theta^t_r)$ be the robot state a time t, and assume it is limited to a bounding box in the three-dimensional pose space, $X^t_r \in \Gamma = \{(x_{min}, x_{max}), (y_{min}, y_{max}), (-\pi, \pi)\}$. The particle filter steps are:

1. Propagation: All particles are propagated using the kinematic model of the robot and the odometric observation.

$$X^t_i = f(X^{t-1}_i, o^t_0); \forall i = 1 \ldots N_p$$

2. Correction: Particle weights are updated according to the likelihood of the particle state given the observations, $k = 1 \ldots N_B$:

$$w^j_i = w^{j-1}_i \prod_{k=1}^{N_B} p(X^t_i | o^t_k); \forall i = 1 \ldots N_p$$

where $p(X^t_i | o^t_k) = L_k(o^t_k, o^g_k(X^t_i))$ and $L_k(o^t_k, o^g_k(X^t_i))$ is the likelihood function between two observations: the current one made by the k-th observer, o^t_k (a ray traced from the laser scanner), and the expected one $o^g_k(Xt)$, computed using the k-th observation and the environment model. This $o^g_k(X^t_i)$ observation is called *synthetic* because is computed using models. We have as many likelihood functions as sensors we have, and they are included in the computation of this likelihood. For example, for the laser scan this function is:

$$L_1(o^t_1, o^q_1(x^t_i)) = \frac{1}{N_L} \sum_{j=1}^{N_L} \text{erfc} \frac{\sigma^t_{1,j} - \sigma^g_{1,i}(X^t_i)}{\sigma_1 \sqrt{2}}$$

where σ_1 is the noise for the laser observation and

$$\sigma^q_{1,i}(X^t_i) = \text{raytr}\left(x^t_i, y^t_i, \theta^t_i - \frac{\Delta\alpha}{2} + j\frac{\Delta\alpha}{N_L}\right)$$

with $\Delta\alpha$ the aperture angle of the laser scan and raytr (X_p) is a function to evaluate the distance of a given position in the map, X_p, with the closest obstacle of the map, following an extended ray casted at the proper heading. The details of the other two steps, set estimate and resampling, can be seen in [30].

Integration of Asynchronous Data

Given the asynchronous nature of the data coming from the various sensors (laser scanners, odometry, and eventually GPS) (see Figure 20a), the particle filter integrates observations taking into account their time stamps. The filter does not propagate the particle set once per iteration as classical approaches do. Instead, the proposed algorithm propagates the particle set only when a new observation arrives with a time stamp greater than the last propagation. At this point, the filter propagates with the kinematic model and the odometric observation and, then, integrates the observation as a delayed one. In order to integrate delayed observations, the proposed approach maintains a delayed history of past estimates and back-propagates particles to compute observation models at those instances where particles were expected to be [32] (see Figure 20b). The experiments using this particle filter in the Barcelona Robot Lab can be seen in Figure 21.

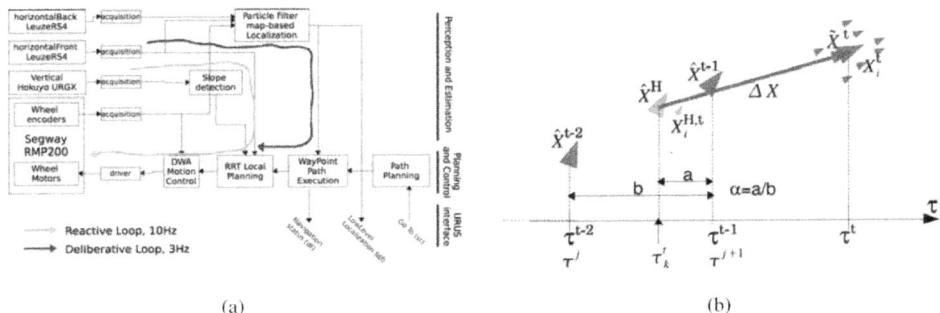

Figure 20. (a) Tibi and Dabo asynchronous navigation. (b) Particle propagation. Backward propagation of the particle X^t_i when integrating observation o^t_k.

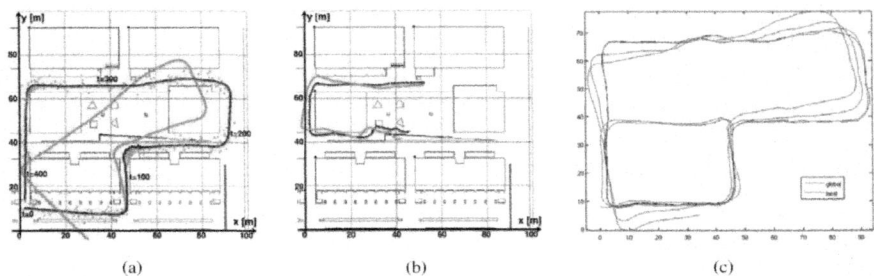

Figure 21. Robot localization. (a) Simulated results with multiple sensors. (b) Robot localization in the real outdoor campus environment with the robot Tibi. The red path indicates ground truth, the green path indicates robot odometry, and the blue path the resulting filter estimate. (c) Map based localization results in the same environment for the robot Romeo. The red path indicates local estimates, and the blue path indicates global localization.

8.2. Tracking People and Detecting Gestures Using the Camera Network of Barcelona Robot Lab

8.2.1. The Method

The fixed cameras cover a wide area of the experiment site and are the basis for the fusion of the other sensors, they are able to track objects of interest both on and across different cameras without explicit calibration. The approach is based on a method proposed by Gilbert and Bowden [33]. Intra camera objects of interest are identified with a Gaussian mixture model [34] and are linked temporally with a Kalman filter to provide intra camera movement trajectories. When the object of interest enters a new camera field of view, the transfer of the object position estimate to the new camera is challenging since cameras have no overlapping fields of view, making many traditional image plane calibration techniques impossible. In addition, handling a large number of cameras means that traditional time consuming calibration is impractical. Therefore, the approach needs to learn the relationships between the cameras automatically. This is achieved by the way of two cues, modeling the color, and the inter-camera motion of objects. These two weak cues, are then combined to allow the technique to determine if objects have been previously tracked on another camera or are new object instances. The approach learns these camera relationships, and unlike previous work, it does not require a priori calibration or explicit training periods. Incrementally learning the cues over time allows for accuracy increase without supervised input.

8.2.2. Forming Temporal Links between Cameras

To learn the temporal links between cameras, we make use of the key assumption that, given time, objects (such as people) will follow similar routes inter-camera and that the repetition of the routes will form marked and consistent trends in the overall data. Initially the system is subdivided so that each camera is a single region. It identifies temporal reappearance links at the camera-to-camera level.

After sufficient evidence has been accumulated, the noise floor level is measured for each link. If the maximum peak of the distribution is found to exceed the noise floor level, this indicates a possible correlation between the two blocks as shown in Figure 22. If a link is found between two regions, they are both subdivided to each create four new equal sized regions providing a higher level of detail. Regions with little or no data are removed to maintain scalability.

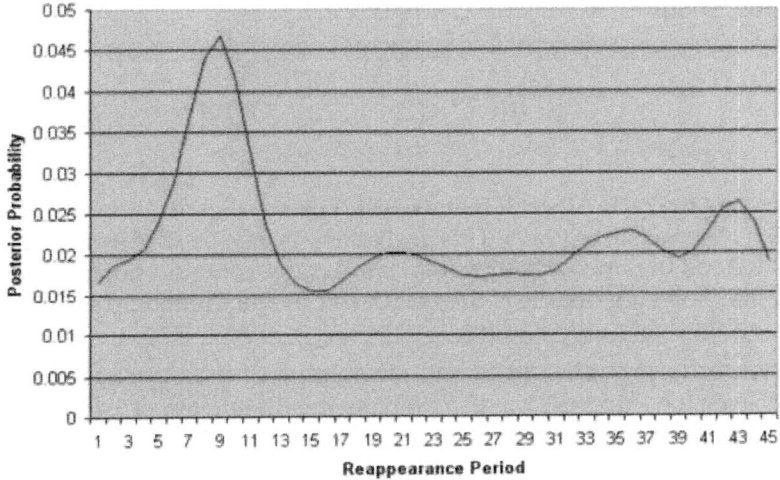

Figure 22. An example of a probability distribution showing a distinct link between two regions.

8.2.3. Modelling Color Variations

The color quantization descriptor, used to form temporal reappearance links, assumes a similar color response between cameras. However this is seldom the case. Therefore, a color calibration of these cameras is proposed that can be learnt incrementally, simultaneously with the temporal relationships discussed in the section above. People tracked inter-camera are automatically used as the calibration objects, and a transformation matrix is formed incrementally to model the color changes between specific cameras.

The transformation matrices for the cameras are initialized as identity matrices assuming a uniform prior of color variation between cameras. When a person is tracked inter-camera and is identified as the same object, the difference between the two color descriptors, is modeled by a transform matrix. The matrix is calculated by computing the transformation that maps the person's descriptor from the previous camera to the person's current descriptor. This transformation is computed via SVD. The matrix is then averaged with the appropriate camera transformation matrix, and repeated with other tracked people to gradually build a color transformation between cameras.

8.2.4. Calculating Posterior Appearance Distributions

With the weak cues learnt, when an object which leaves a region, we can model its reappearance probability over time as

$$P(O_t|O_y) = \sum_{\forall x} w_x P(O_{x,t}|O_y)$$

where the weight w_x at time t is given as

$$w_x = \frac{\sum_{i=0}^{T} f_\Phi^{x|y}}{\sum_{\forall y} \sum_{i=0}^{T} f_\Phi^{x|y}}$$

representing the ratio of weak feature responses in region x given the feature has also been observed in region y. This probability is then used to weight the observation likelihood obtained through color similarity to obtain a posterior probability of a match. Tracking objects is then achieved by maximizing the posterior probability within a set time window.

In order to provide additional information of the objects of interest being tracked over the cameras, we also address the distinction of the objects by classifying them as a person or a robot.

8.2.5. Classification of Objects of Interest as Person or Robot

The intra camera tracking algorithm provides a unique identifier for every object of interest based on color cues. In addition to the identifier, the discrimination of the objects between people and robots provides a useful cue for the fusion procedure. We propose to categorize the objects of interest as human or robots by the differences on their motion patterns, which are obtained from the optical flow [35]. We assume the robots are rigid bodies that produce flow vectors with very similar orientations, while a person produces patterns on different orientations.

The discrimination of the motion patterns relies on: (i) optical flow-based features and (ii) a learning algorithm with temporal consistency. We compare two types of flow features, which have been used previously to detect people: (i) Histogram of gradients (HOG) [36], which computes the histogram of the optic flow orientation weighted by its magnitude and (ii) Motion boundary histogram (MBH) [37], computed from the gradient of the optical flow. Similarly to HOG, this feature is obtained by the weighted histogram of the optical flow's gradient.

In order to classify an object of interest at every frame, a boosting algorithm uses labeled samples of either HOG or MBH features in a binary problem: person vs. robot. We use the GentleBoost algorithm, which provides a framework to sequentially fit additive models (*i.e.*, weak learners) in order to build a final strong classifier. We select a weak learner that considers the temporal evolution of the features, the Temporal Stumps [35], which are an extension of the commonly used decision stumps. The output of the strong classifier at each frame is related to the person/robot likelihood.

8.2.6. Gesture Detection

The key gesture to be recognized by the camera network is people waving. The importance of this specific gesture is due to its "universal" nature: it is used as an attention triggering and emergency indicator by most people independently of their culture. It is worth to point that state-of-the-art, high performance, hand gesture recognition systems still present robustness problems that limit their application to (i) high resolution targets and (ii) uncluttered backgrounds. Thus, the project team concentrated its efforts in developing practical and robust waving detectors that can be applied to low resolution targets and arbitrary background clutter for outdoors environments. We address the robustness under different conditions by implementing two specialized waving detectors: (a) one suited for low resolution targets that relies on the local temporal consistency of optical flow-based features and (ii) the second one suited for arbitrary clutter on the image, which relies on data mining of a very large set of spatio-temporal features. In the following we describe the techniques utilized for each detector.

1. Local temporal consistency of flow-based features [38]. This approach relies on a qualitative representation of body parts' movements in order to build the model of waving patterns. Human activity is modeled using simple motion statistics information, not requiring the (time-consuming) pose reconstruction of parts of the human body. We use focus of attention (FOA) features [39], which compute optical flow statistics with respect to the target's centroid. In order to detect waving activities at every frame, a boosting algorithm uses labeled samples of FOA features in a binary problem: waving vs not waving. We use the Temporal Gentleboost algorithm [40], which improves boosting performance by adding a new parameter to the weak classifier: the (short-term) temporal support of the features. We improve the noise robustness of the boosting classification by defining a waving event, which imposes the occurrence of a minimum number of single-frame waving classifications in a suitably defined temporal window.

2. Scale Invariant Mined Dense Corners Method. The generic human action detector [33] utilizes an over complete set of features, that are data mined to find the optimal subset to represent an action class.

   Space-time features have shown good performance for action recognition [41, 42]. They can provide a compact representation of interest points, with the ability to be invariant to some image transformations. While many are designed to be sparse in occurrence [40], we use dense simple 2D Harris corners [43]. While sparsity makes the problem tractable, it is not necessarily optimal in terms of class separability and classification.

   The features are detected in (x,y), (x,t) and (y,t) channels in the sequences at multiple image scales. This provides information on spatial and temporal image changes but is a far denser detection rate than 3D Harris corners

[44] and encodes both spatial and spatio-temporal aspects of the data. The over complete set of features are then reduced through the levels of mining. Figure 23 shows the large amount of corners detected on two frames.

(a)   (b)

Figure 23. 2D Harris corner detection on two frames.

8.2.7. Neighbourhood Grouping

Given a set of encoded corners, their relationships to other corners are used to build stronger compound features. Our approach is to define neighborhoods centered upon the feature that encode the relative displacement in terms of angle rather than distance hence achieving scale invariance. To do this, each detected interest point forms the centre of a neighborhood. The neighborhood is divided into 8 quadrants in the x, y, t domain which radiate from the centre of the neighborhood out to the borders of the image in x, y and one frame either side either $t - 1$ or $t, t + 1$ (see Figure 24b,c).

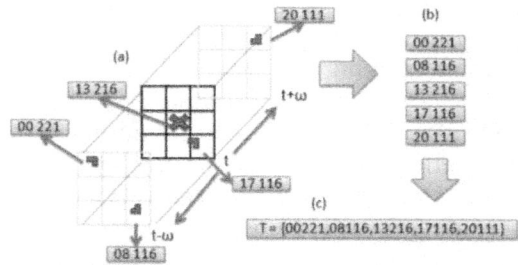

Figure 24. (a) Close-up example of a 3x3x3 neighborhood of an interest point, with size local features shown as corners. (b) Spatial and temporal encoding applied to each local feature. (c) Concatenating the local features into a transaction vector for this interest point.

Each quadrant is given a label, all feature codes found within a unique quadrant are appended with the quadrant label. A vector of these elements is formed for every interest point found in the video sequence and contains the relative spatial encoding to all other features on the frame. This called a Transaction file, these are collected together into a large single file, containing around 2,000,000 transactions, where a single transaction contains around 400 items or encoded features. To condense or summarize this vast amount of information, a priori data mining is employed [45]. This computationally efficient approach, finds the frequent co-occurring configurations of encoded features among the tens of thousands of transaction vectors from the training sequences. Given a new query frame sequence, the features are detected and grouped as outlined above. This forms a new query set of transactions. A global classifier then exhaustively compares frequent mined configurations to the image feature groups in the query transaction set. It works as a voting scheme by accumulating the occurrences of the mined compound features. In order to learn the wave class for the URUS system, the training database is based on the KTH dataset [41]. We used the training setup proposed by the authors, in addition we added training examples from the URUS fixed cameras, and negative examples from other action classes and false positive detections from the URUS cameras.

8.2.8. Fixed Camera Experiments

A series of experiments were performed on the fixed camera system, to illustrate the incrementally learnt cross camera relations, the inter-camera relationships were learnt for a total of 5 days. The number of tracked objects on each camera was 200 per day. This is relatively low and made the region subdivision unsuitable after the second level of subdivision. Figure 25 shows resultant temporal likelihoods for a number of inter-camera links at a single subdivision level.

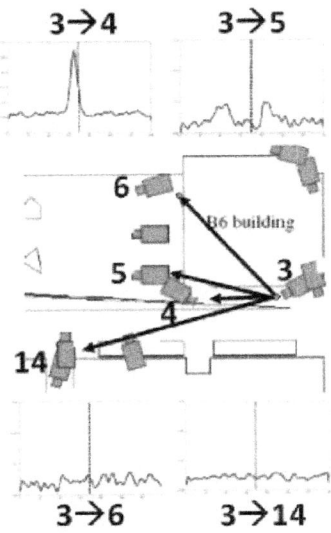

Figure 25. Inter camera temporal likelihoods.

The black vertical line indicates a reappearance of zero seconds, it can be seen that there is strong links between cameras 3 and 4 and between 3 and 5, while there is no visible link between 3 and 6 and between 3 and 14. This is due to the increased distance and people will rarely reappear on cameras 6 and 14 after they were tracked on camera 3.

Table 1 shows the accuracy results of tracking people inter-camera. The inter-camera links were formed over up to 5 days and the test sequence consists of a 1 hour sequence on the cameras, with a total of 50 people tracked inter-camera. A 1 subdivision is a region per camera, 2 subdivision is the where any linked regions are subdivided. All people that moved inter-camera were ground-truthed and a true positive occurred when a person was assigned the same ID as that they were assigned on a previous camera.

Table 1. Accuracy of fixed inter-camera tracking over days.

		Data amount (days)			
Method		0	1	2	5
1	Subdiv	34%	38%	56%	78%
2	Subdiv	34%	10%	60%	83%

(a)
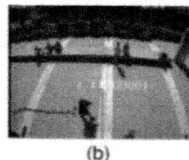
(b)

(c)
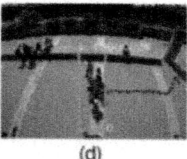
(d)

Figure 26. Cross camera tracking; (a) Person 11000001 on camera 11, (b) Person 11000001 correctly identified on camera 12 (c) Person 13000027 on camera 13 (d) Person 13000027 correctly identified on camera 12.

The column for 0 days indicates performance without learning the camera time and color relationships. It is poor generally due to large color variations inter-camera due to shadow and lighting changes. The 2 level subdivision initially performs poorly as it requires greater data to build relationships. However by 5 days significant improvement is shown for both one and two levels of region subdivision. Little performance is gained from the additional subdivision on this system due to the lower levels of traffic and low level of routes between the cameras due to their closeness. However for a more distributed system the additional detail of the region relationships would aid the tracking performance greater. Figure 26 gives example frames of tracking inter-camera for two separate people.

8.2.9. Classification of Robots and Humans

An initial test of the person *vs.* robot classifier was performed on the ISRobot-Net. We grabbed five groups of sequences, where each one includes images from 10 cameras. One group with a person walking, another group with a different person walking, two groups with the same Pioneer robot moving in two different conditions, and the last group with a third person loitering. The people class videos have a total of 9,500 samples of the optical flow and the robot class videos have a total of 4,100 samples. We follow a cross validation approach to compare the classification result of the GentleBoost algorithm, so we build two different groups of training and testing sets. The best classification result for this setup was 95% recognition rate, provided by the MBH feature [35]. Figure 27 shows examples of correct detections of people and robots.

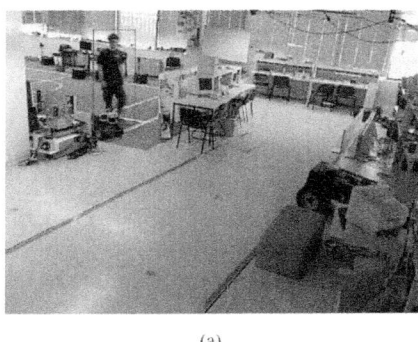 

(a) (b)

Figure 27. Examples of correct detections of (a) a person and (b) a Pioneer robot.

(a) (b)

Figure 28. Examples of correct detections of (a) a person and (b) two robots.

From the results at the ISRobotNet, we select the MBH feature for a second group of tests, at the Barcelona Robot Lab. It is important to mention that the localization of some of the cameras on the Barcelona Robot Lab provide a point of view with large perspective effects on the optical flow features, so it is necessary to perform the training step with the new sequences. We grabbed sequences with people and robots performing naturally (*i.e.*, no given script or pre-defined actions known) for each camera. The actions include walking, loitering and waiting for person. The people class videos have a total of 19,444 samples of the optical flow and the robot class videos have a total of 2,955 samples. We follow the approach of the ISRobotNet test, building two different groups of training and testing sets. The average recognition rate of the two sets is 88.4%. Figure 28 shows examples of correct detections of people and robots.

8.2.10. Gesture Detection with Local Temporal Consistency of Flow-based Features

The learning step of the Temporal Gentleboost algorithm with the FOA features was performed at the ISRobotNet. The training sequences have 4,229 frames (2,303 waving and 1,926 not waving), and the testing sequence has 4,600 frames (1,355 waving and 3,245 not waving). The FOA feature sampling is $\pi/4$. The support window of the Temporal boost algorithm is 20 frames. The event window size is 2s (20 frames), considering a waving event if at least 60% of the single-frame classifications are positive. We follow a cross validation approach to compare the classification result of the GentleBoost algorithm, so we build two different groups of training and testing sets. The average performance of the two sets is 94.4%, on the ISRobotNet.

We assume that the person is not translating on the image while performing the waving action, so the features selected by the Gentleboost algorithm on the ISRobotNet setup, are able to generalize well to different conditions, such as the target distance to the camera and the target size on the image. Thus, we use the parameters of the GentleBoost algorithm learned on the ISRobotNet in order to classify waving events in sequences grabbed at the Barcelona Robot Lab. The testing sequences have 3,288 (1,150 waving and 2,138 not waving). The recognition rate was 86.3%, which shows the generalization capabilities of the method. Figure 29 shows examples of the waving detection on a sequence recorded at the Barcelona Robot Lab.

(a)   (b)

Figure 29. Examples of correct detections of a waving event, marked by the word waving on the bounding box.

It should be noted however that these performance results are preliminary, and a more exhaustive test campaign of the method is envisaged. It is necessary for instance, to validate the method under significant variations of illuminants (severe shadows, cloudy days, different hours of the day, *etc.*). This is the subject of further research.

8.3. Tracking with Mica2 Nodes

The WSN deployed in the environment can be used to estimate the position of mobile nodes of the network from the measured signal strength received by each static node, which is used for person tracking (see Figure 30a).

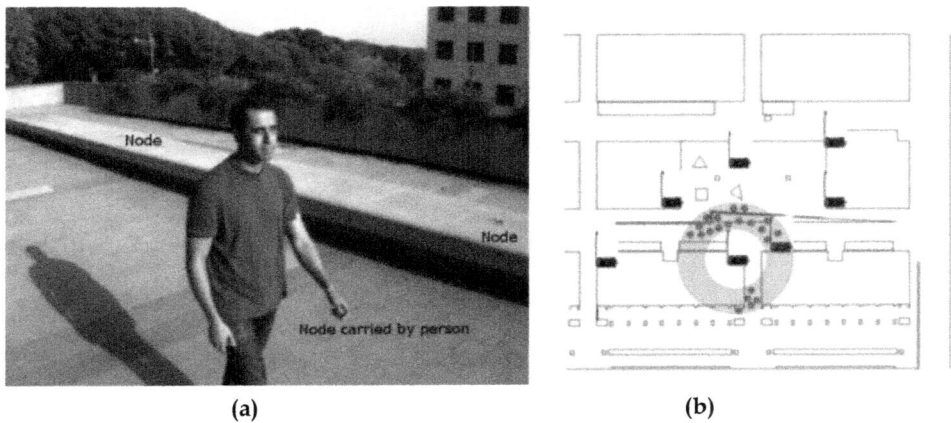

(a) (b)

Figure 30. (a) Setup for person tracking based on radio-signal power. (b) Each power measurement delimits a feasible region within the map.

The algorithm to estimate the mobile node position is, as the robot localization algorithm discussed earlier, also based on particle filtering. In this case however, the current belief represents the position of the mobile node and is described by a set of particles; each particle representing a potential hypothesis about the actual position of the person carrying the node (see Figure 30a). At each filter iteration, kinematic models of the motion of the person and map information are used to predict the future position of the particles. The addition of the map information allows identifying and discarding unlikely motions with simple collision analysis. When the network of static nodes receives new messages from the mobile node, the weight of the different particles is updated with the RSSI indicator. By considering radio propagation models, it is possible to determine the likelihood of receiving a certain signal power as a function of the distance from the particle to the receiving node [46]. Each transmission restricts the position of the particles to an annular shaped area around the receiving node (see Figure 30b).

As a result, the filter provides an estimate of the 3D position of the mobile node with one meter accuracy (depending on the density of nodes in the network). Figure 31 shows the evolution of the particles for a particular person guiding experiment at the Barcelona Robot Lab. Figure 32 shows the estimated position of the person estimated by the WSN, compared to that of the guiding robot which is in average about 3 meters ahead of the tracked person. The red dots indicate individual estimates of the particle filter and are shown to give an idea of the distribution of the localization estimate.

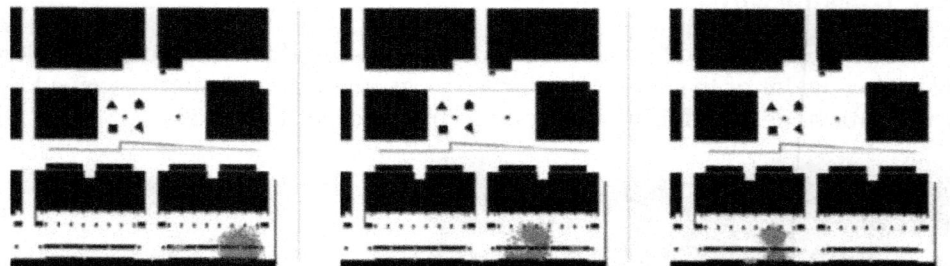

Figure 31. The evolution of the tracking estimates. Particle evolution (in red), and signal strength of each node (in green).

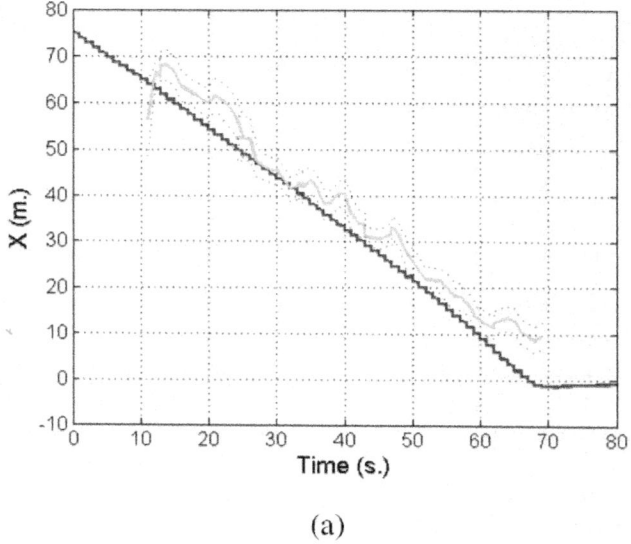

(a)

Figure 32. Estimated position of the person (green) and its standard deviation (red), compared with the position of the robot (blue) during a guiding mission.

Decentralized Tracking with Cameras and Wireless Sensor Network

The full system can take advantage of all the information sources (cameras, robots, WSN) to improve the reliability and accuracy of the different services, as for example in tracking applications. In order to illustrate the benefits of the data fusion process described in Section 6, results from a simple setup are presented here. This setup consists of two fixed cameras and a WSN. The objective was to track one person cooperatively. In the experiment, the person is not always in the field of view of the cameras, appearing and disappearing from the image plane several times.

Three instances of the decentralized algorithm summarized in Section VI (see [26] for a thorough description) are launched, processing each camera's data and the WSN estimations. They communicate to exchange estimated trajectories. The received data is fused with the local estimations (tracks on the image plane), leading to a decentralized tracking of the person.

Figure 33 shows the estimated person position (X and Y coordinates) compared with a centralized estimation (employing a central EKF filter). One important benefit from the system is that, if the communications channels are active, the different elements have nearly the same information. That way, one robot or camera, even not seeing the person, can know where it is. Also, the uncertainty is decreased due to the redundancies in the system.

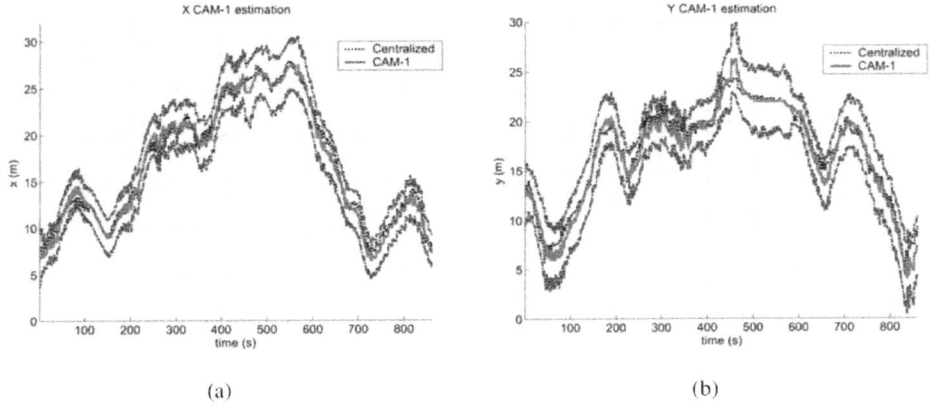

Figure 33. Estimated X and Y by the camera (red) compared to the centralized estimation (black), with the 3-sigma bound. The estimations are nearly the same, except during times when no communications occur.

9. LESSONS LEARNED

Sensor fusion has proven to be a difficult undertaking in the URUS project. The issues that make it difficult include heterogeneity and latency of sensors and systems, the mobility of some of the sensors translating to network connectivity issues, and the challenges of environment perception in outdoor dynamic environments.

With regards to robot localization, we report here on results for a pair of experimental sessions that lasted for nearly three hours of autonomous navigation, and traversing over 3.5 km. The session included 35 navigation requests of which 77% of them were completed successfully. Of the 8 missions that did not reach completion, 1 was due to hardware malfunction of the Wifi interface, and 7 were due to software glitches. Of these, 5 resulted in poor localization, 1 was due to a failure in the path execution module, and 1 was the cause of limited obstacle avoidance capabilities.

The URUS localization module has room for improvement especially for those areas in the environment in which 3D perception is still an issue. This is, on ramps and near stairs. Moreover, path planning should also be improved to more robustly avoid obstacle-occluded intermediate waypoints. Furthermore, the robot sensing capabilities should also be enhanced. Our laser-based localization module cannot deal consistently with glass walls and in situations with extreme illumination conditions. Finally little but dangerous obstacles such as small holes and curbs in the path remain a perception challenge for Segway-type robotic platforms.

For the case of detection and tracking, the use of delayed states and a decentralized system was very important, as the switching between 3G and Wifi, which occurred from time to time in the system, produced significant latencies and communications drop-outs. The system was able to operate in guiding missions under these circumstances. For instance, when communications were re-established, the local estimates at each robot were combined with the accumulated information from other nodes, obtaining a good global estimate (see Section 8).

The point of fusing the data from cameras and the Mica2 WSN together was to increase reliability of the tracking compared to just using the cameras alone, as in cross camera tracking, occlusions and illumination changes will cause the tracking to fail between the uncalibrated non-overlapping cameras. A decentralized approach was used where each sensor had its own processing engine. This meant that they could work independently, and the system was not vulnerable to a sensor failing elsewhere in the system. This improves scalability and robustness of the system. A more detailed account of our decentralized tracking approach that fuses information from the camera network, the Mica sensors, and the onboard robot cameras is presented in [47].

10. CONCLUSIONS

We have presented in this article the architecture design of the URUS sensors for a Network Robot System. This architecture has been deployed in a real urban environment and is used to test and verify people assistant robot services. We have described the sensors used in several of the robots that we have built and the sensors that we have deployed in the urban environment. Moreover, we have described an example how to fuse such heterogeneous sensor information during urban robotic tasks using among other techniques, decentralized data fusion, information-based filtering, and asynchronous data handling. The paper also discussed specific results on some of the application domains of the architecture.

This paper presents significant advances in urban robotic sensor integration. During the course of our research, we have identified several important technological issues that must be solved in order to be able to unleash robots in cities at a commercial level. For instance, the WLAN communication systems and protocols are not appropriate for robot deployment, the laser range sensors must be redesigned to capture 3D data or the HRI sensors must be further developed.

Moreover we have realized that the experimental sites, such as the Barcelona Robot Lab, have allowed us to test the technology in real environments and in contact with people, a main feature required for city robots.

The URUS project is touching upon a very new and exciting technology, that of networked robotics services in urban outdoor environments. The choice of methods and architecture portrayed in this article is of course not the only choice, and perhaps not the most suited in some occasions, but we hope that with this articles and with our research endeavors, we can identify scientific and technological aspects for which the state of the art does not provide compelling solutions, paving the road for future work in the field.

Acknowledgements

This work has been supported by URUS, the Ubiquitous Robotics in Urban Settings project, funded by the European Commission (FP6-IST-045062), by CONET, the Cooperating Objects Network of Excellence (FP7-2007-2-224053), by the Spanish National projects UbRob, Ubiqitous Robotics in Urban Settings (DPI2007-61452) and PAU, Perception and Action under Uncertainty (DPI2008-06022) of the DPI program; and MIPRCV, Multimodal Interaction in Pattern Recognition and Computer Vision of the Consolider Ingenio 2010 program (CSD2007-00018); and by the project SIRE, funded by the Andalusian Government (P06-TEP-01494). The authors would like to express their acknowledgment to the rest of the URUS team, a group of over 50 professionals from several institutions across Europe.

REFERENCES

1. Sanfeliu, A.; Andrade-Cetto, J. Ubiquitous networking robotics in urban settings. In *Proceedings of IROS Workshop Netw. Rob. Syst.*, Beijing, China, October 15, 2006; pp. 14–18.
2. Capita´n, J.; Manteco´n, D.; Soriano, P.; Ollero, A. Autonomous perception techniques for urban and industrial fire scenarios. In *Proceedings of IEEE Int. Workshop Safety, Secur. Rescue Rob.*, Roma, Italy, September 27–29, 2007; pp. 1–6.
3. Grime, S.; Durrant-Whyte, H.F. Data fusion in decentralized sensor networks. *Control Eng. Practice* 1994, 2, 849–863.
4. Sukkarieh, S.; Nettleton, E.; Kim, J.H.; Ridley, M.; Goktogan, A.; Durrant-Whyte, H. The ANSER project: Data fusion across multiple uninhabited air vehicles. *Int. J. Robot. Res.* 2003, 22, 505–539.
5. Barbosa, M.; Ramos, N.; Lima, P. Mermaid - Multiple-robot middleware for intelligent decision-making. In *Proceedings of the 6th IFAC/EURON Sym. Intell. Auton. Vehicles*, Toulouse, France, September 3–5, 2007.
6. Metta, G.; Fitzpatrick, P.; Natale, L. Yarp: Yet another robot platform. *Int. J Adv. Robotic Syst.* 2006, 3, 43–48.
7. Valencia, R.; Teniente, E.; Trulls, E.; Andrade-Cetto, J. 3D Mapping for urban serviece robots. In *Proceedings of IEEE/RSJ Int. Conf. Intell. Robots Syst.*, Saint Louis, MO, USA, October 11–15, 2009; pp. 3076–3081.
8. Thrun, S.; Burgard, W.; Fox, D. *Probabilistic Robotics*; MIT Press: Cambridge, UK, 2005.
9. Ortega, A.; Haddad, I.; Andrade-Cetto, J. Graph-based segmentation of range data with applications to 3D urban mapping. In *Proceedings of European Conf. Mobile Robotics*, Mlini, Croatia, September 23–25, 2009; pp. 193–198.

10. Ortega, A.; Dias, B.; Teniente, E.; Bernardino, A.; Gaspar, J.; Andrade-Cetto, J. Calibrating an Outdoor Distributed Camera Network using Laser Range Finder Data. In *Proceedings of IEEE/RSJ Int. Conf. Intell. Robots Syst.*, Saint Louis, MO, USA, October 11–15, 2009, pp. 303–308.
11. Ila, V.; Porta, J.M.; Andrade-Cetto, J. Information-based compact Pose SLAM. *IEEE Trans. Robot.* 2010, *26*, 78–93.
12. Ila, V.; Porta, J.; Andrade-Cetto, J. Reduced state representation in delayed State SLAM. In *Proceedings of IEEE/RSJ Int. Conf. Intell. Robots Syst.*, Saint Louis, MO, USA, October 11–15, 2009, pp. 4919–4924.
13. Eustice, R.M.; Singh, H.; Leonard, J.J. Exactly sparse delayed-state filters for view-based SLAM. *IEEE Trans. Robot.* 2006, *22*, 1100–1114.
14. Konolige, K.; Agrawal, M.; Sola`, J. Large scale visual odometry for rough terrain. In *Proceedings of the 13th Int. Sym. Robot. Res.*, Hiroshima, Japan, November 26–29, 2007.
15. Ila, V.; Andrade-Cetto, J.; Sanfeliu, A. Outdoor delayed-state visually augmented odometry. In *PProceedings of the 6th IFAC/EURON Sym. Intell. Auton. Vehicles*, Toulouse, France, September 3–5, 2007.
16. Uhlmann, J. Introduction to the algorithmics of data association in Multiple-Target Tracking. In *Handbook of Multisensor Data Fusion*; Liggins, M.E., Hall, D.E., Llinas, J., Eds.; CRC Press: Boca Raton, FL, USA, 2001.
17. Corominas-Murtra, A.; Mirats-Tur, J.; Sanfeliu, A. Action evaluation for mobile robot global localization in cooperative environments. *Robot. Auton. Syst.* 2008, *56*, 807–818.
18. Mirats-Tur, J.; Zinggerling, C.; Corominas-Murtra, A. Geographical information systems for map based navigation in urban environments. *Robot. Auton. Syst.* 2009, *57*, 922–930.
19. Bradski, G. Computer vision face tracking for use in a perceptual user interface. *Intel Techn. J.* 1998, pp. 1–15.
20. Viola, P.; Jones, M. Robust real-time face detection. *Int. J. Comput. Vision* 2004, *57*, 137–154.
21. Gonc¸alves, N.; Sequeira, J. Multirobot task assignment in active surveillance. In *Proceedings of the 14th Portuguese Conf. Artificial Intell.*, Aveiro, Portugal, October 12–15, 2009; Volume 5816, pp. 310–322.
22. Spaan, M.; Gonc¸alves, N.; Sequeira, J. Multirobot Coordination by Auctioning POMDPs. In *Proceedings of IEEE Int. Conf. Robot. Automat.*, Anchorage, AK, USA, May 3–8, 2010, (to appear).
23. Kaelbling, L.; Littman, M.; Cassandra, A. Planning and acting in partially observable stochastic domains. *Artif. Intell.* 1998, *101*, 99–134.
24. Pahliani, A.; Spaan, M.; Lima, P. Decision-theoretic robot guidance for active cooperative perception. In *Proceedings of IEEE/RSJ Int. Conf. Intell. Robots Syst.*, Saint Louis, MO, USA, October 11–15, 2009; pp. 4837–4842.
25. Nettleton, E.; Thrun, S.; Durrant-Whyte, H.; Sukkarieh, S. Decentralised SLAM with low-bandwidth communication for teams of vehicles. In *Field and Service Robots, Recent Advances in Research and Applications*; Springer: Berlin, Germany, 2003; Volume 24, pp. 179–188.
26. Capita´n, J.; Merino, L.; Caballero, F.; Ollero, A. Delayed-state information filter for cooperative decentralized tracking. In *Proceedings of IEEE Int. Conf. Robot. Automat.*, Kobe, Japan, May 12–17, 2009; pp. 3865–3870.
27. Bourgault, F.; Durrant-Whyte, H. Communication in general decentralized filters and the coordinated search strategy. In *Proceedings of the 7th Int. Conf. Information Fusion*, Stockholm, Sweden, June 28–July 1, 2004; pp. 723–730.
28. Lima, P.; Messias, J.; Santos, J.; Estilita, J.; Barbosa, M.; Ahmad, A.; Carreira, J. ISocRob 2009 team description paper. In *Proceedings of Robocup Sym.*, Graz, Austria, June 25–July 5, 2009.
29. Corominas, A.; Mirats, J.; Sandoval, O.; Sanfeliu, A. Real-time software for mobile robot simulation and experimentation in cooperative environments. In *Proceedings of the 1st Int.*

Conf. Simulation, Modelling, Programming Autonomous Robots, Venice, Italy, November 3–7, 2008; Volume 5325, pp. 135–146.

30. Corominas-Murtra, A.; Mirats-Tur, J.; Sanfeliu, A. Efficient active global localization for mobile robots operating in large and cooperative environments. In *Proceedings of IEEE Int. Conf. Robot. Automat.*, Pasadena, CA, USA, May 19–23, 2008; pp. 2758–2763.

31. Fox, D.; Burgard, W.; Kruppa, H.; Thrun, S. A probabilistic approach to collaborative multi-robot localization. *Auton. Robot.* 2000, *8*, 325–344.

32. Corominas-Murtra, A.; Mirats-Tur, J.; Sanfeliu, A. Integrating asynchronous observations for mobile robot position tracking in cooperative environments. In *Proceedings of IEEE/RSJ Int. Conf. Intell. Robots Syst.*, Saint Louis, MO, USA, October 11–15, 2009; pp. 3850–3855.

33. Gilbert, A.; Illingworth, J.; Bowden, R. Scale invariant action recognition using compound features mined from dense spatio-temporal corners. In *Proceedings of the 10th European Conf. Comput. Vision*, Marseille, France, October 12–18, 2008; Volume 5302, pp. 222–233.

34. Kaew-Trakul-Pong, P.; Bowden, R. A real-time adaptive visual surveillance system for tracking low resolution colour targets in dynamically changing scenes. *Image Vision Comput.* 2003, *21*, 913–929.

35. Figueira, D.; Moreno, P.; Bernardino, A.; Gaspar, J.; Santos-Victor, J. Optical flow based detection in mixed human robot environments. In *Proceedings of the 5th Int. Sym. Visual Computing*, Las Vegas, NV, USA, November 30–December 2, 2009; Volume 5875, pp. 223–232.

36. Dalal, N.; Triggs, B. Histograms of oriented gradients for human detection. In *Proceedings of the 19th IEEE Conf. Comput. Vision Pattern Recog.*, San Diego, CA, USA, June 20–25, 2005; pp. 886–893.

37. Dalal, N.; Triggs, B.; Schmid, C. Human detection using oriented histograms of flow and appearance. In *Proceedings of the 9th European Conf. Comput. Vision*, Graz, Austria, May 7–13, 2006; Volume 3951, pp. 428–441.

38. Moreno, P.; Bernardino, A.; Santos-Victor, J. Waving detection using the local temporal consistency of flow-based features for real-time applications. In *Proceedings of the 6th Int. Conf. Image Anal. Recog.*, Halifax, Canada, June 6–8, 2009; Volume 5627, pp. 886–895.

39. Pla, F.; Ribeiro, P.; Santos-Victor, J.; Bernardino, A. Extracting motion features for visual human activity representation. In *Proceedings of the 2nd Iberian Conf. on Pattern Recognition and Image Analysis*, Estoril, Portugal, June 7–9, 2005; Vol. 3522, pp. 537–544.

40. Ribeiro, P.; Moreno, P.; Santos-Victor, J. Boosting with temporal consistent learners: An application to human activity recognition. In *Proceedings of the 3rd Int. Sym. Visual Computing*, Lake Tahoe, NV, November 26–28, 2007; Volume 4841, pp. 464–475.

41. Schuldt, A.; Laptev, I.; Caputo, B. Recognizing human actions: A local SVM approach. In *Proceedings of the 17th IAPR Int. Conf. Pattern Recog.*, Cambridge, UK, August 23–26, 2004; Volume 3; pp. 32–36.

42. Dollar, P.; Rabaud, V.; Cottrell, G.; Belongie, S. Behavior recognition via sparse spatio-temporal features. In *Proceedings of the 14th Int. Conf. Comput. Communications and Networks*, San Diego, CA, USA, October 17–19, 2005; pp. 65–72.

43. Harris, C.G.; Stephens, M. A combined corner edge detector. In *Proceedings of Alvey Vision Conf.*, Manchester, UK, August 31-September 2, 1988; pp. 189-192.

44. Laptev, I.; Marszalek, M.; Schmid, C; .; Rozenfeld, B. Learning realistic human actions from movies. In *Proceedings of the 22nd IEEE Conf. Comput. Vision Pattern Recog.*, Anchorage, AL, USA, June 24-26, 2008; pp. 1–8.

45. Agrawal, R.; Srikant, R. Fast algorithms for mining association rules in large databases. In *Proceedings of the 20th Int. Conf. Very Large Data Bases*, Santiago de Chile, Chile, September 12–15, 1994; pp. 487–499.

46. Caballero, F.; Merino, L.; Gil, P.; Maza, I.; Ollero, A. A probabilistic framework for entire WSN localization using a mobile robot. *Robot. Auton. Syst.* 2008, 56, 798-806.
47. Gilbert, A.; J.Illingworth.; Capitan, J.; Bowden, R.; Merino, L. Accurate fusion of robot, camera and wireless sensors for surveillance applications. In *Proceedings of the 9th IEEE Int. Workshop Visual Surveillance,* Kyoto, Japan, October 3, 2009.

INDEX

A

AC Servomotors, 258
Accidents and Operating Modes, 278
Actuation Devices, 264
Actuation, 94
Adhesive Grippers, 268
Ancient Mythology, 3
Application Challenges, 39
Applications of Vacuum Grippers, 271
Arc Welding Power Sources, 138
Arc Welding Robot, 293
Architecture, 242
Articulated Robot, 73
Artificial Emotions, 105
Artificial Intelligence, 219
Assembly Operation, 293
Attaching Frame to the Object, 60
Automated Assembly, 330
Automated Guided Vehicle, 15
Automated Production Lines, 328
Automatic Battery Swap, 28
Automatic Type Robot, 55
Automation and Society, 339
Autonomously Guided Robot, 12

B

Battery Swap, 27
Behavior Based Control, 216

Biological Analogs, 194
Break-beam Sensors, 196
Brushed DC Servomotors, 259
Bucket of Stuff, 34

C

Cam Actuation, 270
Car Production, 45
Career Training, 109
Cartesian Coordinate Robot, 75
Changing Industrial Landscape, 287
Cleaned, Sprayed and Spatter-Free, 184
Collaborative Robots, 41
Combination Control, 21
Combinations, 232
Communications, 335
Competitions and Prizes, 126
Complexity of Planning, 238
Components, 94
Computational Intelligence, 121
Computer Control, 295
Computer Process Control, 332
Consumables for Robotic Welding, 166
Consumer Products, 339
Continuous Rotation Motors, 210
Controlling Movement, 67
Creativity, 118
Cybernetics and Brain Simulation, 119

D

Decommissioning of Robots, 71
Defining Parameters, 63
Delta Robot, 82
Development of Robotics, 326
Different Types of Robot Joints, 295
Different Types of Sensors, 265
Domestic Robots, 47
During the Machine Age, 313
Dynamics and Kinematics, 108

E

Early AGV-Style Robots, 46
Early Beginnings, 4
Early Developments, 319
Early in the Industrial Revolution, 312
Educational Robot, 41
Elastic Nanotubes, 96
Electric Drive System, 297
Electric Motors – DC Servomotors, 259
Electroactive Polymers, 96
Electrode Tip Dresser, 145
Electromagnets, 272
Electronic Control of Motors, 209
Enchanting, 240
End Effector – Robot's Hand, 274
End-of-arm Tooling, 67

F

Facial Expression, 105
Feedback (Closed Loop) Control, 212
Feedback Controls, 322
Feedback Devices, 264
Fifth Degree Wrist Pitch, 219
Final Considerations, 162
Finished Product Handling, 24
First Degree Shoulder Pitch, 218
Flat Belt, 300

Flexible Manufacturing Systems, 332
Flying and Underwater Robots, 224
Food and Beverage, 26
Force Required to Grip the Object, 187
Formal Definition, 237
Forward Sensing Control, 21
Fourth Degree Elbow Pitch, 219
Frequency Select Mode, 20
Frequent Training Required, 59
Future Development, 52

G

Gameboy Advance (GBA), 231
Gameboy, 231
Gear and Rack Actuation, 270
General Ethical Issues in Science and Technology, 44
General HRI Research, 307
General Intelligence, 118
General Purpose Effectors, 99
Gödel's Incompleteness Theorem, 127
Grand Challenges, 40
Grippers, 274
Guarded Tele-op, 12

H

Haptic Interface Robots, 52
Hardware Design Challenges, 39
Humanoid Robots, 6
Hydraulic Drive System, 297

I

Image Processing, 133
Imaging, 132
Impact on Society, 340
Impact on the Individual, 340
Inductive Transducers, 41
Industrial Robot Control System, 255
Industrial Robot Software, 226

INDEX 353

Inertial (Gyroscopic) Navigation, 17
Inflatable Bladder Type Gripper, 269
Infra Red (IR) Sensors, 198
Inspected and Ready to Weld, 183
Integrating the Approaches, 121
Intelligent Agent Paradigm, 121
Intelligent Robots, 256
Interim AGV-Technologies, 46
Interlocking in Work Cell Control, 257
International Safety Requirements, 278
IR Communication, 198

K

Knowledge Representation, 114
Knowledge-based, 120

L

Languages, 125
Laser Target Navigation, 17
Lead Through, 229
Levels of Processing, 192
Light Sensors, 193
Limitations of a Vision System, 135
Limited Sequence Robots, 255
Linear Actuators, 95
Linear or Translational Movement, 269
Line-following Car, 12
Linkage Actuation, 269
Locomotion, 99
Long-term Goals, 118

M

Machine Load and Unload, 290
Machine Loading, 289
Machine Programming, 323
Machine Unloading, 290
Machine Vision, 200
Main Positions on Roboethics, 43
Manipulation, 98

Manual Data Input Panel, 294
Measuring Direction, 302
Measuring Position, 302
Measuring Rotational Speed, 301
Mechanical Grippers, 273
Mechanics of Robotic Manipulators, 60
Methods for Motion Planning, 308
Methods for Perceiving Humans, 307
Microcontrollers, 230
Military Robots, 42
Mobile Robot, 11
Mobility Limits of the Spider, 224
Modern Autonomous Robots, 6
Modern Developments, 320
Modular Robotics Google Group, 41
Motion and Manipulation, 117

N

Neural Networks, 124
New Functions and Abilities, 53
Next Byte Codes, 243
No Peripheral Decision, 184
Non-market-based Ideas, 317
Normal Operation Mode, 282
Notable Robotic Arms, 190
Numerical Control, 329

O

One-wheeled Balancing Robots, 99
Open Loop Control, 213
Orthogonal Joint, 296
Other Considerations, 170
Overall Trends, 52

P

Packaging, 45
Paint Robot, 75
Pallet Handling, 24
Paper and Print, 26

Parallel Languages, 227
Passive Dynamics, 101
Payback Method, 291
Permanent Magnet AC tachometer, 263
Permanent Magnets, 272
Pivoting Movement, 269
Planning and Control Challenges, 39
Playback Robots, 255
Pneumatic Drive System, 297
Poor Wire Feeding, 160
Poorly Performing Peripherals, 161
Position and Velocity Sensors, 264
Position Sensors, 261
Post-market Theories, 314
Potentiometers, 262
Predictions and Ethics, 128
Premature Cable Failure, 160
Product Revision History, 82
Programmable Controllers, 251
Programmable Logic Controllers, 231
Pronunciation, 241
Proportional Control, 253
Proximity Sensor, 265

Q

Quadrature Shaft Encoding, 197
Quality Control in Work Part Inspection, 257
Quantitative Accomplishment, 38

R

Range Sensor, 265
Raw Materials Handling, 24
Reactive Control, 213
Reconfigurable Robots, 51
Reference Coordinate Frame, 60
Reflective Optosensors, 195
Related Products For Drives and Actuators, 210

Relationship to Unemployment, 44
Relentless March, 286
Remote-controlled Systems, 5
Research Robots, 50
Resistive Position Sensors, 194
Retired Robots and Art, 72
RNA Automation, 86
Robot Competitions, 49
Robot Drive Systems, 296
Robot Kits, 48
Robot Links and Joints, 295
Robot Operating Techniques, 294
Robot Programming and Interfaces, 65
Robot Programming Languages, 235
Robot Programming, 327
Robots and Artificial Intelligence, 129
Rope and Pulley Actuation, 270
Rotational Joint, 296

S

Screw Actuation, 270
Scripting Languages, 227
Search and Optimization, 122
Search and Rescue, 308
Searle's strong AI hypothesis, 127
Second Degree Arm Yaw, 218
Selecting the Best Material, 176
Self-reconfiguring Modular Robot, 30
Semantics of ADL, 238
Sensors used in Robotics, 265
Series Elastic Actuators, 96
Service Industries, 337
Service Robot, 28
Servo Motors, 209
Seven Degrees of Freedom, 218
Seventh Degree Wrist Roll, 219
Shaft Encoding, 196
Shape of the Gripping Surface, 187
Singularities, 68

Sixth Degree Wrist Yaw, 219
Six-wheeled Robots, 100
Sliding Autonomy, 13
Social Intelligence, 118
Soft Robots, 51
Space Exploration, 34
Space Probes, 47
Special Operation Modes, 282
Specular Reflection, 199
Speech Recognition, 104
Spherical Orb Robots, 99
Spot Welding Guns, 144
Spot Welding Robot, 292
Spot Welding Swivel, 146
Steering Control, 18
Stepper Motors, 258
Swarm Robots, 51
Swimming (Piscine), 103
Switch Sensors, 192
Symbolic, 119
Syntax and Semantics, 243
Syntax of ADL, 237

T

Tactile Sensors, 266
Taking Action the Whys and Hows, 179
TCP-Calibration Unit, 142
Teach Method, 228
Technological Trends, 52
Terminology, 248
Things to Remember, 175
Third Degree Shoulder Roll, 218
Tracked Robots, 100
Trailer Loading, 24
Training is Essential, 150
Transformation of Frames, 60
Transportation, 336
Trimming Liners Correctly, 173
Trouble with TCP?, 162

Troubleshooting Robotic Welding, 158
Turing's "Polite Convention", 127
Twisting Joint, 296
Two-wheeled Balancing Robots, 99
Types of Magnetic Grippers, 272
Types of Vacuum Grippers, 271
Typical Programming, 68

U

Ultrasonic Distance Sensing, 198
Using Anti-Spatter Solution, 172
UWA Telerobot, 88

V

Vacuum Cups, 271
Vacuum Grippers, 99
VAL Programming, 248
Variable Assembly Language, 247
Velocity Sensors, 263
VGR Systems Benefits, 135
Vision Guidance, 19
Vision Guided Robotic Systems, 134
Vision Systems for Robot Guidance, 135
Visual Programming Language, 227

W

Warehousing, 26
Weld Timer, 145
Welding Processes, 148
Welding Safety, 148
What are Robots?, 71
What is Your Mode of Welding?, 168
What You Should Know to Improve Performance and Reduce Costs, 166
Wheels, 221
Which Language to Pick?, 233
Why Are Industrial Robots Dangerous?, 277
Wire Feeder, 140

This page left intentionally blank.